Ion Channels

A LABORATORY MANUAL

ALSO FROM COLD SPRING HARBOR LABORATORY PRESS

RELATED TITLES

Addiction

An Introduction to Nervous Systems

Calcium Signaling

Cognition

Cystic Fibrosis: A Trilogy of Biochemistry, Physiology, and Therapy

Epilepsy: The Biology of a Spectrum Disorder

Glia

Learning and Memory

Neurogenesis

Signal Transduction: Principles, Pathways, and Processes

The Synapse

OTHER LABORATORY MANUALS

Antibodies: A Laboratory Manual, Second Edition

Budding Yeast: A Laboratory Manual

Calcium Techniques: A Laboratory Manual

Cell Death Techniques: A Laboratory Manual

CRISPR-Cas: A Laboratory Manual

Fission Yeast: A Laboratory Manual

Manipulating the Mouse Embryo: A Laboratory Manual, Fourth Edition

Molecular Cloning: A Laboratory Manual, Fourth Edition

Molecular Neuroscience: A Laboratory Manual

Mouse Models of Cancer: A Laboratory Manual

Purifying and Culturing Neural Cells: A Laboratory Manual

RNA: A Laboratory Manual

Subcellular Fractionation: A Laboratory Manual

HANDBOOKS

A Bioinformatics Guide for Molecular Biologists

At the Bench: A Laboratory Navigator, Updated Edition

At the Helm: Leading Your Laboratory, Second Edition

Career Options for Biomedical Scientists

Experimental Design for Biologists, Second Edition

Lab Math: A Handbook of Measurements, Calculations, and Other Quantitative Skills for Use at the Bench

Lab Ref: A Handbook of Recipes, Reagents, and Other Reference Tools for Use at the Bench,
    Volume 1 and Volume 2

Next-Generation DNA Sequencing Informatics, Second Edition

Statistics at the Bench: A Step-by-Step Handbook for Biologists

Using R at the Bench: Step-by-Step Data Analytics for Biologists

WEBSITES

www.cshprotocols.org

www.cshperspectives.org

Ion Channels

A LABORATORY MANUAL

EDITED BY

Paul J. Kammermeier

University of Rochester Medical Center

Ian Duguid

Centre for Integrative Physiology
University of Edinburgh

Stephan Brenowitz

Janelia Research Campus
Howard Hughes Medical Institute

CSH PRESS

COLD SPRING HARBOR LABORATORY PRESS
Cold Spring Harbor, New York • www.cshlpress.org

ION CHANNELS
A LABORATORY MANUAL

Publisher	John Inglis
Acquisition Editor	Richard Sever
Managing Editor	Maria Smit
Director of Editorial Services	Jan Argentine
Project Manager	Inez Sialiano
Permissions Coordinator	Carol Brown
Production Editor	Joanne McFadden
Production Manager	Denise Weiss
Director of Product Development & Marketing	Wayne Manos
Cover Designer	Denise Weiss

Front cover image: (*Foreground*) A voltage-dependent potassium channel as viewed from outside of the cell membrane (PDB ID: 2R9R). Each subunit of the tetramer is uniquely colored. A potassium ion is shown as a blue sphere in the channel pore. The voltage-sensing domains are outside of the pore-forming domains. The structures and functions of voltage-gated potassium channels are reviewed in Chapter 2. Image courtesy of Dorothy M. Kim and Crina M. Nimigean. (*Background*) Flourescence image from a hippocampal slice culture. A protocol for preparing slice cultures from rodent hippocampus is included in Chapter 12. Image courtesy of Thomas G. Oertner.

Library of Congress Cataloging-in-Publication Data

Names: Kammermeier, Paul J., editor.
Title: Ion channels : a laboratory manual/edited by Paul J. Kammermeier,
    University of Rochester Medical Center, Ian Duguid, Centre for Integrative
    Physiology, University of Edinburgh, Stephan Brenowitz, Janelia Research
    Campus, Howard Hughes Medical Institute.
Description: Cold Spring Harbor, New York: Cold Spring Harbor Laboratory
  Press, 2016 | Includes bibliographical references and index.
Identifiers: LCCN 2016045412 | ISBN 9781621821205 (hard copy) |
ISBN 9781621821212 (pbk.)
Subjects: LCSH: Ion channels--Laboratory Manuals.
Classification: LCC QH603.I54 I557 2016 | DDC 571.6/4--dc23
LC record available at https://lccn.loc.gov/2016045412

Contents

General Safety and Hazardous Material Information

This manual should be used by laboratory personnel with experience in laboratory and chemical safety or students under the supervision of such trained personnel. The procedures, chemicals, and equipment referenced in this manual are hazardous and can cause serious injury unless performed, handled, and used with care and in a manner consistent with safe laboratory practices. Students and researchers using the procedures in this manual do so at their own risk. It is essential for your safety that you consult the appropriate Material Safety Data Sheets, the manufacturers' manuals accompanying products, and your institution's Environmental Health and Safety Office, as well as the General Safety and Hazardous Material Information Appendix, for proper handling of hazardous materials. Cold Spring Harbor Laboratory makes no representations or warranties with respect to the material set forth in this manual and has no liability in connection with the use of these materials.

All registered trademarks, trade names, and brand names mentioned in this book are the property of the respective owners. Readers should please consult individual manufacturers and other resources for current and specific product information.

Appropriate sources for obtaining safety information and general guidelines for laboratory safety are provided in the General Safety and Hazardous Material Information Appendix.

CHAPTER 1

Ion Channels: History, Diversity, and Impact

Stephan Brenowitz,[1,4] Ian Duguid,[2,4] and Paul J. Kammermeier[3,4]

[1]*Janelia Research Campus, Howard Hughes Medical Institute, Ashburn, Virginia 20147;* [2]*Centre for Integrative Physiology, Edinburgh Medical School: Biomedical Sciences, University of Edinburgh, Edinburgh EH8 9XD, United Kingdom;* [3]*Department of Pharmacology and Physiology, University of Rochester Medical Center, Rochester, New York 14642*

From patch-clamp techniques to recombinant DNA technologies, three-dimensional protein modeling, and optogenetics, diverse and sophisticated methods have been used to study ion channels and how they determine the electrical properties of cells.

THE HISTORY OF ION CHANNELS

The modern era of studying electrical properties of excitable cells began in earnest in the 1930s. It was then that our understanding of how cells produce electrical potentials began to emerge and with it, an understanding of how bioelectric changes could be utilized to perform important cellular functions such as conveying information. As is often the case, technological advances helped drive scientific understanding. By the 1940s, Kenneth Cole and George Mormont had begun to develop the voltage clamp technique, wherein the membrane potential of a (large) cell could be measured and controlled, leading to the earliest descriptions of the electrical properties of membranes and the conductances that underlie neuronal action potentials. Soon after, Alan Hodgkin and Andrew Huxley refined the technique to discover that the action potential was not simply a relaxation of the membrane potential to zero, as had been previously thought, but constituted an overshoot of the membrane potential to positive potentials. Further, they discovered that the depolarizing phase of the action potential was due to sodium flux into the cell, while the repolarization back to the resting membrane potential was due to potassium efflux. It was for these findings that they, along with Sir John Eccles, received the 1963 Nobel Prize in Physiology or Medicine.

Thus began a long exploration of the mechanisms underlying the electrical properties of cells. During this time, competing ideas emerged to explain the discoveries of Hodgkin and Huxley. The first was that the plasma membrane itself changed confirmation to become selectively permeable to sodium, then to potassium. The alternative hypothesis, championed notably by Bertil Hille and Clay Armstrong during the 1960s, was that selective pores were formed by proteins in the membrane to facilitate the passage of ions into and into and out of the cell (T Begenesich, pers. comm.). Further, because of the observation that compounds like tetrodotoxin (TTX) and tetraethylammonium (TEA) could selectively inhibit sodium and potassium conductances, they reasoned that separate proteins most likely underlie these conductances. This was finally confirmed with the biochemical purification

[4]Correspondence: brenowitzs@janelia.hhmi.org; ian.duguid@ed.ac.uk; paul_kammermeier@urmc.rochester.edu

Cite this introduction as *Cold Spring Harb Protoc*; doi:10.1101/pdb.top092288

of the TTX binding protein (the sodium channel) from electroplax membranes by Agnew and colleagues in 1978.

Along the way, new conductances were discovered, including voltage- and ligand-gated conductances, but confirmation of the idea of voltage and ligand-gated pores formed by distinct proteins would require another technological advance, and this came with the development of the patch clamp technique by Bert Sakmann, Erwin Neher, and colleagues, in which they demonstrated that a very high resistance (GΩ) seal could be formed between a glass micropipette and the plasma membrane of a cell. This allowed the voltage clamp technique to be applied to much smaller cells than had previously been possible. Further, the technique could be applied to small patches of membrane, in some cases containing a single channel protein. Sakmann and Neher received the Nobel Prize in Physiology or Medicine in 1991 for the development of this method, which led to an explosion of discoveries in the areas of pharmacology, physiology, and of course ion channel biophysics. Recording of single ion channel currents using the patch-clamp technique remains the only method in biology in which the function of native individual protein molecules can be monitored in real time.

In the ensuing period, continued improvements and innovations have been made to the patch-clamp technique enabling more discoveries in areas previously thought intractable. These include iterative improvements to amplifier designs allowing more accurate and faster voltage clamp, and a multitude of innovations that have enabled patch-clamp recordings from single neurons in vitro and in vivo, patching of organelles (e.g., nuclear and mitochondrial membranes) and subcellular regions of cells (e.g., synaptic terminals, dendrites, cilia). These innovations have transformed our understanding of the role of ion channels, transporters, and pumps in both excitable and nonexcitable cells providing us with the ability, at least in mammalian experimental animals, to describe almost all cellular conductances at the molecular level.

THE ION CHANNEL SUPERFAMILY

The molecular level description of ion channel function was driven in part by the development of electrophysiological techniques but also by the parallel development of "recombinant DNA technologies" in the late 1970s, whereby individual genes could be reliably isolated, sequenced, purified, and cloned. The rapid development of gene sequencing methods, when combined with electrophysiology, created a way to unravel the true extent of mammalian ion channel diversity and to classify ion channels based upon sequence homology, gating properties, and phylogeny. The predicted amino acid sequences uncovered remarkable sequence homologies across groups of ion channels allowing biophysicists to classify "superfamilies" of homologous channel proteins that presumably evolved from common ancestral channels. Although recombinant DNA technologies were being developed in the late 1970s, it was not until the early 1980s that Masaharu Noda and colleagues first managed to clone, sequence, and describe the primary structure of the α-subunit precursor of the nicotinic acetylcholine receptor (nAChR, 1982) and voltage-gated sodium channel (1984) from the electric ray (*Torpedo californica*) and eel (*Electrophorus electricus*), respectively (Noda et al. 1982, 1984). This technical and conceptual enlightenment led to an explosion of interest in unraveling ion channel molecular diversity and to the classification of the two main ion channel superfamilies, voltage- and ligand-gated ion channels.

Voltage-gated ion channels, so called because of the requirement for a change in membrane potential to initiate "gating" or opening of the channel, were found to be tetrameric channels built from four homologous modules comprising a voltage sensor domain and pore-forming domain. Prominent among these are sodium, potassium, and calcium channels that underpin the action potential and calcium signaling cascades present in almost all electrically excitable cells. The biophysical properties and methods with which to isolate voltage-dependent conductances are described in detail in other articles in this collection. In addition to the more prototypic voltage-gated ion channels, several other phylogenetically related ion channels, such as cyclic nucleotide-gated (CNG)

Cite this introduction as *Cold Spring Harb Protoc*; doi:10.1101/pdb.top092288

and transient receptor potential (TRP) channels, were identified due to their weak voltage dependence and retention of a voltage-sensing domain. In parallel, molecular identification and classification of ligand-gated ion channels, so called because of their requirement for extracellular ligand-binding to initiate a conformational change and opening of the ion channel pore, highlighted the presence of several families of ion channels gated by extracellular ligands. Prominent among these are ionotropic glutamate receptors (i.e., AMPA, Kainate, NMDA), cysteine-loop channels (i.e., Ach, GABA, glycine, 5-HT), ATP-gated channels (i.e., P2X) and phosphatidylinositol 4,5-bisphosphate (PIP_2)-gated channels. As the molecular identification of membrane channels gained pace through the 1980s and 1990s, the presence of additional, smaller families of ion channels such as calcium- or light-activated, cyclic nucleotide-gated, and mechanosensitive ion channels added to the increasingly diverse range of membrane proteins expressed by electrically excitable cells. The challenge that faced scientists of the time was to be able to unravel the complex biophysical mechanisms that link ion channel structure to physiological function.

An important feature of developing recombinant DNA technologies became the ability to apply site-directed mutagenesis to directly manipulate the amino acid sequence of a protein to establish a link between structure and function. The application of site-directed mutagenesis led to a number of groundbreaking discoveries including the mechanism of inactivation of Shaker potassium channels (Hoshi et al. 1991), identification of the region of the sodium channel forming the inactivation gate (Stühmer et al. 1989), pore-forming region of the potassium channel (Yellen et al. 1991) and an early description of the conformational change leading to "voltage sensing" in the Shaker potassium channel (Papazian et al. 1995; Larsson et al. 1996; Smith-Maxwell et al. 1998). This powerful genetic manipulation technique combined with advanced single cell electrophysiology heralded a new era for ion channel biophysics.

In parallel to the rapid expansion of recombinant DNA technologies, three-dimensional protein modeling and X-ray crystallography became an invaluable tool in the quest to understand how the amino acid sequence and predicted 3D secondary and tertiary protein structure related to biophysical characteristics of the ion channel. A paradigm shift in our understating of ion channel structure and function came in the early 1990s where Roderick MacKinnon became the first person to crystallize and analyze a potassium channel from the bacterium *Streptomyces lividans* (Doyle et al. 1998; MacKinnon et al. 1998), for which he later received the Nobel Prize in Chemistry in 2003. This powerful approach allowed scientists to visualize the structure of membrane bound proteins at an unprecedented level of resolution, forming 3D models of how sequences of amino acids interact to form a functional ion channel. This novel approach, when combined with emerging recombinant DNA technologies and single cell electrophysiology, generated a new vista of ion channel biophysics that has shaped the way we think about ion channel structure–function relationships to the present day.

RECOMBINANT CHANNEL TECHNOLOGY DRIVES MODERN NEUROBIOLOGY

Knowledge of the protein sequence, molecular structure, and gating behavior of ion channels has driven the development of an array of new methods directed toward understanding their biological function at the cellular, circuit, and behavioral levels. Many of the new approaches to understanding ion channel function rely on the use of light to achieve exquisite spatiotemporal control of channel gating when compared with pharmacological methods. Optical techniques for activating ion channels can be classified into several general approaches: (1) uncaging, the use of light to activate a chemically inert form of an ion channel ligand; (2) photoactivation, the direct activation of a photosensitive conductance such as channelrhodopsin; and (3) photoswitching, the use of light to isomerize a tethered ligand or channel blocker. Progress in each of these areas has resulted from both improvements in optical substrates resulting from molecular engineering and developments in microscopy, illumination, and imaging techniques.

An example of an early method for experimentally controlling ion channel activity is optical "uncaging" of glutamate to activate ionotropic glutamate receptors (Ellis-Davies 2007; Delaney and Shahrezaei 2013). The technique of uncaging requires synthesis of a chemically inert ("caged") form of a neurotransmitter, which upon exposure to light of the appropriate wavelengths (often in the UV portion of the spectrum) releases the biologically active form of the transmitter. Photolytic conversion of the caged to active form of the neurotransmitter can be achieved with brief light pulses (on the order of milliseconds) allowing the ligand to rapidly diffuse and activate its target receptor. Even greater spatial resolution can be achieved using two-photon activation of caged glutamate, which releases glutamate in a diffraction limited point ($\sim$1 μm^3) and can activate glutamate receptors on individually targeted dendritic spines in a manner similar to vesicular release of glutamate (Carter and Sabatini 2004; Svoboda and Yasuda 2006). Glutamate uncaging can be applied to many questions such as determining whether functional glutamate receptors are present on specific target neurons as well as their subcellular distribution, and characterizing the gating properties and kinetics of individual receptors. In addition to glutamate, many other neurotransmitters have been synthesized in a chemically caged form and are commercially available, such as GABA, serotonin, dopamine, and noradrenaline to name but a few.

Another method that has been highly influential in recent years is the optical activation of a cation conductance through activation of the algal photoreceptor channelrhodopsin (ChR2). When this light-gated ion channel is expressed in neurons, its activation produces a depolarizing current that can evoke action potentials. Numerous efforts have been made to increase the conductance and improve the kinetics of ChR2. Since the initial cloning of this receptor (Nagel et al. 2003) it has been widely adopted as a powerful tool for cell type–specific activation. This allows far greater resolution for mapping neural circuits (Petreanu et al. 2007) and the locations of synaptic inputs onto specific target neurons (Petreanu et al. 2009) when compared with electrical stimulation methods. The ability to suppress neuronal activity is also important for determining the roles of specific neurons in circuit function or disease. This has been achieved by optogenetic means by two main classes of protein, an inward chloride pump (halorhodopsin) and an outward proton pump (archaerhodopsin) (Chow et al. 2012).

Although optogenetic photoactivation has been enormously successful in determining circuit-level function, little information can be obtained when using this approach to define the roles of specific ion channels in cellular behavior. For this reason, development of photoswitchable tethered ligands has been a powerful and flexible tool for controlling the activation of specific types of ion channels with spatial, temporal, and molecular specificity (Szobota et al. 2007; Kramer et al. 2013). Several classes of ion channels have been rendered light-sensitive by attaching a channel blocker, such as a quaternary ammonium ion, to a linker such as azobenzene that undergoes a wavelength-dependent conformational change. This linker binds to a cysteine residue inserted near the pore of the Shaker potassium channel. At rest, or in the dark, this azobenzene group remains in the *trans* conformation and the pore is blocked by ammonium. Exposure to 380 nm light induces a change to the shorter *cis* conformation, and the ammonium ion is moved away from the pore, restoring conduction. This optically induced conformational change occurs on a submicrosecond time scale and can be reversed over many cycles. The ability to use light to control activation of ion channels allows far greater spatial and temporal specificity than conventional pharmacological methods.

Application of these widely diverse and sophisticated techniques has arisen to a large extent from the study of ion channels. A direct link can be traced from methods like optogenetics and its emerging dominant role in the broader fields of neuroscience and other arenas in biology, back to the tools initially developed to study channels and to discoveries of channels themselves. Further, analysis of single protein conformational changes in real time obtained with fluorescence techniques such as Förster resonance energy transfer (FRET) may be informed by the development of methods used to study single ion channels. In many ways, all of these techniques owe some debt to those original studies by Marmont, Cole, Hodgkin, and Huxley, and to the many scientists who followed them.

Cite this introduction as *Cold Spring Harb Protoc*; doi:10.1101/pdb.top092288

REFERENCES

Agnew WS, Levinson SR, Brabson JS, Raftery MA. 1978. Purification of the tetrodotoxin-binding component associated with the voltage-sensitive sodium channel from *Electrophorus electricus* electroplax membranes. *Proc Natl Acad Sci* 75: 2606–2610.

Carter AG, Sabatini BL. 2004. State-dependent calcium signaling in dendritic spines of striatal medium spiny neurons. *Neuron* 44: 483–493.

Chow BY, Han X, Boyden ES. 2012. Genetically encoded molecular tools for light-driven silencing of targeted neurons. *Prog Brain Res* 196: 49–61.

Delaney KR, Shahrezaei V. 2013. Uncaging calcium in neurons. *Cold Spring Harb Protoc* 2013: 1115–1124.

Doyle DA, Morais Cabral J, Pfuetzner RA, Kuo A, Gulbis JM, Cohen SL, Chait BT, MacKinnon R. 1998. The structure of the potassium channel: Molecular basis of K+ conduction and selectivity. *Science* 280: 69–77.

Ellis-Davies GC. 2007. Caged compounds: Photorelease technology for control of cellular chemistry and physiology. *Nat Methods* 4: 619–628.

Hoshi T, Zagotta WN, Aldrich RW. 1991. Two types of inactivation in Shaker K+ channels: Effects of alterations in the carboxy-terminal region. *Neuron* 7: 547–556.

Kramer RH, Mourot A, Adesnik H. 2013. Optogenetic pharmacology for control of native neuronal signaling proteins. *Nat Neurosci* 16: 816–823.

Larsson HP, Baker OS, Dhillon DS, Isacoff EY. 1996. Transmembrane movement of the Shaker K+ channel S4. *Neuron* 16: 387–397.

MacKinnon R, Cohen SL, Kuo A, Lee A, Chait BT. 1998. Structural conservation in prokaryotic and eukaryotic potassium channels. *Science* 280: 106–109.

Nagel G, Szellas T, Huhn W, Kateriya S, Adeishvili N, Berthold P, Ollig D, Hegemann P, Bamberg E. 2003. Channelrhodopsin-2, a directly light-gated cation-selective membrane channel. *Proc Natl Acad Sci* 100: 13940–13945.

Noda M, Takahashi H, Tanabe T, Toyosato M, Furutani Y, Hirose T, Asai M, Inayama S, Miyata T, Numa S. 1982. Primary structure of α-subunit precursor of *Torpedo californica* acetylcholine receptor deduced from cDNA sequence. *Nature* 299: 793–797.

Noda M, Shimizu S, Tanabe T, Takai T, Kayano T, Ikeda T, Takahashi H, Nakayama H, Kanaoka Y, Minamino N, et al. 1984. Primary structure of *Electrophorus electricus* sodium channel deduced from cDNA sequence. *Nature* 312: 121–127.

Papazian DM, Shao XM, Seoh SA, Mock AF, Huang Y, Wainstock DH. 1995. Electrostatic interactions of S4 voltage sensor in Shaker K+ channel. *Neuron* 14: 1293–1301.

Petreanu L, Huber D, Sobczyk A, Svoboda K. 2007. Channelrhodopsin-2-assisted circuit mapping of long-range callosal projections. *Nat Neurosci* 10: 663–668.

Petreanu L, Mao T, Sternson SM, Svoboda K. 2009. The subcellular organization of neocortical excitatory connections. *Nature* 457: 1142–1145.

Smith-Maxwell CJ, Ledwell JL, Aldrich RW. 1998. Role of the S4 in cooperativity of voltage-dependent potassium channel activation. *J Gen Physiol* 111: 399–420.

Stühmer W, Conti F, Suzuki H, Wang XD, Noda M, Yahagi N, Kubo H, Numa S. 1989. Structural parts involved in activation and inactivation of the sodium channel. *Nature* 339: 597–603.

Svoboda K, Yasuda R. 2006. Principles of two-photon excitation microscopy and its applications to neuroscience. *Neuron* 50: 823–839.

Szobota S, Gorostiza P, Del Bene F, Wyart C, Fortin DL, Kolstad KD, Tulyathan O, Volgraf M, Numano R, Aaron HL, et al. 2007. Remote control of neuronal activity with a light-gated glutamate receptor. *Neuron* 54: 535–545.

Yellen G, Jurman ME, Abramson T, MacKinnon R. 1991. Mutations affecting internal TEA blockade identify the probable pore-forming region of a K+ channel. *Science* 251: 939–942.

Cite this introduction as *Cold Spring Harb Protoc*; doi:10.1101/pdb.top092288

Voltage-Gated Potassium Channels: A Structural Examination of Selectivity and Gating

Dorothy M. Kim and Crina M. Nimigean

Department of Anesthesiology, Weill Cornell Medical College, New York, New York 10065

Voltage-gated potassium channels play a fundamental role in the generation and propagation of the action potential. The discovery of these channels began with predictions made by early pioneers, and has culminated in their extensive functional and structural characterization by electrophysiological, spectroscopic, and crystallographic studies. With the aid of a variety of crystal structures of these channels, a highly detailed picture emerges of how the voltage-sensing domain reports changes in the membrane electric field and couples this to conformational changes in the activation gate. In addition, high-resolution structural and functional studies of K^+ channel pores, such as KcsA and MthK, offer a comprehensive picture on how selectivity is achieved in K^+ channels. Here, we illustrate the remarkable features of voltage-gated potassium channels and explain the mechanisms used by these machines with experimental data.

INTRODUCTION TO K^+ CHANNELS

Electrical signaling is produced by the interplay of a diverse array of ion channels, which are elegantly orchestrated to respond to a variety of stimuli and propagate this response in an efficient manner. Channel opening and closing is exquisitely balanced among different types of channels to correctly and precisely maintain homeostasis at resting potential and contribute to explosive electrical activity during the action potential. Potassium channels play a role in repolarization of the membrane, which follows membrane depolarization by sodium, and in some cases calcium, channels during the action potential; this is necessary for returning the membrane to a negative resting potential to terminate the action potential signal. The balance required for this interplay is achieved through the extremely high ionic selectivity, high rate of flux, and sophisticated gating mechanisms of potassium channels.

Potassium channels are found in all living organisms. These highly selective channels are conserved across the kingdoms and are present in all cell types including neurons, muscle cells, and other tissues. They have evolved to play different roles in different cells, yet they have retained key features that are conserved throughout the potassium channel family. One universal trait is their high selectivity; some potassium channels show up to 1000-fold preference for K^+ ions, over smaller ions such as Na^+ and Li^+ (Yellen 1984; Neyton and Miller 1988a; Hille 2001; LeMasurier et al. 2001). This high degree of selectivity is necessary for maintaining the resting membrane potential, with high K^+ inside and high Na^+ outside of the cell. This selectivity is achieved by the structure of a highly conserved amino acid sequence that forms a potassium ion selectivity filter within the pore.

Correspondence: dok2004@med.cornell.edu

The membrane permeability of excitable cells to potassium ions was predicted by Julius Bernstein in 1902 (Bernstein 1902). He hypothesized that cells at rest were exclusively permeable to K^+ ions and that the cells were permeable to other ions only during periods of excitation, suggesting that there existed ion-selective components in the membrane. This hypothesis was borne out through the breakthrough studies by Hodgkin and Huxley performed on the squid giant axon in the 1940s and 1950s, which characterized the action potential in terms of the coordinated changes in the cell membrane permeability to Na^+ and K^+ ions and developed a model directly correlating these fluxes with excitation and electrical conduction (Goldman 1943; Hodgkin and Huxley 1945, 1946, 1947, 1952a,b,c,d,e,f; Hodgkin and Katz 1949; Hodgkin et al. 1952; Hodgkin and Keynes 1955). This was correctly hypothesized at the time to result from the orchestrated opening and closing of voltage-dependent ion-selective channels. Their experiments confirmed that the cell was predominantly selective for K^+ ions when it was at rest, resulting in a negative transmembrane potential (measured as the inside of the cell relative to the outside). These studies were key in highlighting the importance of both ion selectivity and regulated gating to generation of the action potential.

In addition to being highly selective, K^+ channels conduct ions at an extremely fast rate, close to the rate of diffusion (Hille 2001). How is this possible? Again, the clever architecture of the selectivity filter provides the answer to this paradox. Structural and functional studies of potassium channels have shown that the placement of ions in the selectivity filter leads to a "knock-on" mechanism (Fig. 1), which exploits the charge-charge repulsion of the K^+ ions that march in single file through the filter (Hodgkin and Keynes 1955; Doyle et al. 1998). Thus, the selectivity filter has evolved to efficiently conduct ions in a highly selective manner, allowing the fast rate of flux required to quickly repolarize the cell.

K^+ channels must respond rapidly on sensing changes in cellular environment. Ion flux is controlled by channel gating; that is, the ions cannot pass through the selectivity filter unless the channel is open, and the flow of ions must also be controlled by the ability of the channel to close. To do this rapidly, potassium channels have evolved gating mechanisms to couple the sensing of environmental cues with physical movements of domains within the channel that control ion flux. A major family of potassium channels is the voltage-gated potassium channels (Kv), which detect changes in transmembrane voltage and couple this detection to channel opening and closing.

Here, we will discuss the mechanisms by which K^+ channels achieve high selectivity while retaining a fast rate of flux, how the channel senses the change in transmembrane voltage, and how this detection of voltage is coupled to gating of the channel.

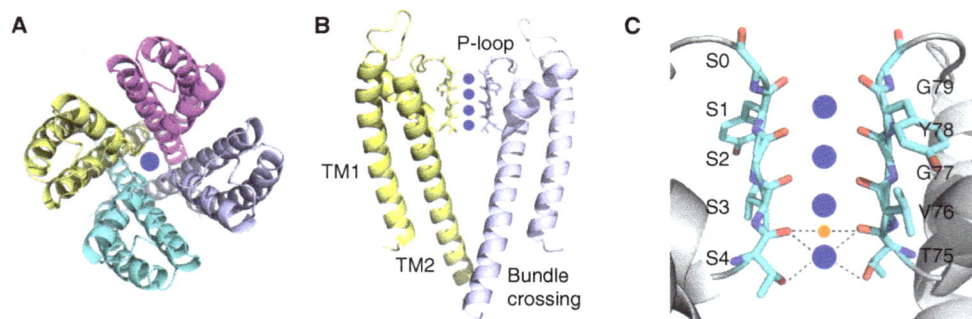

FIGURE 1. Structural details of KcsA. (A) KcsA tetramer as viewed from the top of the membrane (PDBID:14KC). Each subunit is uniquely colored. The central ion conduction pore is shown with a K^+ ion (blue sphere). (B) KcsA as viewed from the side with two opposing subunits removed for clarity. The P-loop contains the signature sequence TVGYG (shown as sticks), and forms the selectivity filter. Four potassium ions are shown in the selectivity filter. Below the selectivity filter is the aqueous cavity, formed by the TM2 helices that form the bundle crossing. (C) A detailed view of the selectivity filter of KcsA. Each binding site, S0–S4, is formed by the oxygen cages originating from backbone carbonyl and side-chain hydroxyls of the selectivity filter signature sequence TVGYG. Dashed lines depict coordination of the K^+ ion in S4 and the oxygens. Sodium (orange sphere) binds in the plane between sites S3 and S4. Dashed lines represent its coordination with carbonyl oxygens.

Cite this introduction as *Cold Spring Harb Perspect Biol* doi:10.1101/cshperspect.a029231

K⁺ CHANNEL SELECTIVITY

The puzzle of ion selectivity was pieced together by ingenious predictions of the pore characteristics by pioneers in the ion channel field. Important findings included pore size, ion dehydration as a requirement for pore entry, and the existence of stabilizing interactions between the pore walls and the ion (Mullins 1959; Eisenman 1962; Bezanilla and Armstrong 1972; Hille 1975a,b). In addition, Bezanilla and Armstrong (1972) and Hille (1973) proposed that K^+ ions were stabilized inside the pore of the K^+ channel by specific binding sites, which were composed of oxygen dipoles forming a "bracelet," and that this region would consist of a cylindrical pore with an inner diameter of 3–3.4 Å. The identification of a signature amino acid sequence, highly conserved among K^+-selective channels, provided a hypothesis for the identity of the K^+ channel selectivity filter, and subsequent functional studies confirmed that mutation of this signature sequence disrupts K^+ selectivity (Heginbotham et al. 1992, 1994). The culmination of our understanding of the selectivity filter arrived with the crystal structure of the KcsA potassium channel, which confirmed all of these predictions based on functional data with structural data (Doyle et al. 1998; Zhou et al. 2001b). The structure shows the selectivity filter to be a narrow pore with dimensions of $\sim$3 Å, with specific potassium binding sites formed by carbonyl oxygen atoms from residues in the signature sequence TVGYG (Fig. 1) (Heginbotham et al. 1994).

The requirement for specific binding sites to stabilize K^+ ions in the K^+ channel selectivity filter seemed to be in direct contrast to the fast flux rates measured in K^+ channels. Notably, Bezanilla and Armstrong suggested that the result of high selectivity would be slow permeation (Bezanilla and Armstrong 1972); however, the large conduction rates of potassium channels contradict this suggestion. This paradox led to the proposal of the multi-ion theory by several groups (Hodgkin and Keynes 1955; Heckmann 1965a,b, 1968, 1972; Hille and Schwarz 1978), which was supported by the following studies. Hodgkin and Keynes explained anomalous unidirectional flux ratios in squid axons by proposing that multiple ions pass through long pores in single file (Hodgkin and Keynes 1955). In addition, changes in permeability ratios in different ion-selective channels in the presence of more than one permeant ion (the anomalous mole fraction effect), could only be explained by the interaction of at least two ions within the pore (Chandler and Meves 1965; Hagiwara and Takahashi 1974; Neher and Sakmann 1975; Cahalan and Begenisich 1976; Sandblom et al. 1977; Hille and Schwarz 1978; Begenisich and Cahalan 1980; Almers and McCleskey 1984; Hess and Tsien 1984; Eisenman et al. 1986). Amazingly, these predictions were later confirmed by structural studies of potassium channels, which showed multiple ions occupying the selectivity filter pore.

Structural studies of potassium channels were made possible by the emergence of recombinant DNA technology in the 1980s, accelerating the ability to study these channels in isolation. Following cloning of the nicotinic acetylcholine receptor channel nAChR from *Torpedo californica* (Noda et al. 1982, 1983) and the voltage-gated sodium channel from the electric eel *Electrophorus electricus* (Noda et al. 1984), the first potassium channel was discovered by the analysis of *Drosophila* mutants. One of these mutants showed uncontrollable shaking, leading to the identification of the *Shaker* locus (Papazian et al. 1987). This gene was the first K^+ channel to be cloned and encodes a member of the voltage-gated potassium channel (Kv) family. Functional studies with Shaker revealed incredible insight into K^+ channel selectivity, including its preference for K^+, Rb^+, and NH_4^+ over Na^+ (Heginbotham and MacKinnon 1993). The cloning of Shaker, along with cloning of additional potassium channels, enabled taxonomic grouping of channels into functional superfamilies by sequence analysis. These analyses led to the discovery of the "signature sequence" of the selectivity filter, which contains the super conserved GYG motif that serves as the basis for ion selectivity in K^+ channels (Heginbotham et al. 1994) (exceptions are the EAG-like K^+ channels and the inwardly rectifying Kir6 family of K^+ channels, which contain the sequence GFG).

The discovery of homologous potassium channels in prokaryotic organisms catapulted the ion channel field into the structural era, with the crystal structure of the bacterial potassium channel KcsA from *Streptomyces lividans*, providing a rich understanding of fundamental properties of ion channels (Doyle et al. 1998; LeMasurier et al. 2001; Zhou et al. 2001b). KcsA was revealed to be a tetramer with fourfold symmetry, containing a pore formed by each subunit along the axis of symmetry (Fig. 1A). The transmembrane region of each subunit consists of two transmembrane domains TM1 and TM2. An extracellular loop, called the turret, followed by a short helix, called the pore helix, link these two helices (Fig. 1B). The carboxyl terminus of this pore helix extends toward the center of the channel pore, and precedes the signature sequence $T_{75}V_{76}G_{77}Y_{78}G_{79}$, located on the P-loop (Fig. 1B,C). The P-loop forms the selectivity filter by extending toward the extracellular end of the channel and creating a narrow pore. Importantly, the oxygens of the backbone carbonyls of residues Y78, G77, V76, and T75 as well as from the side-chain hydroxyl of T75 point directly toward the center of the fourfold symmetric pore. These oxygens from four identical subunits form four cages or K^+ binding sites, numbered S1–S4, which coordinate dehydrated K^+ in single file during permeation. Each K^+ ion is surrounded by eight oxygen atoms, mimicking K^+ hydration in solution (Fig. 1C). This region forms the narrowest part of the pore with dimensions of 3 Å, strikingly similar to the predicted 3–3.4 Å from earlier functional studies (Bezanilla and Armstrong 1972; Hille 1973). The backbone carbonyls of Y78 and G77 form S1, the most extracellular K^+ binding site, whereas the side-chain hydroxyl and backbone carbonyl oxygens of T75 form S4, the most intracellular binding site. Additionally, the crystal structure reveals a fifth binding site for K^+ at the extracellular surface of the channel, S0, which is coordinated by the backbone carbonyl oxygens of Y78 along with four water molecules (Fig. 1C). The structure also led to a hypothesis regarding the high specificity for K^+ binding over Na^+; the oxygen cages were suggested to be unable to properly coordinate smaller ions binding at the same location because the carbonyls would be too far to adequately mimic hydration of Na^+. Structural studies of KcsA and NaK (a channel that conducts both K^+ and Na^+) crystallized in the presence of Na^+ (and Li^+) suggest that instead of binding in the center of the eight-oxygen cage, Na^+ binds in a plane formed by four oxygens, in between two K^+ binding sites (Fig. 1C) (Alam and Jiang 2009b; Thompson et al. 2009; Cheng et al. 2011). Therefore, K^+ channels preferentially pass K^+ over Na^+ because either binding is energetically unfavorable or because it encounters a very high-energy barrier to permeate through the pore (Noskov et al. 2004; Zhou and MacKinnon 2004; Lockless et al. 2007; Alam and Jiang 2009b; Thompson et al. 2009; Liu et al. 2012). With these architectural details from the crystal structure of KcsA, the early predictions about the nature of the potassium channel selectivity filter were confirmed to be impressively accurate. In particular, the hypotheses that oxygen cages would be required to coordinate K^+ ions and that multi-ion occupancy of the selectivity filter would allow for high rate of flux were revealed by the structure to be correct (Bezanilla and Armstrong 1972; Hille and Schwarz 1978; Neyton and Miller 1988a,b).

In addition to revealing the fundamental principles of selectivity, structural studies with KcsA and other prokaryotic K^+ channels also proved to be crucial to understanding the regulation of ion flux in potassium channels. Conductance in potassium channels is regulated by two mechanisms, activation gating, and inactivation gating. In general, activation gating is the opening or closing of the channel in response to stimuli, whereas inactivation gating is a cessation of conduction through a channel in the continued presence of the stimulus that activated the channel. These two regulation mechanisms are elegantly coupled to one another to control conductance in potassium channels. In the next section, we will discuss the gating mechanisms used by potassium channels.

GATING

Activation Gating at the Bundle Crossing

The KcsA channel structure contains an intracellular gate that separates the cytoplasm from a water-filled cavity just below site S4 (Fig. 1B) (Doyle et al. 1998; Zhou et al. 2001b). This gate is formed by the carboxy-terminal ends of the four TM2 helices gathering together into a so-called bundle crossing to constrict access to the pore. In KcsA, this gate appears to be sterically closed to K^+ and is stabilized by a complex hydrogen bond network between multiple ionizable residues on each subunit. The binding of protons to some of these residues disrupts the hydrogen bonds and induces electrostatic forces that favor opening the bundle-crossing gate. KcsA is, therefore, found to open at low pH (high proton concentration) and is completely closed at neutral pH (Doyle et al. 1998; Heginbotham et al. 1999; Cortes et al. 2001; Zhou et al. 2001b; Takeuchi et al. 2007; Thompson et al. 2008).

The second K$^+$ channel structure determined by X-ray crystallography was the MthK channel, a Ca^{2+}-activated K$^+$ channel from the archaeon *Methanothermobacterium thermoautotrophicum* (Jiang et al. 2002). MthK is homologous to KcsA, with two transmembrane segments per subunit, but is activated by binding of intracellular Ca^{2+} to a separate intracellular domain that has been conserved in higher organisms. Because the structure of MthK was solved in the presence of activating Ca^{2+}, it might be expected that the channel would be in an open state. Indeed, in contrast to the closed bundle-crossing gate observed in the KcsA structure, the TM2 helices of MthK are splayed open, without any physical barrier between the intracellular solution and the pore. With these two static pictures, gating can be inferred as a swinging motion at a conserved glycine hinge on the TM2 helix, creating a pivot point to snap the gate shut (open pore conformation of K$^+$ channel) (Fig. 2A) (Jiang et al. 2002). Additional crystal structures of prokaryotic and eukaryotic potassium channels support this hypothesis and provide further details about the mechanism (Jiang et al. 2003a; Long et al. 2005, 2007; Alam and Jiang 2009a; Cuello et al. 2010).

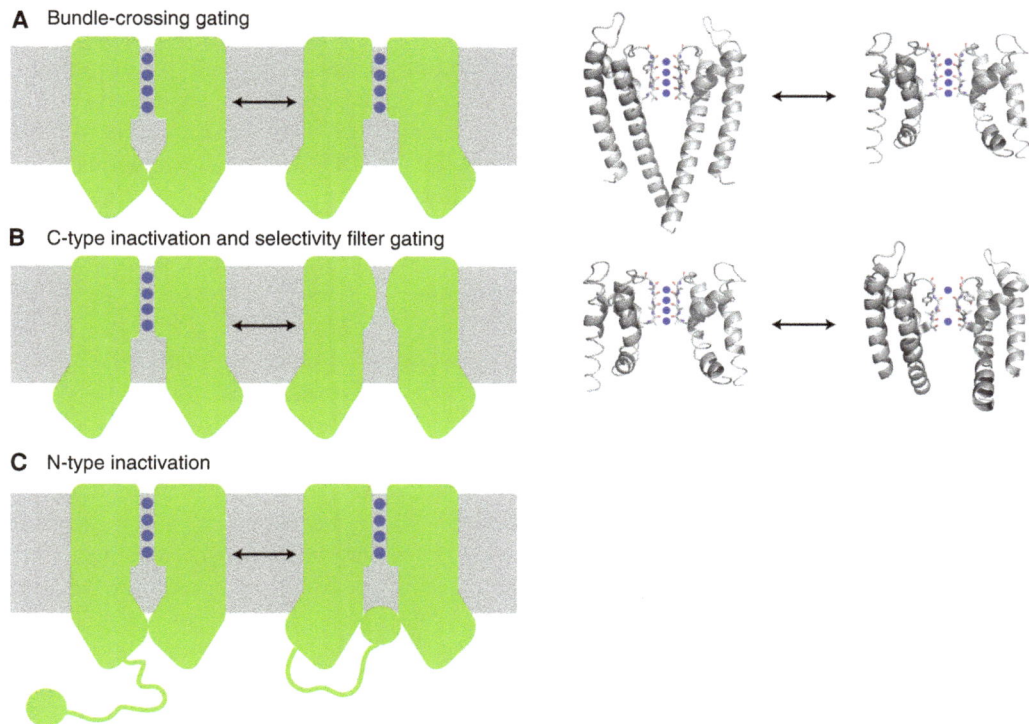

FIGURE 2. Gating mechanisms of potassium channels. (*A*) Gating can occur via conformational changes at the bundle crossing. The inner helices physically block the entry of potassium ions into the aqueous cavity, as shown in the KcsA closed structure (*left* structure, 1K4C). On activation, the inner helices splay open as observed in MthK (*right* structure, 1LNQ) and expose the cavity. Blue spheres denote potassium ions. (*B*) C-type inactivation and selectivity filter gating. In this model, partial collapse of the filter prevents conduction of potassium ions through the pore even if the bundle-crossing gate is open. This model is based on the comparison of the open MthK structure (*left* structure, 1LNQ) to the collapsed filter of a mutant KcsA in the open state (*right* structure, 3F5W). (*C*) N-type inactivation, or ball-and-chain gating, results from binding of an auto-inhibitory peptide to the bundle-crossing gate to physically block entry of ions into the cavity. This peptide can be part of the amino terminus of the channel (as shown) or the amino terminus of an associated β subunit. Currently, no crystal structures exist to illustrate N-type inactivation. The *left* panels show a cartoon representation of the gating mechanism, and the *right* panels show the available crystal structures that represent these conformational states.

Voltage-gated potassium channels are believed to use a bundle-crossing gate analogous to that observed in the KcsA channel (Liu et al. 1997; del Camino and Yellen 2001). The activation gate in many of the eukaryotic Kv channels does not appear to be controlled by a glycine hinge, as in KcsA, but instead appears to result from a hinge formed by the highly conserved amino acid sequence PxP or PxG, in which x is typically a hydrophobic residue (Long et al. 2005, 2007). This region is thought to produce a kink in the gating helix, allowing it to bend or rotate to accommodate the open and closed conformations of the channel, which appears to make a smaller channel opening compared with that observed in the open MthK structure (Doyle et al. 1998; Holmgren et al. 1998; del Camino et al. 2000, 2001; Zhou et al. 2001b; Jiang et al. 2002, 2003a; Long et al. 2005, 2007).

Inactivation Gating: C-Type Inactivation

Inactivation gating can also occur by a mechanism intrinsic to the pore, known as C-type inactivation (also called slow inactivation) (Kurata and Fedida 2006; McCoy and Nimigean 2012). Here, it is hypothesized that the selectivity filter acts as a gate by closing off conductance even in the presence of ongoing stimulus (Fig. 2B). In addition, C-type inactivation is thought to provide a physiological response to high levels of extracellular K^+ (Pardo et al. 1992; Baukrowitz and Yellen 1995); thus, it is probably responsible for regulation of repetitive or prolonged electrical activity.

C-type inactivation is believed to be the result of a conformational change in the selectivity filter. Structural studies of KcsA in low potassium concentrations reveal a collapsed selectivity filter (Zhou et al. 2001b), which was hypothesized to represent the inactivated state. Additionally, functional studies identified E71, located behind the selectivity filter, to be an important residue for the inactivation of KcsA, as the E71A mutant completely removed inactivation gating. The KcsA crystal structure shows that this residue interacts with an aspartate and tryptophan as well as with the GYG sequence from the selectivity filter. Mutation of E71 to an alanine results in disruption of these bonds and a conformational change in this aspartate that disrupts the selectivity filter (Cordero-Morales et al. 2006), preventing entry into the collapsed inactivated state even in the absence of K^+ (Cheng et al. 2011). E71 is not conserved in mammalian Kv channels and, therefore, the mechanism for C-type inactivation in these channels may be different from that of KcsA. However, mutation of equivalent residues in Kv channels also produces a disruption of inactivation (Cordero-Morales et al. 2011), suggesting that the conformational dynamics of the selectivity filter are conserved throughout the K^+ channel superfamily. In addition, studies with Shaker channels show that residues W434 (Perozo et al. 1993; Yang et al. 1997) and Y445 (Hoshi and Armstrong 2013; Armstrong and Hoshi 2014), along with D447 (Molina et al. 1997) and V449 (Lopez-Barneo et al. 1993), play a critical role in C-type inactivation. Nevertheless, other studies show that interfering with collapse of the selectivity filter does not abolish C-type inactivation (Devaraneni et al. 2013) and that water molecules lodged behind the selectivity filter may contribute to the kinetics of inactivation (Ostmeyer et al. 2013); thus, the mechanism of C-type inactivation remains elusive.

Inactivation Gating: N-Type Inactivation

Another way to regulate conductance through the pore is via N-type inactivation, or the ball-and-chain mechanism (also called fast inactivation) (Fig. 2C). This mechanism, similar to the bundle-crossing gating, is based on a physical obstruction to K^+ flux at the intracellular pore entrance. N-type inactivation, however, involves the binding of an autoinhibitory peptide that is either part of the channel (Hoshi et al. 1990) or part of an associated β subunit (Rettig et al. 1994) to the open pore. Interestingly, this type of inactivation can be abolished by the simple removal of the amino-terminal peptide (Hoshi et al. 1990), and the phenotype can then be rescued by addition of a soluble peptide of the same sequence (Zagotta et al. 1990). This mechanism is distinct from C-type inactivation, which persists in Shaker channels on deletion of the amino-terminal peptide (Hoshi et al. 1991). Mutational analysis provides evidence that the peptide binds at the cavity site (Zhou et al. 2001a). This mechanism of inactivation is found in several voltage-gated potassium channels. For example, the Shaker channel contains an amino-terminal region that is purported to bind directly to the cavity to block conductance (Demo and Yellen 1991). No direct crystallographic evidence of this mechanism is available to date, but a plausible model was proposed by MacKinnon and coworkers (Long et al. 2005).

Activation Gating at the Selectivity Filter

Several K^+ channels, such as the Ca^{2+}-activated K^+ channels (BK, IK, SK) have been proposed to gate at the selectivity filter (Bruening-Wright et al. 2002; Wilkens and Aldrich 2006; Klein et al. 2007; Tang et al. 2009; Zhou et al. 2011; Thompson and Begenisich 2012) instead of the bundle crossing, unlike KcsA and the voltage-gated K^+ channels mentioned above (Fig. 2B). Removal of the stimulus—Ca^{2+} in this case—presumably gates Ca^{2+}-activated K^+ channels by pinching the selectivity filter shut via a conformational change similar to that occurring during C-type inactivation. A constriction at the bundle crossing, insufficient for blocking K^+ flux through the channel, has been proposed to accompany the selectivity filter closing in BK channels, which are both Ca^{2+} and voltage-activated (Marty 1981; Latorre and Miller 1983; Latorre et al. 1989). Because of complexities in the gating of BK channels, the location of the gate is still uncertain. However, reinforcement in support of the BK-gating mechanism comes from MthK, a prokaryotic homolog of BK channels (Jiang et al. 2003a; Posson et al. 2013), which is gated solely by Ca^{2+}. In functional studies (blocker accessibility with single-channel recording and stopped flow fluorescence assays), MthK was found to gate at the selectivity filter with concomitant movement of intracellular helices (Posson et al. 2013, 2015), lending support to this type of gating mechanism. Conversely, a recent cryo-electron microscopy (cryo-EM) structure of slo2.2, a Na^+-activated K^+ channel that is a member of the *slo* family of channels that contain RCK domains, displays a bundle-crossing gate that is sterically closed, which supports the bundle-crossing gate model (Hite et al. 2015).

These gating mechanisms at the bundle crossing and the selectivity filter serve to modulate conductance in response to environmental stimuli. Therefore, the channels must be able to sense this stimulus and allosterically couple it to a change in gating. Voltage-gated K^+ channels sense changes in the transmembrane voltage and open or close in response. In the next section, we will explore how voltage-gated K^+ channels sense shifts in voltage across the cell membrane and how this is coupled to channel gating.

VOLTAGE SENSING IN K^+ CHANNELS

Voltage-gated ion channels have evolved to provide feedback loops to enhance or diminish electrical activity following a deviation from the resting potential in excitable cells. For example, initial changes in membrane potential occur as a result of opening of voltage-gated Na^+ or Ca^{2+} channels to allow cation influx, resulting in cell depolarization. This depolarization opens more Na^+ channels, allowing more Na^+ influx, which leads to more depolarization. This positive feedback loop leads to the rapid and massive membrane depolarization during the rising phase of the action potential, giving it its "all-or-none" character. Depolarization also opens voltage-gated K^+ channels, allowing K^+ efflux, which, together with rapid Na^+ channel inactivation, quickly repolarizes the membrane during the falling phase of the action potential. This enables the cell to propagate electrical signals quickly and with high efficiency. The sensitivity of these channels to changes in voltage led early researchers to hypothesize that activation of this conductance must be the result of the movement of charged particles across the membrane electrical field (Hodgkin and Huxley 1952f). Today, we know these "gating charges" are positively charged amino acids on the voltage-sensor domain of voltage-gated channels (Fig. 3A), and that movement of 12–14 e_o gating charges (where e_o is the charge of an electron) move across the membrane electric field during gating of the Shaker Kv channel (Schoppa et al. 1992; Aggarwal and MacKinnon 1996; Seoh et al. 1996). Extensive electrophysiological and structural studies have examined the voltage sensor of Kv channels and shed light on the mechanistic details of how these channels couple voltage sensing to gating.

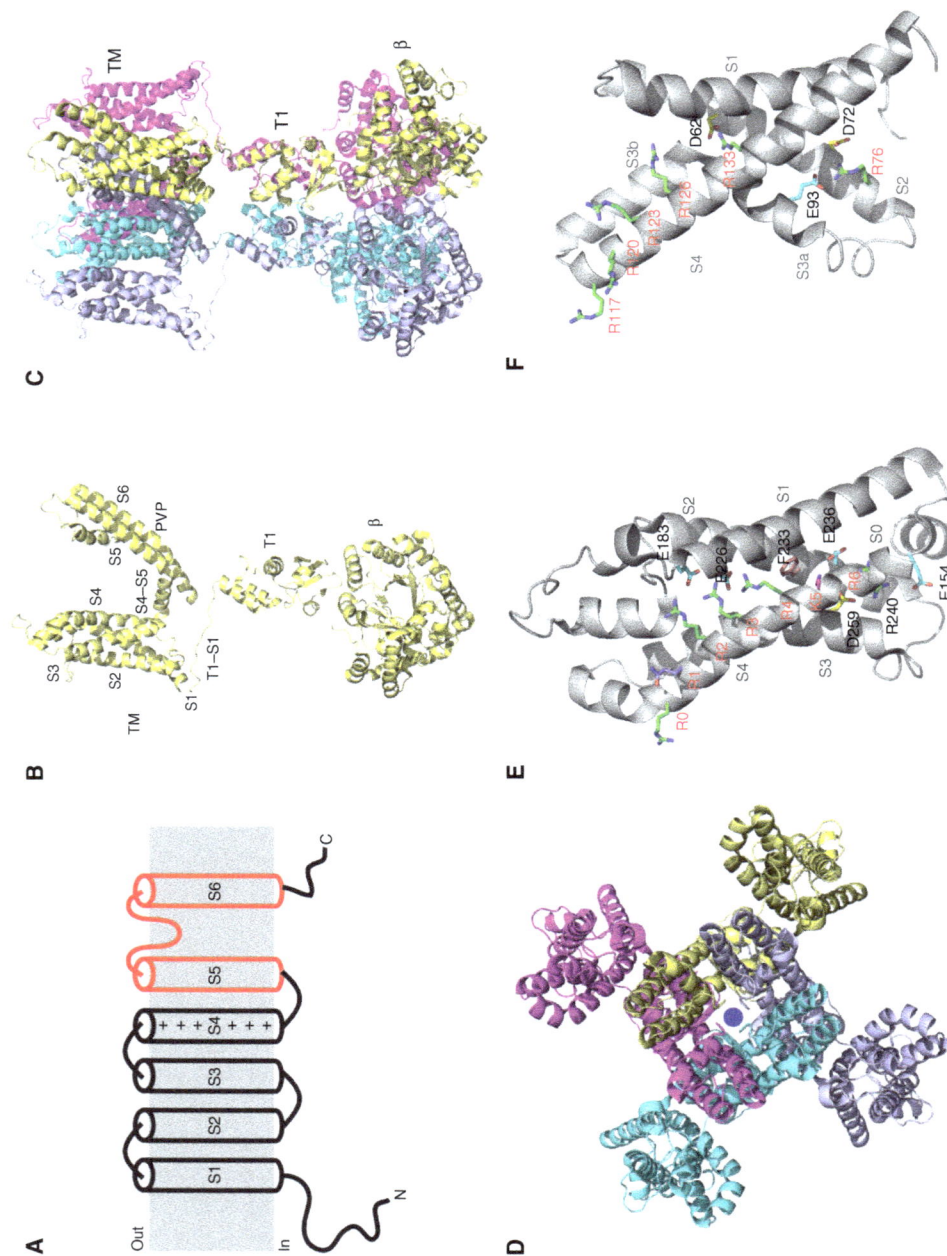

FIGURE 3. Structure of the voltage-dependent potassium channel. (*A*) Membrane topology of Kv channels consists of six transmembrane helices S1–S6, where both the amino and carboxyl termini are intracellular. Helices S1–S4 form the voltage sensor, whereas helices S5 and S6 comprise the channel pore and are analogous to TM1 and TM2 of KcsA and MthK. The S4 helix contains the conserved positively charged residues (plus signs) that are critical to the voltage-sensing mechanism. (*B*) The crystal structure of a chimeric channel comprised of the Kv1.2 with the voltage-sensor paddle of Kv2.1, crystallized in the presence of detergent and lipids (PDBID:2R9R). The β subunit is bound to the channel via the T1 intracellular domain. One subunit is depicted here and helices S1–S6, the S4–S5 linker, the PVP hinge, and the T1–S1 linker are indicated. (*C*) *Side* view of the channel tetramer β subunit complex. (*D*) *Top* view of the channel tetramer. Each subunit is differently colored, and the channel pore is visible in the *center* of the structure. The voltage-sensing domains are located *outside* of the channel pore. A potassium ion is shown as a blue sphere. (*E*) A detailed view of the voltage-sensor domain of the chimeric channel. Critical arginine residues are shown as green sticks and labeled as the putative gating charges R0–R4, followed by K5 and R6 on the S4 helix. Two clusters of negatively charged residues serve as countercharges to the gating charges on the external side (E154, E183, and E226) and the internal side (E154, E236, and D259) and are separated by the F233 on S2. (*F*) The 1.9 Å resolution structure of the isolated voltage-sensor domain of KvAP (PDBID:1ORQ). Critical arginine residues on the S4 helix are shown in green sticks. Gating charge R133 forms a salt bridge with D62 on the S2 helix, whereas R76, D72, and E93 form a salt-bridge network to bridge S2 with S3a.

 Cite this introduction as *Cold Spring Harb Perspect Biol* doi:10.1101/cshperspect.a029231

Before the elucidation of Kv channel crystal structures, many of the fundamental principles about voltage gating were established. Kv channels, like KcsA, are tetrameric but contain six transmembrane helices S1–S6 (Fig. 3A). Helices S1–S4 were found to form the voltage sensors, whereas S5–S6, homologous to TM1 and TM2 in KcsA and MthK, form the pore (Tempel et al. 1987; Kamb et al. 1988; Pongs et al. 1988; MacKinnon 1991). The prediction that the voltage-sensor region would contain multiple positively charged residues (Hodgkin and Huxley 1952f) enabled the identification of this region by sequence analysis (Noda et al. 1984). One region was found to be highly conserved among all voltage-gated ion channels (Na^+, Ca^{2+}, and K^+ channels) in the fourth transmembrane helix, S4 (Noda et al. 1984). This sequence motif contains an arginine or lysine at every third position, and mutagenesis studies have largely supported the hypothesis that these residues comprise the gating charges that move across the membrane to gate the channels (Aggarwal and MacKinnon 1996; Seoh et al. 1996). Indeed, the movement of these gating charges can be measured as transient electrical currents that are distinct from ion conduction (Armstrong and Bezanilla 1973; Schneider and Chandler 1973; Keynes and Rojas 1974). Voltage-dependent translocation of these charges was confirmed by cysteine-scanning mutagenesis and subsequent testing for voltage-dependent accessibility of cysteine-modifying agents (Larsson et al. 1996; Yang et al. 1996). Studies using fluorescently labeled channels also support conformational changes in S4 during gating (Mannuzzu et al. 1996; Cha and Bezanilla 1997; Bezanilla 2000). Collectively, these studies provided a model in which the S4 helices moved within the transmembrane region in response to changes in voltage. Most of these functional predictions were confirmed with the elucidation of the crystal structures of Kv channels at atomic resolution.

Several voltage-gated K^+ channel structures have been solved to date: KvAP from the thermophilic archaebacteria *Aeropyrum pernix* at 3.2 Å (Jiang et al. 2003a), Kv1.2 channel from rat (Long et al. 2005) to 2.9 Å, and a chimeric Kv1.2 channel containing a section of the voltage sensor from Kv2.1 (called the S3b–S4 voltage-sensor paddle) (Long et al. 2007) at 2.4 Å. These structures share certain features, including the K^+ channel selectivity filter, which is identical to that of KcsA, and the overall architecture of the pore. The pore is in an apparent open conformation because of the similarity of the bundle crossing to the open MthK structure rather than to the constricted KcsA structure (Fig. 3B–D, compare with Fig. 2, right panels). The voltage-sensor domains of the Kv1.2 and chimera Kv1.2/2.1 are very similar and are also similar to the isolated voltage sensor of KvAP (Fig. 3E,F). However, because the KvAP was crystallized in the absence of lipids and with a bound antibody fragment, its voltage sensor appears to be in a distorted conformation compared with Kv channels crystallized with lipids (Jiang et al. 2003a; Lee et al. 2005; Long et al. 2005, 2007). Our discussion will include the structural features of KvAP, Kv1.2, and the chimeric eukaryotic Kv1.2/2.1, but we will mainly refer to the structure of the chimera (shown in Fig. 3B–E) because of its high resolution.

The Structure of the Voltage Sensor in Voltage-Gated K^+ Channels

Structural studies of KvAP, Kv1.2, and Kv1.2/2.1 chimera show that the voltage-sensor domain (S1–S4) is conserved between eukaryotes and prokaryotes (Fig. 3E,F). S3 and S4 adopt an antiparallel arrangement forming a paddle that may move as a unit, and the voltage sensors are in a conformation that exposes the gating charges to the extracellular solution, providing further support for the channel being in the activated open state. The voltage sensors are unexpectedly located on the outside of the channel pore and are not tightly packed against the pore (Fig. 3D). This arrangement led to the hypothesis that the voltage sensors are independent and that the shielding of gating charges from the lipid as well as interactions with countercharges may occur within the voltage-sensing domain itself (Papazian et al. 1987; Long et al. 2007).

The high resolution of the voltage-sensor domain allows for detailed mapping of previous functional data onto the structure (Fig. 3E). The structure clearly shows the interactions between the conserved gating charge residues R1–R4 with their acidic countercharges on helices S0–S3a. The arginine residues make contacts with conserved Glu and Asp residues in two regions, at the intracellular and extracellular sides, separated by 15 Å and by a hydrophobic patch at F233 on helix S2 (Fig.

3E). This Phe residue is a highly conserved residue in Kv channels and is located near the midpoint of the membrane, but is absent in the KvAP voltage-sensing domain (Fig. 3F). In addition, the structure also shows a 3_{10}-helix hydrogen-bonding pattern in the lower half of the S4 helix, which elongates S4 by 5 Å compared with an α-helix, thus bringing it closer to the extracellular side and also directing the Arg and Lys residues at every third position to occupy the same face of the helix (Long et al. 2007).

This presumed open state structure shows that the first five arginines on the S4 helix (R0–R4) are close to the extracellular side and reside in an aqueous cleft (Fig 3E). The structure is consistent with the 12–14 e_o gating charges per channel previously calculated for Shaker (Schoppa et al. 1992; Aggarwal and MacKinnon 1996; Seoh et al. 1996), with R1–R4 each contributing a gating charge. Movement of R1–R4 toward the intracellular side would result in closing of the channel, as evidenced by cysteine accessibility experiments (Larsson et al. 1996), requiring a large translation of S4 toward the internal solution. Further support for this mechanism was provided by avidin–biotin accessibility studies on KvAP, which suggested that a 15 Å downward movement of S4 is required for closing (Ruta et al. 2005). However, other studies suggest that movement of the gating charges requires a smaller translation of S4 (Starace and Bezanilla 2001, 2004; Posson et al. 2005). The structural and functional data on Kv channels provide the starting point for understanding the molecular mechanism for how the S4 helix may move downward, as much as ∼15 Å, in the presence of a negative membrane voltage to close the channel. To date, we have no crystal structure of a closed Kv channel, with gating charges in contact with the intracellular solution. However, models of the resting state have emerged from both experimentally and computationally derived evidence (Yarov-Yarovoy et al. 2006, 2012; Campos et al. 2007; Pathak et al. 2007; DeCaen et al. 2008, 2009, 2011; Delemotte et al. 2010; Henrion et al. 2012; Jensen et al. 2012; Vargas et al. 2012), and advances have been made in obtaining closed state structures of the VSD alone (Li et al. 2014a,b) as discussed below.

Coupling between the Voltage Sensor and the Channel Gate

These structural studies suggest a hypothesis for how ion channel gating is coupled to the voltage-sensing mechanism (Fig. 4) (Long et al. 2007). The investigators propose that coupling is made possible by the PVP hinge region in eukaryotes (glycine in prokaryotes), which produces curvature in the S6 inner helix. This allows for the S4–S5 linker to cross over the top of the S6 inner helix. A large inward displacement of the S4 voltage sensor would push down on the S4–S5 linker, ultimately causing the S6 helix to constrict the pore and close the channel (Fig. 4B). Therefore, the interaction between the S4–S5 linker and the S6 helix appears to be crucial to coupling of the movements in the voltage sensor to gating. The critical arginine residues were proposed to move between two clusters of countercharges, separated by ∼15 Å, which are strategically placed on S0–S3 during gating. This mechanism is likened to a switch transition, requiring the arginines to cross a hydrophobic patch at F233 to close the channel. With these mechanistic details, the chimeric channel structure along with the KvAP and Kv1.2 structures could provide an emerging model to correlate functional data with the structural properties of a Kv channel.

Voltage-Sensor Movement between Up and Down Conformations

The observation that the voltage sensor is in a sense "floating" as a separate domain from the pore in the structure suggests that these domains can move independently of the remainder of the channel. Gating is retained when the voltage sensor is separated from the pore by cutting the S4–S5 linker in the KCNH family of Kv channels (Lorinczi et al. 2015). However, this is in direct contradiction to the putative role of the linker in the coupling mechanism suggested by the chimera structure (Long et al. 2007). KCNH channels lack the PVP motif, and, therefore, the coupling mechanism may differ in this channel family. The independent nature of the voltage-sensing domain allows its functional transfer to a nonvoltage dependent K^+ channel (Lu et al. 2001) or from one channel to another (Alabi et al. 2007). In addition, this explains the ability to express the voltage-sensing domain in the absence of an associated channel pore (Jiang et al. 2003b). Furthermore, Hv1 proton channels lack a canonical pore domain but retain a voltage-gating mechanism (Ramsey et al. 2006; Sasaki et al. 2006; Lee et al.

Cite this introduction as *Cold Spring Harb Perspect Biol* doi:10.1101/cshperspect.a029231

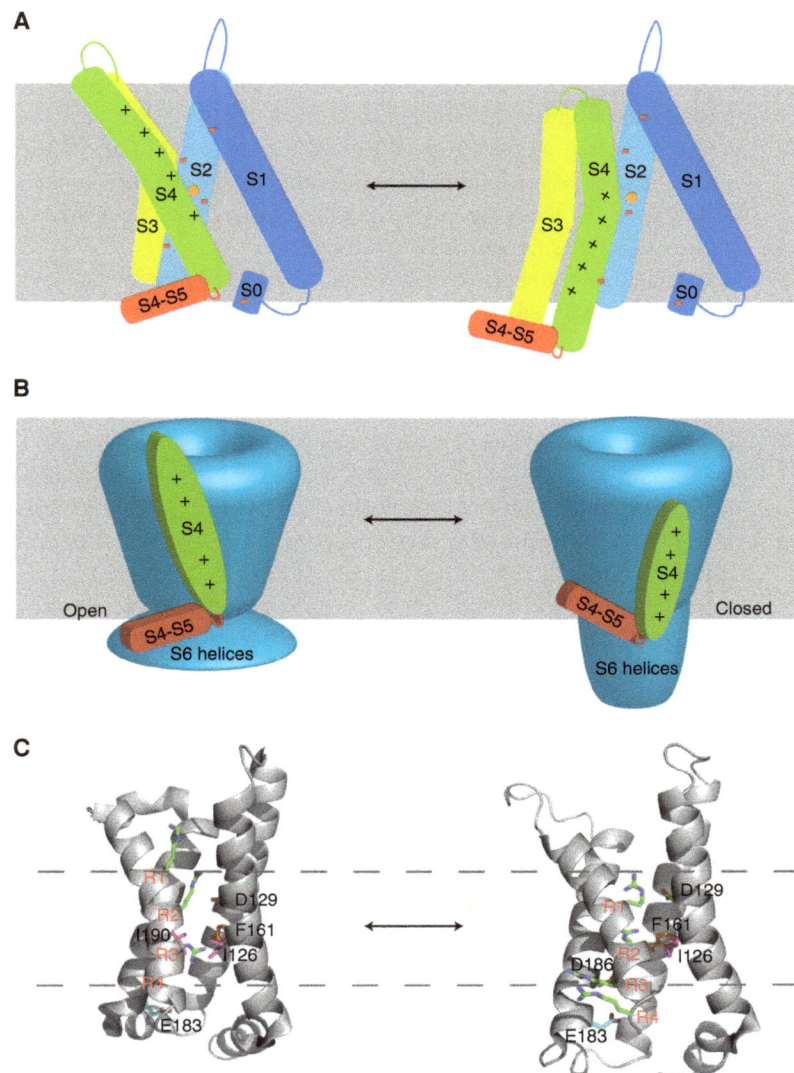

FIGURE 4. A model for coupling of voltage sensing to channel opening in Kv channels. (*A*) A cartoon depiction of the paddle model (Long et al. 2007) voltage-sensing mechanism. Helices S1–S4 and the S4–S5 linker are shown as cylinders of different colors. Gating charges are located on the S4 helix and represented by a black + sign. Counter-charges on S0 (a short α-helical region before S1), S1, and S2 are represented by a red dash. In the up or open conformation (*left*), S3 and S4 are located within the membrane and the gating charges are closer to the extracellular side, interacting with the external cluster of countercharges. The hydrophobic plug is represented by an orange circle on the S2 helix. A change in membrane potential would cause the S4 helix to move into the down conformation (*right*) with the gating charges now closer to the intracellular side and interacting with the internal cluster of countercharges. This displacement of S4 pushes down on the S4–S5 linker, tilting it toward the intracellular side, poising it to interact with the S6 helices to close the pore. (*B*) The proposed mechanism for coupling of voltage sensing to gating in Kv channels (Long et al. 2007). The channel pore and bundle crossing are represented in blue. S4 and the S4–S5 helix of the voltage-sensing domain are depicted in green and red, respectively. Based on the crystal structure, S4 is in the up conformation (*left*), and the S4–S5 linker rests on the S6 helices in the bundle crossing in the open state. Transition to the hypothetical closed state of the channel requires movement of S4 into the down conformation (*right*). This downward movement of the S4 helix pushes on the amino-terminal end of the S4–S5 helix, which tilts toward the intracellular side and pushes the S6 helices down into the closed state. (*C*) A comparison of the conformational states of the voltage-sensing phosphatase Ci-VSP. R217E Ci-VSP (*left*, PDBID:4G7V) shows the voltage sensor in the up conformation, whereas wild-type Ci-VSP (*right*, PDBID:4G80) shows the voltage sensor in the down conformation. The structures were aligned using the S1 helix for reference. Critical Arg residues R1–R4 are shown as green sticks. Countercharges are shown along with residues comprising the hydrophobic plug region. The gray dotted lines denote the position of R1 (*top*) and R4 (*bottom*) on the up conformation of R217E Ci-VSP for comparison to their positions on the down conformation of wild-type.

2009), providing further support for the independent nature of the voltage-sensing domain. Finally, this feature also enables voltage-sensing capabilities in nonchannel proteins, such as the voltage-dependent phosphatase enzyme, which contains the S1–S4 helices but does not contain a pore or function as a channel (Murata et al. 2005; Kohout et al. 2008). Therefore, the voltage-sensing domain has evolved to couple voltage-dependent conformational changes to a variety of physiological processes. Studies of both chimeric channels and the voltage-sensing phosphatases have provided considerable advancements in understanding the detailed mechanism of voltage-dependent conformational changes in Kv channels.

Because all the reported Kv crystal structures were in the open state, one obvious missing piece of this puzzle was a high-resolution structure of a Kv channel in the closed state to verify this mechanism and further expand the model. Recent studies of the lipid-dependence of KvAP and voltage-dependent phosphatases have filled in this gap (Li et al. 2014a,b). The difficulty in capturing the closed state of a Kv channel may be because this conformation requires an electric field with negative voltage, and the crystallization process occurs at 0 mV. All crystal structures may, thus, be in the activated/inactivated state with the voltage sensors in the up conformation. Therefore, it is necessary to bias toward the closed conformation using other methods. Previous experiments with voltage-gated channels showed that lipid–protein interactions could change the energetic landscape of membrane-reconstituted Kv channels, suggesting that transitions between the up and down conformations could require gating–charge interactions with surrounding phospholipids (Ramu et al. 2006; Schmidt et al. 2006; Xu et al. 2008; Zheng et al. 2011). In the absence of phospholipids, voltage sensors are "trapped" in the down conformation (Zheng et al. 2011). This was exploited in studies of the KvAP voltage-sensing domain, which used site-directed spin labeling and EPR to show that non-phosphate-containing lipids could trigger the reorientation of the S4 helix, producing a sensor in the down state at 0 mV (Li et al. 2014b). This study showed that in DOTAP lipids, S4 is displaced downward by only 2 Å (compared with the 15 Å predicted by Long et al. 2007) and also tilts away from S1 and S2 by 3 Å. This suggests a new "tilt-shift" model, which is supported by previous studies (Posson et al. 2005) and joins the sliding helix–helical screw model (Catterall 1986; Guy and Seetharamulu 1986), the tethered hydrophobic cation sensor (Jiang et al. 2003a,b; Ruta et al. 2005), and refocusing of the electric field around gating charges (Yang et al. 1996; Chanda and Bezanilla 2008) as possible models for voltage sensing. The KvAP study with DOTAP lipids shows that the down state of the sensor, either a closed state or an intermediate state not fully closed, can be analyzed in the absence of an electric field by biasing the conformational equilibrium in a lipid-dependent manner.

The crystal structure of a voltage-sensing domain in the down state was elucidated using a similar approach to bias the domain toward one state. The discovery of voltage-sensing phosphatases (VSPs) revealed that voltage-sensing domains are not limited to ion channels (Murata et al. 2005). The VSP from *Ciona intenstinalis* (Ci-VSP) contains a PTEN-related phosphoinositide phosphatase (Matsuda et al. 2011), which is under the control of a canonical voltage-sensing domain that has high sequence similarity to S1–S4 of Kv channels (Li et al. 2014a). These enzymes have a Q-V curve that is shifted to more positive potentials with half-maximal voltage near +60 mV (Murata et al. 2005); therefore, at 0 mV, the VSP should be in the down conformation. In addition, a mutation was identified, which results in a negative shift of 120 mV with a half-maximal voltage near −60 mV (Villalba-Galea et al. 2008; Li et al. 2014a), providing a simple strategy to obtain the active or up conformation at 0 mV. Therefore, crystal structures could be obtained of the same protein in the active and resting states with little manipulation. The structures show that, in contrast to Kv channels, there are no countercharges in S2 of Ci-VSP (Fig. 4C) (Li et al. 2014a). However, the hydrophobic patch is conserved and formed by a pair of isoleucines and a phenylalanine (Fig. 4C), similar to Kv channels, and functions as separating partition between the two sides of the membrane as well as a dielectric barrier to the permeation of water and ions. Alignment of the activated (voltage sensor up) and resting (voltage sensor down) structures shows displacement of the S4 helix by 5 Å downward with every gating charge in a one-click down fashion, along with a 60° clockwise rotation, differing from the 15 Å movement of S4 in the paddle model but supporting the sliding helix–helical screw model (Fig. 4C). Comparison of these structures to the Kv channel structures (Fig. 3E,F) suggests that there are mechanistic variances in voltage sensing owing to the different positions of the gating charges and also their interactions with the coevolved countercharges. These studies with the voltage-dependent phosphatases, in conjunction with studies of the lipid-dependent conformational states of KvAP (Li et al. 2014a,b), foster a greater understanding of the structural and mechanistic details of the active and resting states.

Molecular Dynamics Simulations of the Voltage-Sensor Movement

To fully understand how voltage sensors work, it is imperative to elucidate the myriad transition states that connect these two conformations. In silico studies using molecular dynamics (MDs) and Rosetta modeling techniques fill this gap by theorizing the conformational states that have not been elucidated by crystallography, using functional data as constraints to model voltage-sensor movement and channel gating. Available MD simulations to date using the known crystal structures as starting points are in agreement with most of the functional data (Jensen et al. 2012; Vargas et al. 2012). These studies tend to dispute the paddle mechanism although providing support for the classical sliding helix–helical screw model. A major caveat of these studies, however, is the requirement to use large hyperpolarizing potentials (750 mV) to allow for the voltage-dependent transition toward the resting state to occur within accessible computing time. The resulting resting state models show the S4 helix to be rotated and translated inward along the main axis relative to the up conformation observed in the crystal structures, whereas the S1 and S2 helices retain their position (Jensen et al. 2012). In the resting state, R1 is positioned between the countercharges E1 and E2 on the S2 helix, above the Phe residue (Fig. 3E). This positions positive gating charges to form salt bridges with countercharges on S1–S3. Kinetic models of voltage gating in Shaker (Islas and Sigworth 2001; Asamoah et al. 2003; Starace and Bezanilla 2004) are also supported by MD simulations showing that the ion conducting pore closes before any of the VSDs have transitioned to the most stable resting state conformation (Jensen et al. 2012), which is further validated by crystal structures of the bacterial Nav channel (Payandeh et al. 2011, 2012; Zhang et al. 2012). The agreement of these MD studies with structural and functional data provides a sense of validation; however, no experimental data has been obtained at 750 mV, and, therefore, dynamic properties of gating could be somewhat skewed in these studies.

CONCLUSION

The evolution of a relatively small helical domain that can sense changes in voltage across the cell membrane and relay that information to control gating of potassium channels is a feat of nature. Understanding how the voltage-sensing domain can move in response to a change in membrane potential and couple this to conformational changes at the pore is crucial to deciphering the sophisticated mechanisms of gating in Kv channels. Experimental studies with a wide variety of techniques, such as electrophysiology and crystallography, together with in silico MD simulations and computational modeling studies, provide us with an enriched understanding of how these machines work. Although a clearer picture of the mechanistic details of voltage-dependent gating is emerging, many questions linger about these fascinating ion channels.

REFERENCES

Aggarwal SK, MacKinnon R. 1996. Contribution of the S4 segment to gating charge in the Shaker K+ channel. *Neuron* **16:** 1169–1177.

Alabi AA, Bahamonde MI, Jung HJ, Kim JI, Swartz KJ. 2007. Portability of paddle motif function and pharmacology in voltage sensors. *Nature* **450:** 370–375.

Alam A, Jiang Y. 2009a. High-resolution structure of the open NaK channel. *Nat Struct Mol Biol* **16:** 30–34.

Alam A, Jiang Y. 2009b. Structural analysis of ion selectivity in the NaK channel. *Nat Struct Mol Biol* **16:** 35–41.

Almers W, McCleskey EW. 1984. Non-selective conductance in calcium channels of frog muscle: Calcium selectivity in a single-file pore. *J Physiol* **353:** 585–608.

Armstrong CM, Bezanilla F. 1973. Currents related to movement of the gating particles of the sodium channels. *Nature* **242:** 459–461.

Armstrong CM, Hoshi T. 2014. K$^+$ channel gating: C-type inactivation is enhanced by calcium or lanthanum outside. *J Gen Physiol* **144**: 221–230.

Asamoah OK, Wuskell JP, Loew LM, Bezanilla F. 2003. A fluorometric approach to local electric field measurements in a voltage-gated ion channel. *Neuron* **37**: 85–97.

Baukrowitz T, Yellen G. 1995. Modulation of K$^+$ current by frequency and external [K$^+$]: A tale of two inactivation mechanisms. *Neuron* **15**: 951–960.

Begenisich TB, Cahalan MD. 1980. Sodium channel permeation in squid axons. II: Non-independence and current-voltage relations. *J Physiol* **307**: 243–257.

Bernstein J. 1902. Untersuchungen zur thermodynamik der bioelektrischen Ströme [Studies on thermodynamics of bioelectrical currents]. *Pflügers Arch* **92**: 521–562.

Bezanilla F. 2000. The voltage sensor in voltage-dependent ion channels. *Physiol Rev* **80**: 555–592.

Bezanilla F, Armstrong CM. 1972. Negative conductance caused by entry of sodium and cesium ions into the potassium channels of squid axons. *J Gen Physiol* **60**: 588–608.

Bruening-Wright A, Schumacher MA, Adelman JP, Maylie J. 2002. Localization of the activation gate for small conductance Ca^{2+}-activated K$^+$ channels. *J Neurosci* **22**: 6499–6506.

Cahalan M, Begenisich T. 1976. Sodium channel selectivity. Dependence on internal permeant ion concentration. *J Gen Physiol* **68**: 111–125.

Campos FV, Chanda B, Roux B, Bezanilla F. 2007. Two atomic constraints unambiguously position the S4 segment relative to S1 and S2 segments in the closed state of Shaker K channel. *Proc Natl Acad Sci* **104**: 7904–7909.

Catterall WA. 1986. Molecular properties of voltage-sensitive sodium channels. *Annu Rev Biochem* **55**: 953–985.

Cha A, Bezanilla F. 1997. Characterizing voltage-dependent conformational changes in the Shaker K$^+$ channel with fluorescence. *Neuron* **19**: 1127–1140.

Chanda B, Bezanilla F. 2008. A common pathway for charge transport through voltage-sensing domains. *Neuron* **57**: 345–351.

Chandler WK, Meves H. 1965. Voltage clamp experiments on internally perfused giant axons. *J Physiol* **180**: 788–820.

Cheng WW, McCoy JG, Thompson AN, Nichols CG, Nimigean CM. 2011. Mechanism for selectivity-inactivation coupling in KcsA potassium channels. *Proc Natl Acad Sci* **108**: 5272–5277.

Cordero-Morales JF, Cuello LG, Zhao Y, Jogini V, Cortes DM, Roux B, Perozo E. 2006. Molecular determinants of gating at the potassium-channel selectivity filter. *Nat Struct Mol Biol* **13**: 311–318.

Cordero-Morales JF, Jogini V, Chakrapani S, Perozo E. 2011. A multipoint hydrogen-bond network underlying KcsA C-type inactivation. *Biophys J* **100**: 2387–2393.

Cortes DM, Cuello LG, Perozo E. 2001. Molecular architecture of full-length KcsA: Role of cytoplasmic domains in ion permeation and activation gating. *J Gen Physiol* **117**: 165–180.

Cuello LG, Jogini V, Cortes DM, Pan AC, Gagnon DG, Dalmas O, Cordero-Morales JF, Chakrapani S, Roux B, Perozo E. 2010. Structural basis for the coupling between activation and inactivation gates in K$^+$ channels. *Nature* **466**: 272–275.

DeCaen PG, Yarov-Yarovoy V, Zhao Y, Scheuer T, Catterall WA. 2008. Disulfide locking a sodium channel voltage sensor reveals ion pair formation during activation. *Proc Natl Acad Sci* **105**: 15142–15147.

DeCaen PG, Yarov-Yarovoy V, Sharp EM, Scheuer T, Catterall WA. 2009. Sequential formation of ion pairs during activation of a sodium channel voltage sensor. *Proc Natl Acad Sci* **106**: 22498–22503.

DeCaen PG, Yarov-Yarovoy V, Scheuer T, Catterall WA. 2011. Gating charge interactions with the S1 segment during activation of a Na$^+$ channel voltage sensor. *Proc Natl Acad Sci* **108**: 18825–18830.

del Camino D, Yellen G. 2001. Tight steric closure at the intracellular activation gate of a voltage-gated K$^+$ channel. *Neuron* **32**: 649–656.

del Camino D, Holmgren M, Liu Y, Yellen G. 2000. Blocker protection in the pore of a voltage-gated K$^+$ channel and its structural implications. *Nature* **403**: 321–325.

Delemotte L, Treptow W, Klein ML, Tarek M. 2010. Effect of sensor domain mutations on the properties of voltage-gated ion channels: Molecular dynamics studies of the potassium channel Kv1.2. *Biophys J* **99**: L72–L74.

Demo SD, Yellen G. 1991. The inactivation gate of the Shaker K$^+$ channel behaves like an open-channel blocker. *Neuron* **7**: 743–753.

Devaraneni PK, Komarov AG, Costantino CA, Devereaux JJ, Matulef K, Valiyaveetil FI. 2013. Semisynthetic K$^+$ channels show that the constricted conformation of the selectivity filter is not the C-type inactivated state. *Proc Natl Acad Sci* **110**: 15698–15703.

Doyle DA, Morais Cabral J, Pfuetzner RA, Kuo A, Gulbis JM, Cohen SL, Chait BT, MacKinnon R. 1998. The structure of the potassium channel: Molecular basis of K$^+$ conduction and selectivity. *Science* **280**: 69–77.

Eisenman W. 1962. A two-way affair. *Science* **136**: 182.

Eisenman G, Latorre R, Miller C. 1986. Multi-ion conduction and selectivity in the high-conductance Ca^{2+}-activated K$^+$ channel from skeletal muscle. *Biophys J* **50**: 1025–1034.

Goldman DE. 1943. Potential, impedance, and rectification in membranes. *J Gen Physiol* **27**: 37–60.

Guy HR, Seetharamulu P. 1986. Molecular model of the action potential sodium channel. *Proc Natl Acad Sci* **83**: 508–512.

Hagiwara S, Takahashi K. 1974. Mechanism of anion permeation through the muscle fibre membrane of an elasmobranch fish, *Taeniura lymma*. *J Physiol* **238**: 109–127.

Heckmann K. 1965a. Zur theorie der "single file"-diffusion. Part I. *Z Phys Chem* **44**: 184–203.

Heckmann K. 1965b. Zur theorie der "single file"-diffusion. Part II. *Z Phys Chem* **46**: 1–25.

Heckmann K. 1968. Zur theorie der "single file"-diffusion. Part III: Sigmoide Konzantratonsabhangigkeit unidirectionaler flusse bei "single file" diffusion. *Z Phys Chem* **58**: 210–219.

Heckmann K. 1972. *Single-file diffusion*. Plenum, New York.

Heginbotham L, MacKinnon R. 1993. Conduction properties of the cloned *Shaker* K$^+$ channel. *Biophys J* **65**: 2089–2096.

Heginbotham L, Abramson T, MacKinnon R. 1992. A functional connection between the pores of distantly related ion channels as revealed by mutant K$^+$ channels. *Science* **258**: 1152–1155.

Heginbotham L, Lu Z, Abramson T, MacKinnon R. 1994. Mutations in the K$^+$ channel signature sequence. *Biophys J* **66**: 1061–1067.

Heginbotham L, LeMasurier M, Kolmakova-Partensky L, Miller C. 1999. Single streptomyces lividans K$^+$ channels: Functional asymmetries and sidedness of proton activation. *J Gen Physiol* **114**: 551–560.

Henrion U, Zumhagen S, Steinke K, Strutz-Seebohm N, Stallmeyer B, Lang F, Schulze-Bahr E, Seebohm G. 2012. Overlapping cardiac phenotype associated with a familial mutation in the voltage sensor of the KCNQ1 channel. *Cell Physiol Biochem* **29**: 809–818.

Hess P, Tsien RW. 1984. Mechanism of ion permeation through calcium channels. *Nature* **309**: 453–456.

Hille B. 1973. Potassium channels in myelinated nerve. Selective permeability to small cations. *J Gen Physiol* **61**: 669–686.

Hille B. 1975a. Ionic selectivity of Na and K channels of nerve membranes. *Membranes* **3**: 255–323.

Hille B. 1975b. Ionic selectivity, saturation, and block in sodium channels. A four-barrier model. *J Gen Physiol* **66**: 535–560.

Hille B. 2001. *Ion channels of excitable membranes*. Sinauer Associates, Sunderland, MA.

Hille B, Schwarz W. 1978. Potassium channels as multi-ion single-file pores. *J Gen Physiol* **72**: 409–442.

Hite RK, Yuan P, Li Z, Hsuing Y, Walz T, MacKinnon R. 2015. Cryo-electron microscopy structure of the Slo2.2 Na-activated K channel. *Nature* **527**: 198–203.

Hodgkin AL, Huxley AF. 1945. Resting and action potentials in single nerve fibres. *J Physiol* **104**: 176–195.

Hodgkin AL, Huxley AF. 1946. Potassium leakage from an active nerve fibre. *Nature* **158**: 376.

Hodgkin AL, Huxley AF. 1947. Potassium leakage from an active nerve fibre. *J Physiol* **106**: 341–367.

Hodgkin AL, Huxley AF. 1952a. The components of membrane conductance in the giant axon of *Loligo*. *J Physiol* **116**: 473–496.

Hodgkin AL, Huxley AF. 1952b. Currents carried by sodium and potassium ions through the membrane of the giant axon of *Loligo*. *J Physiol* **116**: 449–472.

Hodgkin AL, Huxley AF. 1952c. The dual effect of membrane potential on sodium conductance in the giant axon of *Loligo*. *J Physiol* **116**: 497–506.

Hodgkin AL, Huxley AF. 1952d. Movement of sodium and potassium ions during nervous activity. *Cold Spring Harb Symp Quant Biol* **17**: 43–52.

Hodgkin AL, Huxley AF. 1952e. Propagation of electrical signals along giant nerve fibers. *Proc R Soc Lond B Biol Sci* **140**: 177–183.

Hodgkin AL, Huxley AF. 1952f. A quantitative description of membrane current and its application to conduction and excitation in nerve. *J Physiol* **117**: 500–544.

Hodgkin AL, Katz B. 1949. The effect of sodium ions on the electrical activity of giant axon of the squid. *J Physiol* **108**: 37–77.

Hodgkin AL, Keynes RD. 1955. The potassium permeability of a giant nerve fibre. *J Physiol* **128**: 61–88.

Hodgkin AL, Huxley AF, Katz B. 1952. Measurement of current-voltage relations in the membrane of the giant axon of *Loligo*. *J Physiol* **116**: 424–448.

Holmgren M, Shin KS, Yellen G. 1998. The activation gate of a voltage-gated K^+ channel can be trapped in the open state by an intersubunit metal bridge. *Neuron* **21**: 617–621.

Hoshi T, Armstrong CM. 2013. C-type inactivation of voltage-gated K^+ channels: Pore constriction or dilation? *J Gen Physiol* **141**: 151–160.

Hoshi T, Zagotta WN, Aldrich RW. 1990. Biophysical and molecular mechanisms of *Shaker* potassium channel inactivation. *Science* **250**: 533–538.

Hoshi T, Zagotta WN, Aldrich RW. 1991. Two types of inactivation in Shaker K^+ channels: Effects of alterations in the carboxy-terminal region. *Neuron* **7**: 547–556.

Islas LD, Sigworth FJ. 2001. Electrostatics and the gating pore of Shaker potassium channels. *J Gen Physiol* **117**: 69–89.

Jensen MO, Jogini V, Borhani DW, Leffler AE, Dror RO, Shaw DE. 2012. Mechanism of voltage gating in potassium channels. *Science* **336**: 229–233.

Jiang Y, Lee A, Chen J, Cadene M, Chait BT, MacKinnon R. 2002. Crystal structure and mechanism of a calcium-gated potassium channel. *Nature* **417**: 515–522.

Jiang Y, Lee A, Chen J, Ruta V, Cadene M, Chait BT, MacKinnon R. 2003a. X-ray structure of a voltage-dependent K^+ channel. *Nature* **423**: 33–41.

Jiang Y, Ruta V, Chen J, Lee A, MacKinnon R. 2003b. The principle of gating charge movement in a voltage-dependent K^+ channel. *Nature* **423**: 42–48.

Kamb A, Tseng-Crank J, Tanouye MA. 1988. Multiple products of the *Drosophila Shaker* gene may contribute to potassium channel diversity. *Neuron* **1**: 421–430.

Keynes RD, Rojas E. 1974. Kinetics and steady-state properties of the charged system controlling sodium conductance in the squid giant axon. *J Physiol* **239**: 393–434.

Klein H, Garneau L, Banderali U, Simoes M, Parent L, Sauve R. 2007. Structural determinants of the closed KCa3.1 channel pore in relation to channel gating: Results from a substituted cysteine accessibility analysis. *J Gen Physiol* **129**: 299–315.

Kohout SC, Ulbrich MH, Bell SC, Isacoff EY. 2008. Subunit organization and functional transitions in Ci-VSP. *Nat Struct Mol Biology* **15**: 106–108.

Kurata HT, Fedida D. 2006. A structural interpretation of voltage-gated potassium channel inactivation. *Prog Biophys Mol Biol* **92**: 185–208.

Larsson HP, Baker OS, Dhillon DS, Isacoff EY. 1996. Transmembrane movement of the *Shaker* K^+ channel S4. *Neuron* **16**: 387–397.

Latorre R, Miller C. 1983. Conduction and selectivity in potassium channels. *J Membr Biol* **71**: 11–30.

Latorre R, Oberhauser A, Labarca P, Alvarez O. 1989. Varieties of calcium-activated potassium channels. *Annu Rev Physiol* **51**: 385–399.

Lee SY, Lee A, Chen J, MacKinnon R. 2005. Structure of the KvAP voltage-dependent K^+ channel and its dependence on the lipid membrane. *Proc Natl Acad Sci* **102**: 15441–15446.

Lee SY, Letts JA, MacKinnon R. 2009. Functional reconstitution of purified human Hv1 H^+ channels. *J Mol Biol* **387**: 1055–1060.

LeMasurier M, Heginbotham L, Miller C. 2001. KcsA: It's a potassium channel. *J Gen Physiol* **118**: 303–314.

Li Q, Wanderling S, Paduch M, Medovoy D, Singharoy A, McGreevy R, Villalba-Galea CA, Hulse RE, Roux B, Schulten K, et al. 2014a. Structural mechanism of voltage-dependent gating in an isolated voltage-sensing domain. *Nat Struct Mol Biol* **21**: 244–252.

Li Q, Wanderling S, Sompornpisut P, Perozo E. 2014b. Structural basis of lipid-driven conformational transitions in the KvAP voltage-sensing domain. *Nat Struct Mol Biol* **21**: 160–166.

Liu Y, Holmgren M, Jurman ME, Yellen G. 1997. Gated access to the pore of a voltage-dependent K^+ channel. *Neuron* **19**: 175–184.

Liu S, Bian X, Lockless SW. 2012. Preferential binding of K^+ ions in the selectivity filter at equilibrium explains high selectivity of K^+ channels. *J Gen Physiol* **140**: 671–679.

Lockless SW, Zhou M, MacKinnon R. 2007. Structural and thermodynamic properties of selective ion binding in a K^+ channel. *PLoS Biol* **5**: e121.

Long SB, Campbell EB, Mackinnon R. 2005. Crystal structure of a mammalian voltage-dependent *Shaker* family K^+ channel. *Science* **309**: 897–903.

Long SB, Tao X, Campbell EB, MacKinnon R. 2007. Atomic structure of a voltage-dependent K^+ channel in a lipid membrane-like environment. *Nature* **450**: 376–382.

Lopez-Barneo J, Hoshi T, Heinemann SH, Aldrich RW. 1993. Effects of external cations and mutations in the pore region on C-type inactivation of Shaker potassium channels. *Receptor Channel* **1**: 61–71.

Lorinczi E, Gomez-Posada JC, de la Pena P, Tomczak AP, Fernandez-Trillo J, Leipscher U, Stuhmer W, Barros F, Pardo LA. 2015. Voltage-dependent gating of KCNH potassium channels lacking a covalent link between voltage-sensing and pore domains. *Nat Commun* **6**: 1–14.

Lu Z, Klem AM, Ramu Y. 2001. Ion conduction pore is conserved among potassium channels. *Nature* **413**: 809–813.

MacKinnon R. 1991. Determination of the subunit stoichiometry of a voltage-activated potassium channel. *Nature* **350**: 232–235.

Mannuzzu LM, Moronne MM, Isacoff EY. 1996. Direct physical measure of conformational rearrangement underlying potassium channel gating. *Science* **271**: 213–216.

Marty A. 1981. Ca-dependent K channels with large unitary conductance in chromaffin cell membranes. *Nature* **291**: 497–500.

Matsuda M, Takeshita K, Kurokawa T, Sakata S, Suzuki M, Yamashita E, Okamura Y, Nakagawa A. 2011. Crystal structure of the cytoplasmic phosphatase and tensin homolog (PTEN)-like region of Ciona intestinalis voltage-sensing phosphatase provides insight into substrate specificity and redox regulation of the phosphoinositide phosphatase activity. *J Biol Chem* **286**: 23368–23377.

McCoy JG, Nimigean CM. 2012. Structural correlates of selectivity and inactivation in potassium channels. *Biochim Biophys Acta* **1818**: 272–285.

Molina A, Castellano AG, Lopez-Barneo J. 1997. Pore mutations in *Shaker* K^+ channels distinguish between the sites of tetraethylammonium blockade and C-type inactivation. *J Physiol* **499**: 361–367.

Mullins LJ. 1959. An analysis of conductance changes in squid axon. *J Gen Physiol* **42**: 1013–1035.

Murata Y, Iwasaki H, Sasaki M, Inaba K, Okamura Y. 2005. Phosphoinositide phosphatase activity coupled to an intrinsic voltage sensor. *Nature* **435**: 1239–1243.

Neher E, Sakmann B. 1975. Voltage-dependence of drug-induced conductance in frog neuromuscular junction. *Proc Natl Acad Sci* **72**: 2140–2144.

Neyton J, Miller C. 1988a. Discrete Ba^{2+} block as a probe of ion occupancy and pore structure in the high-conductance Ca^{2+}-activated K^+ channel. *J Gen Physiol* **92**: 569–586.

Neyton J, Miller C. 1988b. Potassium blocks barium permeation through a calcium-activated potassium channel. *J Gen Physiol* **92**: 549–567.

Noda M, Takahashi H, Tanabe T, Toyosato M, Furutani Y, Hirose T, Asai M, Inayama S, Miyata T, Numa S. 1982. Primary structure of α-subunit precursor of *Torpedo californica* acetylcholine receptor deduced from cDNA sequence. *Nature* **299**: 793–797.

Noda M, Takahashi H, Tanabe T, Toyosato M, Kikyotani S, Hirose T, Asai M, Takashima H, Inayama S, Miyata T, et al. 1983. Primary structures of β- and δ-subunit precursors of *Torpedo californica* acetylcholine receptor deduced from cDNA sequences. *Nature* **301**: 251–255.

Noda M, Shimizu S, Tanabe T, Takai T, Kayano T, Ikeda T, Takahashi H, Nakayama H, Kanaoka Y, Minamino N, et al. 1984. Primary structure of *Electrophorus electricus* sodium channel deduced from cDNA sequence. *Nature* **312**: 121–127.

Noskov SY, Berneche S, Roux B. 2004. Control of ion selectivity in potassium channels by electrostatic and dynamic properties of carbonyl ligands. *Nature* **431**: 830–834.

Ostmeyer J, Chakrapani S, Pan AC, Perozo E, Roux B. 2013. Recovery from slow inactivation in K^+ channels is controlled by water molecules. *Nature* **501**: 121–124.

Papazian DM, Schwarz TL, Tempel BL, Jan YN, Jan LY. 1987. Cloning of genomic and complementary DNA from *Shaker*, a putative potassium channel gene from *Drosophila*. *Science* **237**: 749–753.

Pardo LA, Heinemann SH, Terlau H, Ludewig U, Lorra C, Pongs O, Stuhmer W. 1992. Extracellular K$^+$ specifically modulates a rat brain K$^+$ channel. *Proc Natl Acad Sci* **89**: 2466–2470.

Pathak MM, Yarov-Yarovoy V, Agarwal G, Roux B, Barth P, Kohout S, Tombola F, Isacoff EY. 2007. Closing in on the resting state of the Shaker K$^+$ channel. *Neuron* **56**: 124–140.

Payandeh J, Scheuer T, Zheng N, Catterall WA. 2011. The crystal structure of a voltage-gated sodium channel. *Nature* **475**: 353–358.

Payandeh J, Gamal El-Din TM, Scheuer T, Zheng N, Catterall WA. 2012. Crystal structure of a voltage-gated sodium channel in two potentially inactivated states. *Nature* **486**: 135–139.

Perozo E, MacKinnon R, Bezanilla F, Stefani E. 1993. Gating currents from a nonconducting mutant reveal open-closed conformations in *Shaker* K$^+$ channels. *Neuron* **11**: 353–358.

Pongs O, Kecskemethy N, Muller R, Krah-Jentgens I, Baumann A, Kiltz HH, Canal I, Llamazares S, Ferrus A. 1988. Shaker encodes a family of putative potassium channel proteins in the nervous system of *Drosophila*. *EMBO J* **7**: 1087–1096.

Posson DJ, Ge P, Miller C, Bezanilla F, Selvin PR. 2005. Small vertical movement of a K$^+$ channel voltage sensor measured with luminescence energy transfer. *Nature* **436**: 848–851.

Posson DJ, McCoy JG, Nimigean CM. 2013. The voltage-dependent gate in MthK potassium channels is located at the selectivity filter. *Nat Struct Mol Biol* **20**: 159–166.

Posson DJ, Rusinova R, Andersen OS, Nimigean CM. 2015. Calcium ions open a selectivity filter gate during activation of the MthK potassium channel. *Nat Commun* **6**: 8342.

Ramsey IS, Moran MM, Chong JA, Clapham DE. 2006. A voltage-gated proton-selective channel lacking the pore domain. *Nature* **440**: 1213–1216.

Ramu Y, Xu Y, Lu Z. 2006. Enzymatic activation of voltage-gated potassium channels. *Nature* **442**: 696–699.

Rettig J, Heinemann SH, Wunder F, Lorra C, Parcej DN, Dolly JO, Pongs O. 1994. Inactivation properties of voltage-gated K$^+$ channels altered by presence of β-subunit. *Nature* **369**: 289–294.

Ruta V, Chen J, MacKinnon R. 2005. Calibrated measurement of gating-charge arginine displacement in the KvAP voltage-dependent K$^+$ channel. *Cell* **123**: 463–475.

Sandblom J, Eisenman G, Neher E. 1977. Ionic selectivity, saturation and block in gramicidin A channels. I: Theory for the electrical properties of ion selective channels having two pairs of binding sites and multiple conductance states. *J Membr Biol* **31**: 383–347.

Sasaki M, Takagi M, Okamura Y. 2006. A voltage sensor-domain protein is a voltage-gated proton channel. *Science* **312**: 589–592.

Schmidt D, Jiang QX, MacKinnon R. 2006. Phospholipids and the origin of cationic gating charges in voltage sensors. *Nature* **444**: 775–779.

Schneider MF, Chandler WK. 1973. Voltage dependent charge movement of skeletal muscle: A possible step in excitation-contraction coupling. *Nature* **242**: 244–246.

Schoppa NE, McCormack K, Tanouye MA, Sigworth FJ. 1992. The size of gating charge in wild-type and mutant Shaker potassium channels. *Science* **255**: 1712–1715.

Seoh SA, Sigg D, Papazian DM, Bezanilla F. 1996. Voltage-sensing residues in the S2 and S4 segments of the *Shaker* K$^+$ channel. *Neuron* **16**: 1159–1167.

Starace DM, Bezanilla F. 2001. Histidine scanning mutagenesis of basic residues of the S4 segment of the *Shaker* K$^+$ channel. *J Gen Physiol* **117**: 469–490.

Starace DM, Bezanilla F. 2004. A proton pore in a potassium channel voltage sensor reveals a focused electric field. *Nature* **427**: 548–553.

Takeuchi K, Takahashi H, Kawano S, Shimada I. 2007. Identification and characterization of the slowly exchanging pH-dependent conformational rearrangement in KcsA. *J Biol Chem* **282**: 15179–15186.

Tang QY, Zeng XH, Lingle CJ. 2009. Closed-channel block of BK potassium channels by bbTBA requires partial activation. *J Gen Physiol* **134**: 409–436.

Tempel BL, Papazian DM, Schwarz TL, Jan YN, Jan LY. 1987. Sequence of a probable potassium channel component encoded at *Shaker* locus of *Drosophila*. *Science* **237**: 770–775.

Thompson J, Begenisich T. 2012. Selectivity filter gating in large-conductance Ca^{2+}-activated K$^+$ channels. *J Gen Physiol* **139**: 235–244.

Thompson AN, Posson DJ, Parsa PV, Nimigean CM. 2008. Molecular mechanism of pH sensing in KcsA potassium channels. *Proc Natl Acad Sci* **105**: 6900–6905.

Thompson AN, Kim I, Panosian TD, Iverson TM, Allen TW, Nimigean CM. 2009. Mechanism of potassium-channel selectivity revealed by Na$^+$ and Li$^+$ binding sites within the KcsA pore. *Nat Struct Mol Biol* **16**: 1317–1324.

Vargas E, Yarov-Yarovoy V, Khalili-Araghi F, Catterall WA, Klein ML, Tarek M, Lindahl E, Schulten K, Perozo E, Bezanilla F, et al. 2012. An emerging consensus on voltage-dependent gating from computational modeling and molecular dynamics simulations. *J Gen Physiol* **140**: 587–594.

Villalba-Galea CA, Sandtner W, Starace DM, Bezanilla F. 2008. S4-based voltage sensors have three major conformations. *Proc Natl Acad Sci* **105**: 17600–17607.

Wilkens CM, Aldrich RW. 2006. State-independent block of BK channels by an intracellular quaternary ammonium. *J Gen Physiol* **128**: 347–364.

Xu Y, Ramu Y, Lu Z. 2008. Removal of phospho-head groups of membrane lipids immobilizes voltage sensors of K$^+$ channels. *Nature* **451**: 826–829.

Yang N, George AL Jr, Horn R. 1996. Molecular basis of charge movement in voltage-gated sodium channels. *Neuron* **16**: 113–122.

Yang Y, Yan Y, Sigworth FJ. 1997. How does the W434F mutation block current in *Shaker* potassium channels? *J Gen Physiol* **109**: 779–789.

Yarov-Yarovoy V, Baker D, Catterall WA. 2006. Voltage sensor conformations in the open and closed states in ROSETTA structural models of K$^+$ channels. *Proc Natl Acad Sci* **103**: 7292–7297.

Yarov-Yarovoy V, DeCaen PG, Westenbroek RE, Pan CY, Scheuer T, Baker D, Catterall WA. 2012. Structural basis for gating charge movement in the voltage sensor of a sodium channel. *Proc Natl Acad Sci* **109**: E93–E102.

Yellen G. 1984. Ionic permeation and blockade in Ca^{2+}-activated K$^+$ channels of bovine chromaffin cells. *J Gen Physiol* **84**: 157–186.

Zagotta WN, Hoshi T, Aldrich RW. 1990. Restoration of inactivation in mutants of Shaker potassium channels by a peptide derived from ShB. *Science* **250**: 568–571.

Zhang X, Ren W, DeCaen P, Yan C, Tao X, Tang L, Wang J, Hasegawa K, Kumasaka T, He J, et al. 2012. Crystal structure of an orthologue of the NaChBac voltage-gated sodium channel. *Nature* **486**: 130–134.

Zheng H, Liu W, Anderson LY, Jiang QX. 2011. Lipid-dependent gating of a voltage-gated potassium channel. *Nat Commun* **2**: 250.

Zhou Y, MacKinnon R. 2004. Ion binding affinity in the cavity of the KcsA potassium channel. *Biochemstry* **43**: 4978–4982.

Zhou M, Morais-Cabral JH, Mann S, MacKinnon R. 2001a. Potassium channel receptor site for the inactivation gate and quaternary amine inhibitors. *Nature* **411**: 657–661.

Zhou Y, Morais-Cabral JH, Kaufman A, MacKinnon R. 2001b. Chemistry of ion coordination and hydration revealed by a K$^+$ channel-Fab complex at 2.0 Å resolution. *Nature* **414**: 43–48.

Zhou Y, Xia XM, Lingle CJ. 2011. Cysteine scanning and modification reveal major differences between BK channels and Kv channels in the inner pore region. *Proc Natl Acad Sci* **108**: 12161–12166.

CHAPTER 3

Voltage-Gated Na⁺ Channels: Not Just for Conduction

Larisa C. Kruger and Lori L. Isom

Department of Pharmacology, University of Michigan, Ann Arbor, Michigan 48109-5632

Voltage-gated sodium channels (VGSCs), composed of a pore-forming α subunit and up to two associated β subunits, are critical for the initiation of the action potential (AP) in excitable tissues. Building on the monumental discovery and description of sodium current in 1952, intrepid researchers described the voltage-dependent gating mechanism, selectivity of the channel, and general structure of the VGSC channel. Recently, crystal structures of bacterial VGSC α subunits have confirmed many of these studies and provided new insights into VGSC function. VGSC β subunits, first cloned in 1992, modulate sodium current but also have nonconducting roles as cell-adhesion molecules and function in neurite outgrowth and neuronal pathfinding. Mutations in VGSC α and β genes are associated with diseases caused by dysfunction of excitable tissues such as epilepsy. Because of the multigenic and drug-resistant nature of some of these diseases, induced pluripotent stem cells and other novel approaches are being used to screen for new drugs and further understand how mutations in VGSC genes contribute to pathophysiology.

Voltage-gated sodium channels (VGSCs) conduct inward current that depolarizes the plasma membrane and initiates the action potential (AP) in excitable cells, including neurons, cardiomyocytes, and skeletal muscle cells. Because of the intrinsic link between VGSCs and cellular excitability, it is not surprising that mutations in VGSC genes are linked with epilepsy, cardiac arrhythmia, neuropathic pain, migraine, and neuromuscular disorders. The goal of this review is to provide an overview of critical discoveries in VGSC physiology and discuss the challenges of studying VGSCs in disease, including some of the exciting techniques to address these challenges.

VGSC DISCOVERY AND STRUCTURE

In 1952, the Nobel Laureates Alan Lloyd Hodgkin and Andrew Fielding Huxley first recorded sodium current (I_{Na}) using their voltage-clamp technique on the squid giant axon. Their experiments showed three key features of I_{Na}—selective Na⁺ conductance, voltage-dependent activation, and rapid inactivation (Hodgkin and Huxley 1952). The Hodgkin–Huxley model mathematically described voltage-dependent initiation of the AP by inward I_{Na}, followed by fast inactivation of I_{Na}, and simultaneous activation of outward I_K. Outward I_K was postulated to reestablish the charge balance across the plasma membrane (Hodgkin and Huxley 1952). These results were supported by studies of peripheral motor neurons by Sir John Carew Eccles, who shared the Nobel Prize with Hodgkin and Huxley. Among other contributions, he identified deviations in membrane potential and APs because of injections of sodium into motor neurons (Coombs et al. 1955). In addition to this monumental discovery, refinement of the voltage-clamp technique opened the door for a plethora of research on membrane potential physiology.

Correspondence: lisom@umich.edu

In the 1960s, Bertil Hille and Clay Armstrong proposed the idea that I_{Na} and I_K are conducted through specific ion channels. At that time, many studies centered on performing voltage-clamp recordings of squid axon with sodium-free solutions containing organic cations (e.g., ammonium) to determine the size of the ion-conducting pore. The axon membrane became transiently permeable to these ions following depolarization, and this permeability was abolished by tetrodotoxin (TTX) (Larramendi et al. 1956; Lorente De No et al. 1957; Tasaki et al. 1965, 1966; Tasaki and Singer 1966; Binstock and Lecar 1969). Hille added to this body of work by using additional metal and many organic cations to develop a model of the narrowest region of the ion-conducting pore of the channel (Hille 1971, 1972). Hille's model proposed a partial dehydration of Na^+ through interaction with a high-field-strength site at the extracellular end of the pore followed by rehydration in the lumen of the pore. This model is astonishingly close to what we now know to be true about VGSCs from crystal structure information (Payandeh et al. 2011; McCusker et al. 2012; Zhang et al. 2012). Armstrong and Bezanilla developed signal-averaging techniques to detect the movement of "gating charges" (Armstrong and Bezanilla 1973, 1974) corresponding to what we now understand to be the movement of voltage sensors, which respond to changes in membrane potential. Remarkably, without having any of the knowledge that we possess today about nucleotide/amino acid sequence and crystal structures, these early studies made many accurate predictions about VGSC structure and the functions of various channel domains.

Although early physiological studies provided evidence for the existence of a VGSC, the biochemical proof that tetrodotoxin/saxitoxin (STX) receptors were also ion-conducting did not yet exist. Membrane protein purification techniques were being established, but they required an extraordinarily large amount of time and biological starting material compared to today's techniques. Photoaffinity labeling of STX receptors with a scorpion toxin derivative by William Catterall's group identified a family of VGSC proteins that were designated α and β subunits (Beneski and Catterall 1980). Purification of VGSC protein to theoretical homogeneity was challenging because of the difficulty of solubilizing a high molecular weight, highly lipophilic membrane protein and the large amount of biological starting material required. Again, the Catterall laboratory was the first to overcome these challenges and purify VGSCs from rat brain (Hartshorne and Catterall 1981, 1984; Hartshorne et al. 1982). This work was closely followed by purification of VGSCs from rat and rabbit skeletal muscle (Barchi 1983; Kraner et al. 1985), and from chicken heart (Lombet and Lazdunski 1984). Reconstitution of purified VGSCs in a lipid bilayer membrane allowed observation of Na^+ flux and confirmed that a functional VGSC protein had been purified (Hartshorne et al. 1985). The evolution of VGSC purification methods is discussed in detail in Catterall (1992). Purification from rat brain revealed that VGSCs are heterotrimers, composed of a single α subunit and two non-pore-forming β subunits, β1 and β2 (Hartshorne and Catterall 1981). We now know that there are more than two VGSC β subunits. Subsequent homology cloning and heterologous expression studies showed that each VGSC α is associated at the plasma membrane with a noncovalently linked β1 or β3 subunit and a covalently linked β2 or β4 subunit (Morgan et al. 2000; Yu et al. 2003).

To clone the first VGSC α subunit complementary DNA (cDNA), researchers from Shosaku Numa's group took on the arduous task of purifying brain VGSCs and obtaining partial amino acid sequence by amino-terminal Edman degradation as well as through cleaved peptides from the α subunit. They then generated and screened multiple cDNA libraries based first on predicted degenerate cDNA sequences and then on cloned sequences. Assembly of the cloned VGSC cDNA fragments revealed what we now know to be *Scn1a* (Noda et al. 1984). Heterologous expression of the cloned cDNAs gave a functional channel, which we now recognize as the VGSC α subunit $Na_v1.1$.

VGSC β SUBUNITS ARE REQUIRED TO RECAPITULATE PHYSIOLOGICAL EXPRESSION OF I_{Na}

Early cloning of $Na_v1.2$ cDNA and its expression in oocytes resulted in I_{Na} that inactivated more slowly than I_{Na} recorded from neurons. Co-injection of low-molecular-weight rat brain messenger RNA (mRNA) was required for recapitulation of physiological I_{Na} in terms of rates of activation and inactivation, and voltage dependence (Auld et al. 1988; Krafte 1990). Cloning and expression of the VGSC β1 and β2 subunits showed that they mimicked the effects of low-molecular-weight mRNA on α subunit expression in oocytes (Isom et al. 1992, 1995b). Coexpression of $Na_v1.2 + β1$ in oocytes resulted in a larger I_{Na} that activated and inactivated more rapidly and had a negative shift in the voltage-dependence of inactivation compared to α alone. Coexpression of $Na_v1.2$ with both β1 and β2 further increased I_{Na} density (Isom et al. 1995b). As discussed by Calhoun and Isom, when expressed in heterologous systems, β subunits in general increase I_{Na} density, shift the voltage-dependence of current activation and inactivation, and accelerate the rates of activation and inactivation. However, the magnitude of current increase and direction of the shifts in activation and inactivation are dependent on the particular heterologous cell line and the identities of the β and α cDNAs expressed (Calhoun and Isom 2014). Importantly, heterologous expression systems cannot replicate the native cellular milieu, especially the multiprotein VGSC complexes that are known to form in specific subcellular domains of neurons and cardiac myocytes (Calhoun and Isom 2014). For example, in *Scn1b* null ventricular myocytes, peak and persistent sodium current is increased, mediated by increased $Na_v1.5$ expression, with no effect on channel kinetics and voltage dependence (Lopez-Santiago et al. 2007). Furthermore, loss of *Scn1b* also results in increased TTX-S current and *Scn3a* expression in ventricular myocytes (Lin et al. 2015). However, in the dorsal root ganglia (DRG), critical neurons for peripheral pain sensation, *Scn1b* deletion causes a depolarizing shift in the voltage dependence of VGSC inactivation and decreased persistent I_{Na} (Lopez-Santiago et al. 2011). As heterologous expression systems cannot accurately replicate these cell types, in vivo and transgenic mouse models offer more appropriate methods for studying the functional roles of VGSC physiology.

MECHANISTIC INSIGHTS ON VGSC STRUCTURE AND FUNCTION

VGSC α subunits are highly evolutionarily conserved. Each α subunit is composed of four homologous domains, each containing six transmembrane helices (S1–S6), which come together in pseudotetrafold symmetry to form the ion-conducting pore (Fig. 1). As discussed earlier, long before the emergence of crystal structures, much of VGSC structure and function was inferred through studies modeling gating and using site-specific antibodies and toxins during electrophysiological recordings.

The sliding-helix model of VGSC voltage-dependent activation was proposed in the 1980s by Catterall (1986a,b). The S4 segments of each domain that serve as voltage sensors contain arginine residues at every third position, which pair with negatively charged residues in nearby transmembrane segments. On membrane depolarization, these residues change ion-pair partners, causing each S4 helix to rotate or slide outward, pushing a single arginine residue in each of the four segments out, and generating gating current that activates, or opens, the channel pore (Payandeh et al. 2011). Data from potassium channel structures prompted the paddle model as an alternate mechanism of voltage-dependent activation (Jiang et al. 2003); however, the solving of multiple prokaryotic α subunit crystal structures showed the validity of the sliding-helix model for VGSCs (Payandeh et al. 2011; Vargas et al. 2012).

Another critical property of VGSCs is voltage-dependent inactivation, which occurs despite an ongoing depolarizing pulse. Early studies identified the intracellular loop between domains III and IV as the inactivation gate (Fig. 1). This was proposed to "swing" into the pore in a voltage-dependent manner, physically blocking ion conduction, and effectively inactivating the pore (Vassilev et al. 1988, 1989). A short amino acid sequence in the inactivation gate, isoleucine, phenylalanine, and methionine (IFM), was identified as critical to the process of inactivation. Mutation of these three residues to glutamine resulted in a noninactivating channel (West et al. 1992; Kellenberger 1997). These studies, combined with a three-dimensional nuclear magnetic resonance (NMR) structure of the inactivation gate (Rohl et al. 1999), helped form the current understanding of the molecular basis for fast inactivation of VGSCs.

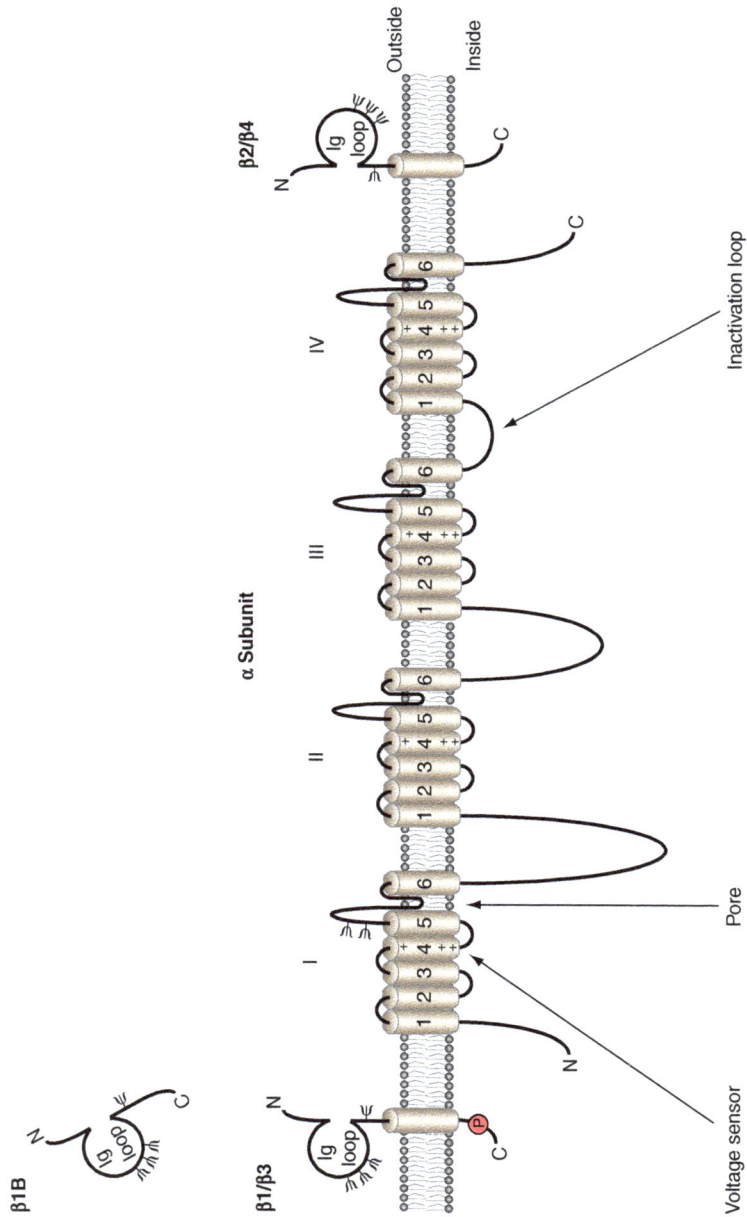

FIGURE 1. Voltage-gated sodium channel (VGSC) α and β subunit topology. The pore-forming α subunit is composed of four homologous domains (I–IV), each with six transmembrane segments. The fourth segment in each domain contains the voltage-sensing domain of the channel. The inactivation loop between domains III and IV contains the isoleucine, phenylalanine, and methionine (IFM) domain required for channel inactivation. Up to two β subunits may be associated with VGSCs. β1 and β3 subunits associate noncovalently, whereas β2 and β4 subunits associate via a disulfide link. (From Brackenbury and Isom 2011; reprinted, with permission, © Frontiers Media SD.)

Recent crystal structures of bacterial VGSCs have led to a more detailed understanding of the pore and selectivity filter. The membrane reentrant P-loop, located between transmembrane helices S5 and S6 of each domain that line the pore, was identified as the target for the pore-blocking TTX, suggesting its close location to the pore (Noda et al. 1989; Terlau et al. 1991). The crystal structure of a bacterial VGSC showed that the four P-loops together form a ring of glutamates, located near the extracellular end of the pore (Payandeh et al. 2011). As predicted in 1992, these highly conserved glutamates form a high-field-strength site critical in determining ion selectivity of the channel (Heinemann et al. 1992). These residues stabilize multiple ionic occupancy states, preferentially conducting hydrated Na$^+$ through the pore (Payandeh et al. 2011). In contrast, potassium channels conduct dehydrated K$^+$ (Doyle et al. 1998).

Capturing crystal structures of bacterial VGSCs in different activation states has provided new insights and biochemical modeling tools. A closed-pore conformation of the *Arcobacter* channel (Na$_v$Ab), crystallized within a lipid-based bicelle, offered the first VGSC 3D structure (Payandeh et al. 2011). A putatively inactive Na$_v$Rh channel from the marine bacterium *Rickettsiales* showed significant differences from Na$_v$Ab and supplied new information about conformational rearrangements required for inactivation (Zhang et al. 2012). Finally, the apparently open conformation of the Na$_v$M structure from the marine bacterium *Magnetococcus* presented further insights into channel gating and selectivity (McCusker et al. 2012).

Despite the wealth of information provided by these prokaryotic VGSC structures, crystallization of a eukaryotic VGSC is the essential next step. Although prokaryotic VGSCs are homotetramers, mammalian VGSC α subunits are single polypeptides containing four nonidentical domains. Further, the extensive extracellular and intracellular loops of mammalian channels are not present in their prokaryotic orthologs. Thus, major gaps in the field include the crystal structure of a mammalian α subunit and the co-crystal structure of an α subunit associated with β subunits. This information will be critical in fully understanding mammalian VGSC structure.

VGSC Diversity

Originally, VGSC α subunit proteins were named according to the tissue from which they were purified. For example, brain type II, rat II, and R-II are all outdated terms for Na$_v$1.2 (Catterall et al. 2005). To date, nine VGSC α proteins (Na$_v$1.1–1.9) and five β proteins (β1–4 and β1B), encoded by *SCN(X)A* and *SCN1B-SCN4B* genes, respectively, have been identified in mammalian genomes.

A BLAST search of the cDNAs encoding β1 and β2 in 1995 revealed important, and unexpected, new information. Both proteins contained areas of homology with known CAMs of the immuno-globulin (Ig) superfamily, for example, contactin and myelin P0. This led to the hypothesis that β1 and β2 function as CAMs in addition to channel modulation (Isom et al. 1995b). We now know that all five β subunit proteins contain an extracellular Ig domain and have CAM function, as discussed later in this review. β1–4 (encoded by *SCN1B-SCN4B*) are type 1 transmembrane proteins with an extracellular amino terminus (containing the Ig domain) and an intracellular carboxyl terminus (Isom et al. 1992, 1995b; Morgan et al. 2000). In contrast, β1B, a splice variant of *SCN1B*, is soluble and, thus, a portion is secreted. Retention of, intron 3 results in generation of an alternate carboxy-terminal domain that does not contain a transmembrane region (Patino et al. 2011). Recently, the three-dimensional crystal structures of the human Ig domains of β3 and β4 were solved. β3 formed a trimer when crystalized (Namadurai et al. 2014). In contrast, the β4 Ig domain was monomeric in crystal form (Gilchrist et al. 2013). Although this could represent a physiological difference in homophilic interactions, it could also be because of the removal of 31 amino acids at the β4 amino terminus to facilitate crystallization (Gilchrist et al. 2013). Comparisons of these structures will give new insights into the function of β subunits, but the co-crystal structure of α and β subunits remains the next frontier.

VGSC α and β subunits are highly expressed in central and peripheral neurons, cardiac myocytes, skeletal muscle, and some nonexcitable cells (Maier et al. 2004; Patino and Isom 2010; Brunklaus et al.

TABLE 1. VGSC genes, expression patterns, and disease associations

Gene symbol	Type	Tissue distribution	Consequence of mutations	TTX sensitivity
SCN1A	Na$_v$1.1	CNS, PNS, heart, DRG	DS, familial autism, FHM3, FS[+], GEFS[+], SUDEP	+
SCN2A	Na$_v$1.2	CNS, PNS, DRG	BFNIS, DS, EOEE, familial autism, GEFS, OS	+
SCN3A	Na$_v$1.3	CNS, PNS, heart, DRG	Unclear	+
SCN4A	Na$_v$1.4	Skeletal muscle, heart, DRG	PAM, PMC, HyperPP, HypoPP, SNEL	+
SCN5A	Na$_v$1.5	Skeletal muscle, heart, CNS, DRG	AF, AS, BS, DCM, LQTS, PCCD, SIDS, SSS, SUDEP	-
SCN8A	Na$_v$1.6	CNS, PNS, heart, DRG	EOEE; cognitive impairment, paralysis, ataxia, dystonia	+
SCN9A	Na$_v$1.7	DRG	CIP, IEM, PEPD, PPN	+
SCN10A	Na$_v$1.8	DRG	PPN	-
SCN11A	Na$_v$1.9	DRG	PPN	-
SCN1B	β1	CNS, PNS, heart, DRG, glia (oligodendrocytes, Schwann cells, astrocytes, radial glia)	AF, BS, DS, GEFS[+], LQTS, PCCD, TLE	N/A
SCN2B	β2	CNS, PNS, heart, DRG	AF, BS	N/A
SCN3B	β3	CNS, PNS, heart, DRG	AF, BS, PCCD, SIDS, ventricular fibrillation	N/A
SCN4B	β4	CNS, PNS, heart, DRG	LQTS, SIDS	N/A
SCN1B	β1B	Fetal CNS, PNS, heart, DRG	BS, PCCD, epilepsy	N/A

Modified, with permission, from Patino and Isom 2010, © Elsevier Ireland; Catterall 2012, © Wiley; Brunklaus et al. 2014, © BMJ.
AF, Atrial fibrillation; AS, atrial standstill; BFNIS, benign familial neonatal-infantile seizures; BS, Brugada syndrome; CIP, channelopathy-associated insensitivity to pain; CNS, central nervous system; DCM, dilated cardiomyopathy; DRG, dorsal root ganglia; DS, Dravet syndrome; EOEE, early-onset epileptic encephalopathy; FHM3, familial hemiplegic migraine type 3; FS[+], febrile seizures plus; GEFS[+], genetic epilepsy with febrile seizures plus; HyperPP, hyperkalemic periodic paralysis; HypoPP, hypokalemic periodic paralysis; IEM, inherited erythromelaglia; LQTS, long QT syndrome; OS, Ohtahara syndrome; PAM, potassium-aggravated myotonia; PCCD, progressive cardiac conduction disease; PEPD, paroxysmal extreme pain disorder formally known as familial rectal pain syndrome; PMC, paramyotonia congenital; PNS, peripheral nervous system; PPN, painful peripheral neuropathies; SIDS, sudden infant death syndrome; SNEL, severe neonatal episodic laryngospasm; SSS, sick sinus syndrome; SUDEP, sudden unexplained death in epilepsy; TLE, temporal lobe epilepsy.

2014), including breast cancer cells (Fraser et al. 2005), astrocytes, and oligodendrocytes. Table 1 shows the gene name, known tissue expression, and diseases associated with mutations for each VGSC subunit (adapted from Patino and Isom 2010; Catterall 2012; Brunklaus et al. 2014). VGSC expression is developmentally regulated. For example, Na$_v$1.3, β3, and β1B are most prevalent in embryonic and neonatal rodent brain. This expression profile then changes such that Na$_v$1.1, Na$_v$1.2, Na$_v$1.6, β1, β2, and β4 are predominant in the adult brain, albeit distribution is not equivalent across brain regions (Kazen-Gillespie et al. 2000; Catterall et al. 2005; Patino et al. 2011).

VGSC subcellular localization is critical for normal physiological functions. In neurons, VGSCs are concentrated at the axon initial segment (AIS), where the AP is initiated, and nodes of Ranvier in myelinated axons, which are critical for saltatory conduction. In cardiomyocytes, VGSCs are differentially localized at intercalated discs and transverse tubules (see Bao and Isom 2014; Calhoun and Isom 2014 for reviews that discuss this in more detail).

Despite their high sequence similarity, each VGSC α subunit has subtle differences in biophysical properties and pharmacological sensitivities (as summarized in Catterall et al. 2005; Kwong and Carr 2015). Many of the agents that target α subunits are toxins, including TTX, STX, μ-conotoxins, and their derivatives, such as 4,9-anhydro-TTX, which preferentially blocks Na$_v$1.6 (Rosker et al. 2007; Kwong and Carr 2015). VGSCs are canonically separated into TTX-sensitive (TTX-S) and TTX-resistant categories (TTX-R), which are blocked by nanomolar or micromolar concentrations of TTX, respectively (Catterall et al. 2005). Interestingly, mutation of a single residue changes TTX sensitivity (Noda et al. 1989). Modulation of I_{Na} by these toxins (reviewed in Gilchrist et al. 2014) can be affected by β subunit expression. For example, expression of β1 and β3, but not β2 or β4, tend to increase the rate of association and binding affinity of μ-conotoxins for VGSCs. However, modulation of I_{Na} by STX or TTX was unaffected by β subunit expression (Isom et al. 1995a; Zhang et al. 2013). In vivo, a combination of electrophysiology and toxicology are used to infer which VGSCs contribute to aspects of endogenous I_{Na}. Although challenging, in vivo studies are crucial, as heterologous results can be difficult to translate to in vivo physiological and pathophysiological mechanisms.

Posttranslational Modifications of α and β Subunits

VGSC α subunits are heavily glycosylated, although the extent of glycosylation varies among subunits (Bennett 2002). 40%–50% of these added carbohydrate residues are sialic acid moieties such that each channel contains ∼100 sialic acid residues (Miller et al. 1983; Roberts and Barchi 1987). Differential glycosylation of VGSC subunits can affect channel gating, putatively because of the local negative charge of sialic acid residues. In general, when α subunits are less glycosylated, channel gating occurs at more depolarized voltages (Recio-Pinto et al. 1990; Bennett et al. 1997; Zhang et al. 1999; Tyrrell et al. 2001). For example, Na$_v$1.9 expressed in neonatal DRG neurons is more glycosylated than in adult DRG, suggesting developmental regulation of glycosylation. This may account for developmental differences in gating of persistent I_{Na} attributed to Na$_v$1.9 (Tyrrell et al. 2001). Voltage-dependent activation and inactivation of Na$_v$1.4, but not Na$_v$1.5, shifts when expressed in cells incapable of sialylation, indicating that not all VGSCs are affected equally (Bennett 2002). Glycosylation accounts for about one-third of the molecular weights of β subunits. Mature β subunits contain three to four N-linked glycosylation sites in the Ig domain (Isom et al. 1992; McCormick et al. 1998) all of which are exposed in the β3 crystal structure (Namadurai et al. 2014). Sialic acid residues in these sites affect β1- and β2-mediated modulation of I_{Na} and channel surface expression (Johnson et al. 2004; Johnson and Bennett 2006). Critically, VGSC glycosylation varies developmentally and among cell types; for example, neonatal and adult myocytes have distinct differences in glycogen expression, as do atrial and ventricular myocytes (Stocker and Bennett 2006). Therefore, tissue-specific variations in β subunit processing likely include altered sialylation/glycosylation and may contribute to differential modulation of I_{Na} by β subunits among cell types (Zhang et al. 1999; Bennett 2002).

β1 can be phosphorylated on intracellular tyrosine residue Y-181 (Malhotra et al. 2002, 2004). Tyrosine phosphorylation regulates β1-mediated recruitment of ankyrin to points of cell–cell contact (Malhotra et al. 2002) and likely occurs via fyn kinase, an src family kinase (Brackenbury et al. 2008; Nelson et al. 2014). Receptor tyrosine phosphatase β interacts with the intracellular domain of β1 and may modulate β1 phosphorylation (Ratcliffe et al. 2000). In cardiac myocytes, nonphosphorylated β1 localizes at the T-tubules with ankyrinB, whereas phosphorylated β1 localizes at intercalated discs with connexin-43, N-cadherin, and Na$_v$1.5 (Malhotra et al. 2004). Therefore, the phosphorylation state Y-181 regulates β1 association with cytoskeletal proteins, as well as its subcellular localization.

β subunits are substrates for sequential proteolytic cleavage by BACE1 (β-secretase) and γ-secretase, similar to amyloid precursor protein. BACE1 cleavage releases the extracellular domain (ECD), followed by a γ-secretase-mediated cleavage event to release the carboxy-terminal fragment (CTF) (Wong et al. 2005). These cleaved peptides may have important physiological functions. For example, the β2-CTF translocates to the nucleus and promotes *Scn1a* mRNA and protein expression when heterologously expressed in cultured neurons (Kim et al. 2005, 2007). Consistent with this proposed mechanism, total and surface expression of Na$_v$1.1 in acutely dissociated hippocampal slices is significantly reduced in BACE1 null mice compared to wild-type (Kim et al. 2011). The physiological importance of BACE1 and γ-secretase cleavage of the other β subunits is not understood. As β1B can act as a CAM, it is possible that the β1-ECD, which contains the identical Ig loop domain, functions as a ligand for cell adhesion (Patino et al. 2011). γ-Secretase activity and, thus, formation of the β1-CTF is critical for β1-mediated neurite outgrowth in vitro (Brackenbury and Isom 2011). Determining whether these functions occur in vivo represents a critical next step in our understanding of posttranslational processing of VGSCs.

β Subunits Are Multifunctional

Effects on Surface Expression

Coexpression of β subunits with α results in increased I_{Na} density, in part, as a result of promotion of α cell-surface expression. Following posttranslational processing, α subunits are stored in an intracellular pool associated with the plasma membrane (Schmidt et al. 1985; Schmidt and Catterall 1986). Concomitant with plasma membrane insertion, α and β2 covalently associate via a disulfide bond (Chen et al. 2012). *Scn2b* null mice on the C57BL/6×129SV background have a 50% reduction in hippocampal I_{Na} density and level of surface α subunits (Chen et al. 2002). However, a second study of *Scn2b* null mice congenic on the C57BL/6 background saw no change in hippocampal I_{Na} (Uebachs et al. 2010), suggesting genetic background influences on β2 function. The effect of β2 on channel surface expression may also depend on the particular α subunit. *Scn2b* null small-fast DRG neurons have reduced TTX-S I_{Na}, but left unchanged TTX-R I_{Na} (Lopez-Santiago et al. 2006). An *Scn2b*-linked Brugada syndrome mutation results in decreased surface $Na_v1.5$ protein compared to wild-type (Riuró et al. 2013). Thus, modulation of α subunit surface expression by β2 may contribute to disease mechanisms. This is not unique to β2; β1 and β1B also promote VGSC surface expression (reviewed in Calhoun and Isom 2014).

Effects of β4 on Resurgent I_Na

Resurgent I_{Na} is caused by the transient opening of VGSCs during recovery from inactivation following the AP. This specialized current is an important adaptation for high-frequency firing neurons, such as cerebellar Purkinje neurons (Raman and Bean 1997). For this to occur, a "blocking protein," putatively the ICD of β4, must bind to open channels during depolarization and unbind on repolarization, shortening the typical refractory period and producing a resurgent I_{Na}. Without β4 expression, resurgent current is lost from cerebellar Purkinje neurons, but is rescued by a β4 ICD peptide (Grieco et al. 2005). This same β4 peptide is sufficient to invoke resurgent I_{Na} in $Na_v1.7$-expressing HEK cells (Theile and Cummins 2011). With β4 knockdown in cerebellar granule neurons, resurgent I_{Na} is decreased, but is rescued by the β4 peptide (Bant and Raman 2010). Other β subunits may also contribute to modulation of resurgent I_{Na}, as *Scn1b* null cerebellar granule neurons have reduced resurgent I_{Na} (Brackenbury et al. 2010).

Cell Adhesion and Neurite Outgrowth

β Subunits are multifunctional. They play critical roles in neurite outgrowth, migration, and maintenance of nodes of Ranvier. β Subunits are CAMs and, as such, recruit other signaling proteins to the VGSC complex. Most is known about the CAM interactions of β1 and β2, which participate in *trans*-homophilic adhesion, via the Ig loop, with β1 or β2 molecules on adjacent cells (Malhotra et al. 2000). β1 or β2 *trans*-homophilic adhesion recruits the cytoskeletal protein ankyrin to points of cell–cell contact (Malhotra et al. 2000). Ankyrin recruitment is abolished by phosphorylation of β1-Y181 (Malhotra et al. 2002). β1–β1 *trans*-homophilic adhesion promotes neurite outgrowth via a pathway requiring fyn kinase, the CAM contactin, γ-secretase activity, and localized I_{Na} (Brackenbury et al. 2008, 2010; Brackenbury and Isom 2011). Because β1B contains the identical Ig loop, it similarly promotes neurite outgrowth of neurons expressing β1 (Patino et al. 2011). Together, these properties may contribute to neuronal development in vivo; *Scn1b* null mice have neuronal pathfinding dysfunction and differences in subcellular localization of α subunits at AIS (Chen et al. 2004; Brackenbury et al. 2008).

β1, and likely β1B, interacts heterophilically with other CAMs and extracellular matrix proteins, including contactin, neurofascin-186, NrCAM, N-cadherin, and tenascin-R (Kazarinova-Noyes et al. 2001; Ratcliffe et al. 2001; Malhotra et al. 2004; McEwen and Isom 2004; Patino et al. 2011). Some of these interactions affect β1-modulation of α surface expression. For instance, coexpression of NF186 or contactin with $Na_v1.2$ and β1 increases $Na_v1.2$ surface expression compared with $Na_v1.2$ and β1 alone, although NF186 and contactin have no impact on $Na_v1.2$ expression in the absence of β1 (Kazarinova-Noyes et al. 2001; McEwen and Isom 2004). Heterophilic interactions with tenascin-R, an extracellular matrix protein secreted by oligodendrocytes during myelination, repels cells expressing β1 or β2 and, therefore, may play a role in restricting VGSCs to nodes of Ranvier in myelinated axons (Xiao et al. 1999). The cell-adhesive properties of β1 and β1B have likely clinical

Cite this introduction as *Cold Spring Harb Perspect Boil* doi:10.1101/cshperspect.a029264

relevance, as most known *SCN1B* epilepsy mutations are located in or near the Ig loop domain (reviewed in O'Malley and Isom 2015).

Much less is known about the CAM functions of β3 and β4. Studies of β3 homophilic adhesion give conflicting results: the β3 Ig domain appears to interact with full-length β3 when expressed heterologously (Yereddi et al. 2013), yet another group reported that β3 expression could not induce aggregation of *Drosophila* S2 cells, unlike β1 and β2 (McEwen et al. 2009). β3 and β1 likely do not associate, as β3 does not interact with a peptide containing the β1-ECD (McEwen and Isom 2004). β4 did not promote neurite outgrowth of cerebellar granule neurons, but this observation does not preclude other CAM interactions (Davis et al. 2004).

PATHOPHYSIOLOGICAL ROLES OF VGSCs

Mutations in VGSC genes are associated with many types of genetic epilepsy, peripheral neuropathy, long QT syndrome, neuromuscular disorders, and other diseases associated with dysfunction of excitable tissues. Recent reviews discuss the complex pathophysiological mechanisms of sodium channelopathies associated with pain and cardiac disease (Dib-Hajj et al. 2010; Adsit et al. 2013; Bao and Isom 2014; O'Malley and Isom 2015). Here, we will focus on the role of VGSCs in the genetic epilepsies and cancer as examples of disease resulting from aberrant or loss of VGSC function in mammals.

Epilepsy

Epilepsy results from an imbalance between excitation and inhibition in the brain. More than two million Americans have epilepsy, which can be caused by genetic mutations or brain injury (Hirtz et al. 2007). Epilepsy significantly impacts the quality of life for both patients and caregivers because of the unpredictability of seizures and the risk of comorbidities including cognitive decline, intellectual disability, developmental delay, and sudden unexpected death in epilepsy (SUDEP).

Mutations in genes encoding VGSC α and β subunits are associated with epilepsies with a wide range of phenotypic severities, including genetic epilepsy with febrile seizures plus (GEFS⁺), an inherited epilepsy with a wide range of phenotypes, and Dravet syndrome (DS), one of the most devastating pediatric epileptic encephalopathies. An intriguing discussion of the complex mechanisms underlying DS was published recently (Chopra and Isom 2014). In 1998, the first VGSC mutation associated with epilepsy was identified in *SCN1B* in a patient with GEFS⁺. However, GEFS⁺ is incompletely penetrant and family members with the same *SCN1B* mutation present with a wide range of epilepsy severities (Wallace et al. 1998), a now familiar characteristic of the genetic epilepsies. Since this initial discovery, a large number of mutations in *SCN1A*, *SCN2A*, *SCN8A*, and *SCN1B* (Escayg et al. 2000; O'Malley and Isom 2015) have been implicated in multiple types of epilepsy (see Table 1) (reviewed in part by Shi et al. 2012; Steinlein 2014; Wagnon and Meisler 2015). More than 1200 mutations have been identified in *SCN1A* alone (Meng et al. 2015). Identifying the pathophysiological mechanisms underlying DS is of particular importance, as traditional antiepileptic drugs often aggravate seizures in DS patients and development of new therapeutics to help these patients is especially critical. Although 70%–80% of DS patients have an identified heterozygous de novo *SCN1A* mutation (Marini et al. 2011), homozygous mutations in *SCN1B* have also been reported (Patino et al. 2009; Ogiwara et al. 2012).

Studies of DS *SCN1A* mutations have shown that a majority result in loss-of-function (Meng et al. 2015). Studies of $Scn1a^{+/-}$ null mice and knockin mice expressing human mutations have led to the "interneuron hypothesis," suggesting that selective I_{Na} loss in GABAergic interneurons causes epilepsy via loss of inhibitory tone (Yu et al. 2006). However, this has become controversial (Chopra and Isom 2014; Isom 2014). More recent studies of DS patient-derived induced pluripotent stem cell (iPSC) neurons showed either that both inhibitory and excitatory (Liu et al. 2013) or excitatory neurons (Jiao et al. 2013) are hyperexcitable compared to nonepileptic controls (Mistry et al. 2014). These studies suggest that *SCN1A* haploinsufficiency may result in compensatory up-regulation of other VGSC α subunits. Hyperexcitability of both inhibitory and excitatory neurons may lead to seizures in DS via increased network excitability or synchronization of firing.

Scn1b null mice also model DS; they experience spontaneous seizures and early lethality. At least one DS *SCN1B* mutation causes β1 to be retained intracellularly, preventing surface expression and, thus, function (Patino et al. 2009). *Scn1b* null mice have differences in neuronal pathfinding that are observed before seizure onset (Brackenbury et al. 2013). Thus, *SCN1B*-linked DS mutations may alter neuronal excitability by affecting α subunit surface expression/function as well as by altering neuronal development via changes in cell adhesion.

A major concern for epilepsy patients and their families is the risk of SUDEP. Research suggests that a "perfect storm" of seizures, cardiac arrhythmias, respiratory dysfunction, and autonomic and/or parasympathetic dysfunction contributes to SUDEP (Auerbach et al. 2013; Kalume 2013; Massey et al. 2014). Heterozygous *Scn1a-R1407X* knockin GEFS$^+$/DS mice, as well as *Scn1b* null DS mice, have increased transient and persistent I_{Na} in cardiac myocytes, which may provide substrates for cardiac arrhythmias and contribute to SUDEP (Lopez-Santiago et al. 2007; Auerbach et al. 2013). Fully understanding the mechanisms underlying SUDEP will be essential to epilepsy treatment.

One of the greatest challenges in epilepsy research is the need for novel antiepileptic drugs. Even with optimal treatment, 20%–30% of all epilepsy patients are pharmacoresistant, defined as being unresponsive to at least two different, tolerated antiepileptic drugs (Kwan et al. 2010, 2011). Even in nonrefractory patients, antiepileptic drugs treat disease symptoms rather than preventing disease progression and often have intolerable side effects. A major goal of studying epilepsy-associated VGSC mutations is to further understand epileptogenesis (how a brain becomes epileptic) to inform new antiepileptic drug discovery.

Emerging Roles for VGSCs in Cancer

An exciting area of research is the role of VGSCs in nonexcitable cells, including multiple cancer cell types. Almost two decades ago, researchers observed that cancer cell lines with higher VGSC expression have increased cell motility and metastatic potential. Furthermore, the invasive capacity of these cells could be reduced by incubation with TTX (Grimes et al. 1995; Laniado et al. 1997). Since then, both α and β subunits have been detected in many cancer cell lines and in some patient biopsies. VGSC α subunits are expressed in some cervical, ovarian, breast, and colon tumors in vivo (Fraser et al. 2005; Gao et al. 2010; House et al. 2010; Hernandez-Plata et al. 2012), although some prostate, breast, cervical, and lung cancers express β subunits (reviewed in Patel and Brackenbury 2015). Studies have identified both positive and negative associations between VGSC expression and metastatic potential. Na$_v$1.5 expression in breast cancer cells correlates positively with increased risk of recurrence and metastasis (Fraser et al. 2005; Yang et al. 2012), and a similar trend has been described for colon, prostate, and ovarian cancers. However, there is an inverse correlation between α expression and clinical grade in glioma and no correlation in lung cancer cell lines (Schrey et al. 2002; Onganer and Djamgoz 2005; Roger et al. 2007). Expression of α subunits can potentiate lateral motility, adhesion, process extension, and other cellular behaviors associated with metastasis (Brackenbury 2012). It is not clear how increased expression of the pore-forming α subunits potentiates invasive changes in the tumor cells. Three potential models have been proposed: increased Na$^+$ influx enhances H$^+$ efflux, thus activating pH-dependent extracellular matrix degradation and invasion; regulation of an "invasion gene network," particularly by *SCN5A*, and increased intracellular Ca^{2+}, which enhances formation of invasive projections (reviewed in Brackenbury 2012).

SCN1B and *SCN2B* expression are associated with metastatic potential in prostate cancer cells (Diss et al. 2008; Jansson et al. 2012). However, studies of the invasive potential of *SCN1B*-expressing breast cancer in vitro have been contradicted by in vivo studies. β1 expression is associated with less invasive breast cancer cell lines (Chioni et al. 2009); however, in a mouse model of breast cancer, overexpression of β1 increased tumor growth, metastasis, and angiogenesis (Nelson et al. 2014). β1 *trans*-homophilic adhesion mediates process outgrowth from breast cancer cells (Nelson et al. 2014),

Cite this introduction as *Cold Spring Harb Perspect Boil* doi:10.1101/cshperspect.a029264

thus it is proposed that β1 promotes metastasis by a mechanism similar to β1-mediated neurite outgrowth.

NEW FRONTIERS AND NOVEL TECHNIQUES

The list of pathogenic VGSC gene mutations is growing rapidly. The explosion of whole-genome sequencing has led to multi-institution efforts, such as the NINDS-funded Epi4K Project (www .ninds.nih.gov/news_and_events/news_articles/pressrelease_childhood_epilepsy_genes_08112013.htm), with the goal of sequencing the genomes of 4000 patients with epilepsy, and the England Department of Health funded 100,000 Genomes Project (genomicsengland.co.uk) (Kearney 2014; Siva 2015). However, even with familial genetic information in hand, determining whether mutations are causative and understanding their pathophysiology are not simple tasks. A given patient can have mutations in several genes, which may be causative, benign, or modify another mutation. A single causative mutation may result in a wide range of phenotypes among individuals, for example, DS and GEFS⁺ patients may express the same *SCN1A* or *SCN1B* mutation. Genetic background is critical. $Scn1a^{+/-}$ mouse models of DS have phenotypes of varying severity depending on genetic background (Yu et al. 2006; Miller et al. 2014). This issue can now be addressed directly with the induced-pluripotent stem cell (iPSC) technique. Patient iPSCs, usually derived from a skin cell biopsy, can theoretically be differentiated into any cell type, including cardiomyocytes, and specific neuronal subtypes (Parent and Anderson 2015). With this remarkable technique, we now have the opportunity to study disease-causing mutations in human cells within the context of the patient's unique genomic background (Takahashi et al. 2007). Furthermore, endogenous genes that may contribute to pathology remain present. Brain organoids made from human iPSC-derived cells with a mutation in a gene involved in pluripotent cell maintenance were used to model microcephaly, providing proof-of-concept for use of organoids to model other neurological diseases (Lancaster et al. 2013). Another advantage of iPSCs is the opportunity to perform drug screening on differentiated human cells, alongside animal models, thus providing insight into toxicity and species-specific effects before clinical trials (reviewed in Ko and Gelb 2014). Patient-derived iPSC cells are critical for precision medicine approaches. For instance, antiepileptic drug effectiveness or toxicity may be optimized on patient-derived iPSC-neurons before patient treatment. As the iPSC technique is further refined and developed, it will provide additional, critical tools in therapeutic development.

Another challenge to the field is the non-specificity of clinically available VGSC-targeted drugs. There are six different potential sites for small molecule targeting on VGSCs, as shown by the wide variety of toxin binding sites. However, most of the current antiarrhythmic, antiepileptic, and analgesic VGSC drugs on the market, for example, lamotrigine, phenytoin, and lidocaine, target the "local anesthetic" site located in domain IVS6 (Kwong and Carr 2015). This site is highly conserved among α subunits; thus, these drugs have little selectivity for specific VGSCs and often cause adverse effects. Development of VGSC agents that target other binding sites remains an important task. Although, in most cases, toxins cannot be used directly for therapeutics, they can greatly inform drug design; small molecules can be developed that preserve beneficial effects while reducing harm. Drugs targeting specific toxin binding sites may allow more selective blockade of specific α subunits to treat disease, such as neuropathic pain, cardiac arrhythmias, epilepsies, and perhaps even cancer (Stevens et al. 2011; Xiao et al. 2014).

Transgenic zebrafish are an economical and rapid vertebrate system that can be used to screen small molecule libraries and FDA-approved drugs for novel applications. Zebrafish with mutations in *scn1Lab*, the gene orthologous to *SCN1A*, have spontaneous seizures that are resistant to many antiepileptic drugs and appear to model DS. Peter de Witte's group showed that DS zebrafish responded to fenfluramine (Zhang et al. 2015). Scott Baraban used locomotion tracking in mutant zebrafish to monitor behavioral seizures in a drug screen, which was validated using diazepam, valproate, and stiripentol, antiepileptic drugs effective in some DS patients. Out of 320 compounds tested, clemizole, an FDA-approved antihistamine, suppressed spontaneous seizures in vivo (Baraban et al. 2013). This approach can be adapted for screening potential therapeutics for any monogenic epilepsy, especially if combined with the CRISPR/Cas9 technique. Discovered in 2010 as a bacterial defense mechanism against phages (Deveau et al. 2010), the CRISPR/Cas9 system has since been adapted to easily and quickly introduce heritable genetic mutations in mice, rats, zebrafish, and rabbits (Chang et al. 2013; Li et al. 2013; Yang et al. 2014a). Compared to zinc-finger nucleases and transcription activator-like effector nucleases (TALENs), CRISPR/Cas creates double-stranded DNA breaks in a sequence-specific manner at sites complementary to a "single-guide RNA." Because these are easy to design, the CRISPR/Cas9 system vastly improves the precision, speed, and accuracy of creating transgenic cell lines (including iPSCs) and animal models for single gene mutations (Yang et al. 2014b). These and other cutting edge techniques will allow us to discover novel disease mechanisms and therapeutics for diseases linked to VGSC mutations for another six decades (Catterall 2012).

REFERENCES

Adsit GS, Vaidyanathan R, Galler CM, Kyle JW, Makielski JC. 2013. Channelopathies from mutations in the cardiac sodium channel protein complex. *J Mol Cell Cardiol* **61**: 34–43.

Armstrong CM, Bezanilla F. 1973. Currents related to movement of the gating particles of the sodium channels. *Nature* **242**: 459–461.

Armstrong CM, Bezanilla F. 1974. Charge movement associated with the opening and closing of the activation gates of the Na channels. *J Gen Physiol* **63**: 533–552.

Auerbach DS, Jones J, Clawson BC, Offord J, Lenk GM, Ogiwara I, Yamakawa K, Meisler MH, Parent JM, Isom LL. 2013. Altered cardiac electrophysiology and SUDEP in a model of Dravet syndrome. *PLoS ONE* **8**: 1–15.

Auld VJ, Goldin AL, Krafte DS, Marshall J, Dunn JM, Catterall WA, Lester HA, Davidson N, Dunn RJ. 1988. A rat brain Na$^+$ channel α subunit with novel gating properties. *Neuron* **1**: 449–461.

Bant JS, Raman IM. 2010. Control of transient, resurgent, and persistent current by open-channel block by Na channel β4 in cultured cerebellar granule neurons. *Proc Natl Acad Sci* **107**: 12357–12362.

Bao Y, Isom LL. 2014. Na$_v$1.5 and regulatory β subunits in cardiac sodium channelopathies. *Card Electrophysiol Clin* **6**: 679–694.

Baraban SC, Dinday MT, Hortopan GA. 2013. Drug screening in *Scn1a* zebrafish mutant identifies clemizole as a potential Dravet syndrome treatment. *Nat Commun* **4**: 2410.

Barchi RL. 1983. Protein components of the purified sodium channel from rat skeletal muscle sarcolemma. *J Neurochem* **40**: 1377–1385.

Beneski DA, Catterall WA. 1980. Covalent labeling of protein components of the sodium channel with a photoactivable derivative of scorpion toxin. *Proc Natl Acad Sci* **77**: 639–643.

Bennett ES. 2002. Isoform-specific effects of sialic acid on voltage-dependent Na$^+$ channel gating: Functional sialic acids are localized to the S5-S6 loop of domain I. *J Physiol* **538**: 675–690.

Bennett E, Urcan MS, Tinkle SS, Koszowski AG, Levinson SR. 1997. Contribution of sialic acid to the voltage dependence of sodium channel gating. A possible electrostatic mechanism. *J Gen Physiol* **109**: 327–343.

Binstock L, Lecar H. 1969. Ammonium ion currents in the squid giant axon. *J Gen Physiol* **53**: 342–361.

Brackenbury WJ. 2012. Voltage-gated sodium channels and metastatic disease. *Channels (Austin)* **6**: 352–361.

Brackenbury WJ, Isom LL. 2011. Na Channel β subunits: Overachievers of the ion channel family. *Front Pharmacol* **2**: 1–11.

Brackenbury WJ, Davis TH, Chen C, Slat EA, Detrow MJ, Dickendesher TL, Ranscht B, Isom LL. 2008. Voltage-gated Na$^+$ channel β1 subunit-mediated neurite outgrowth requires Fyn kinase and contributes to postnatal CNS development in vivo. *J Neurosci* **28**: 3246–3256.

Brackenbury WJ, Calhoun JD, Chen C, Miyazaki H, Nukina N, Oyama F, Ranscht B, Isom LL. 2010. Functional reciprocity between Na$^+$ channel Na$_v$1.6 and β1 subunits in the coordinated regulation of excitability and neurite outgrowth. *Proc Natl Acad Sci* **107**: 2283–2288.

Brackenbury WJ, Yuan Y, O'Malley HA, Parent JM, Isom LL. 2013. Abnormal neuronal patterning occurs during early postnatal brain development of *Scn1b*-null mice and precedes hyperexcitability. *Proc Natl Acad Sci* **110**: 1089–1094.

Brunklaus A, Ellis R, Reavey E, Semsarian C, Zuberi SM. 2014. Genotype phenotype associations across the voltage-gated sodium channel family. *J Med Genet* **51**: 650–658.

Calhoun JD, Isom LL. 2014. The role of non-pore-forming β subunits in physiology and pathophysiology of voltage-gated sodium channels. In *Voltage gated sodium channels* (ed. Ruben PC), pp. 51–89. Springer, Berlin.

Catterall WA. 1986a. Molecular properties of voltage-sensitive sodium channels. *Annu Rev Biochem* **55**: 953–985.

Catterall WA. 1986b. Voltage-dependent gating of sodium channels: Correlating structure and function. *Trends Neurosci* **9**: 7–10.

Catterall WA. 1992. Cellular and molecular biology of voltage-gated sodium channels. *Physiol Rev* **72**: S15–S48.

Catterall WA. 2012. Voltage-gated sodium channels at 60: Structure, function and pathophysiology. *J Physiol* **590**: 2577–2589.

Catterall WA, Goldin AL, Waxman SG. 2005. International Union of Pharmacology. XLVII: Nomenclature and structure–function relationships of voltage-gated sodium channels. *Pharmacol Rev* **57**: 397–409.

Chang N, Sun C, Gao L, Zhu D, Xu X, Zhu X, Xiong J-W, Xi JJ. 2013. Genome editing with RNA-guided Cas9 nuclease in zebrafish embryos. *Cell Res* **23**: 465–472.

Chen C, Bharucha V, Chen Y, Westenbroek RE, Brown A, Malhotra JD, Jones D, Avery C, Gillespie PJ, Kazen-Gillespie KA, et al. 2002. Reduced sodium channel density, altered voltage dependence of inactivation, and increased susceptibility to seizures in mice lacking sodium channel β2-subunits. *Proc Natl Acad Sci* **99**: 17072–17077.

Chen C, Westenbroek RE, Xu X, Edwards CA, Sorenson DR, Chen Y, McEwen DP, O'Malley Ha, Bharucha V, Meadows LS, et al. 2004. Mice lacking sodium channel β1 subunits display defects in neuronal excitability, sodium channel expression, and nodal architecture. *J Neurosci* **24**: 4030–4042.

Chen C, Calhoun JD, Zhang Y, Lopez-Santiago L, Zhou N, Davis TH, Salzer JL, Isom LL. 2012. Identification of the cysteine residue responsible for disulfide linkage of Na$^+$ channel α and β2 subunits. *J Biol Chem* **287**: 39061–39069.

Chioni A-MM, Brackenbury WJ, Calhoun JD, Isom LL, Djamgoz MBA. 2009. A novel adhesion molecule in human breast cancer cells: Voltage-gated Na$^+$ channel β1 subunit. *Int J Biochem Cell Biol* **41**: 1216–1227.

Chopra R, Isom LL. 2014. Untangling the dravet syndrome seizure network: The changing face of a rare genetic epilepsy. *Epilepsy Curr* **14**: 86–89.

Coombs J, Eccles J, Fatt P. 1955. The electrical properties of the motoneurone membrane. *J Physiol* **130**: 291–325.

Davis TH, Chen C, Isom LL. 2004. Sodium channel β1 subunits promote neurite outgrowth in cerebellar granule neurons. *J Biol Chem* **279**: 51424–51432.

Deveau H, Garneau JE, Moineau S. 2010. CRISPR/Cas system and its role in phage-bacteria interactions. *Annu Rev Microbiol* **64:** 475–493.

Dib-Hajj SD, Cummins TR, Black JA, Waxman SG. 2010. Sodium channels in normal and pathological pain. *Annu Rev Neurosci* **33:** 325–347.

Diss JKJ, Fraser SP, Walker MM, Patel A, Latchman DS, Djamgoz MBA. 2008. β-subunits of voltage-gated sodium channels in human prostate cancer: Quantitative in vitro and in vivo analyses of mRNA expression. *Prostate Cancer Prostatic Dis* **11:** 325–333.

Doyle DA, Morais Cabral J, Pfuetzner RA, Kuo A, Gulbis JM, Cohen SL, Chait BT, MacKinnon R. 1998. The structure of the potassium channel: Molecular basis of K⁺ conduction and selectivity. *Science* **280:** 69–77.

Escayg A, MacDonald BT, Meisler MH, Baulac S, Huberfeld G, An-Gourfinkel I, Brice A, LeGuern E, Moulard B, Chaigne D, et al. 2000. Mutations of *SCN1A*, encoding a neuronal sodium channel, in two families with GEFS⁺². *Nat Genet* **24:** 343–345.

Fraser S, Diss J, Chioni A, Mycielska M, Pan H, Yamaci R, Pani F, Siwy Z, Krasowska M, Grzywna Z, et al. 2005. Voltage-gated sodium channel expression and potentiation of human breast cancer metastasis. *Clin Cancer Res* **11:** 5381–5389.

Gao R, Shen Y, Cai J, Lei M, Wang Z. 2010. Expression of voltage-gated sodium channel α subunit in human ovarian cancer. *Oncol Rep* **23:** 1293–1299.

Gilchrist J, Das S, Van Petegem F, Bosmans F. 2013. Crystallographic insights into sodium-channel modulation by the β4 subunit. *Proc Natl Acad Sci* **110:** E5016–E5024.

Gilchrist J, Olivera BM, Bosmans F. 2014. Animal toxins influence voltage-gated sodium channel function. *Handb Exp Pharmacol* **221:** 203–229.

Grieco TM, Malhotra JD, Chen C, Isom LL, Raman IM. 2005. Open-channel block by the cytoplasmic tail of sodium channel β4 as a mechanism for resurgent sodium current. *Neuron* **45:** 233–244.

Grimes JA, Fraser SP, Stephens GJ, Downing JE, Laniado ME, Foster CS, Abel PD, Djamgoz MB. 1995. Differential expression of voltage-activated Na⁺ currents in two prostatic tumour cell lines: Contribution to invasiveness in vitro. *FEBS Lett* **369:** 290–294.

Hartshorne RP, Catterall WA. 1981. Purification of the saxitoxin receptor of the sodium channel from rat brain. *Proc Natl Acad Sci* **78:** 4620–4624.

Hartshorne RP, Catterall WA. 1984. The sodium channel from rat brain. Purification and subunit composition. *J Biol Chem* **259:** 1667–1675.

Hartshorne RP, Messner DJ, Coppersmith JC, Catterall WA. 1982. The saxitoxin receptor of the sodium channel from rat brain. Evidence for two nonidentical β subunits. *J Biol Chem* **257:** 13888–13891.

Hartshorne RP, Keller BU, Talvenheimo JA, Catterall WA, Montal M. 1985. Functional reconstitution of the purified brain sodium channel in planar lipid bilayers. *Proc Natl Acad Sci* **82:** 240–244.

Heinemann SH, Terlau H, Stühmer W, Imoto K, Numa S. 1992. Calcium channel characteristics conferred on the sodium channel by single mutations. *Nature* **356:** 441–443.

Hernandez-Plata E, Ortiz CS, Marquina-Castillo B, Medina-Martinez I, Alfaro A, Berumen J, Rivera M, Gomora JC. 2012. Overexpression of Na_v1.6 channels is associated with the invasion capacity of human cervical cancer. *Int J Cancer* **130:** 2013–2023.

Hille B. 1971. The permeability of the sodium channel to organic cations in myelinated nerve. *J Gen Physiol* **58:** 599–619.

Hille B. 1972. The permeability of the sodium channel to metal cations in myelinated nerve. *J Gen Physiol* **59:** 637–658.

Hirtz D, Thurman DJ, Gwinn-Hardy K, Mohamed M, Chaudhuri AR, Zalutsky R. 2007. How common are the "common" neurologic disorders? *Neurology* **68:** 326–337.

Hodgkin AL, Huxley AF. 1952. A quantitative description of membrane current and its application to conduction and excitation in nerves. *J Physiol* **117:** 500–544.

House CD, Vaske CJ, Schwartz AM, Obias V, Frank B, Luu T, Sarvazyan N, Irby R, Strausberg RL, Hales TG, et al. 2010. Voltage-gated Na⁺ channel *SCN5A* is a key regulator of a gene transcriptional network that controls colon cancer invasion. *Cancer Res* **70:** 6957–6967.

Isom LL. 2014. "It was the interneuron with the parvalbumin in the hippocampus!" "No, it was the pyramidal cell with the glutamate in the cortex!" searching for clues to the mechanism of dravet syndrome—The plot thickens. *Epilepsy Curr* **14:** 350–352.

Isom LL, De Jongh KS, Patton DE, Reber BF, Offord J, Charbonneau H, Walsh K, Goldin AL, Catterall WA. 1992. Primary structure and functional expression of the β1 subunit of the rat brain sodium channel. *Science* **256:** 839–842.

Isom LL, Scheuer T, Brownstein AB, Ragsdale DS, Murphy BJ, Catterall WA. 1995a. Functional co-expression of the β1 and type IIA α subunits of sodium channels in a mammalian cell line. *J Biol Chem* **270:** 3306–3312.

Isom LL, Ragsdale DS, De Jongh KS, Westenbroek RE, Reber BF, Scheuer T, Catterall WA. 1995b. Structure and function of the β2 subunit of brain sodium channels, a transmembrane glycoprotein with a CAM motif. *Cell* **83:** 433–442.

Jansson KH, Lynch JE, Lepori-Bui N, Czymmek KJ, Duncan RL, Sikes RA. 2012. Overexpression of the VSSC-associated CAM, β-2, enhances LNCaP cell metastasis associated behavior. *Prostate* **72:** 1080–1092.

Jiang Y, Lee A, Chen J, Ruta V, Cadene M, Chait BT, MacKinnon R. 2003. X-ray structure of a voltage-dependent K⁺ channel. *Nature* **423:** 33–41.

Jiao J, Yang Y, Shi Y, Chen J, Gao R, Fan Y, Yao H, Liao W, Sun X-F, Gao S. 2013. Modeling Dravet syndrome using induced pluripotent stem cells (iPSCs) and directly converted neurons. *Hum Mol Genet* **22:** 1–12.

Johnson D, Bennett ES. 2006. Isoform-specific effects of the β2 subunit on voltage-gated sodium channel gating. *J Biol Chem* **281:** 25875–25881.

Johnson D, Montpetit ML, Stocker PJ, Bennett ES. 2004. The sialic acid component of the β1 subunit modulates voltage-gated sodium channel function. *J Biol Chem* **279:** 44303–44310.

Kalume F. 2013. Sudden unexpected death in Dravet syndrome: Respiratory and other physiological dysfunctions. *Respir Physiol Neurobiol* **189:** 324–328.

Kazarinova-Noyes K, Malhotra JD, McEwen DP, Mattei LN, Berglund EO, Ranscht B, Levinson SR, Schachner M, Shrager P, Isom LL, et al. 2001. Contactin associates with Na⁺ channels and increases their functional expression. *J Neurosci* **21:** 7517–7525.

Kazen-Gillespie KA, Ragsdale DS, D'Andrea MR, Mattei LN, Rogers KE, Isom LL. 2000. Cloning, localization, and functional expression of sodium channel β1A subunits. *J Biol Chem* **275:** 1079–1088.

Kearney JA. 2014. Epi4K phase I: Gene discovery in epileptic encephalopathies by exome sequencing. *Epilepsy Curr* **14:** 208–210.

Kellenberger S. 1997. Molecular analysis of the putative inactivation particle in the inactivation gate of brain type IIA Na⁺ channels. *J Gen Physiol* **109:** 589–605.

Kim DY, Ingano LAM, Carey BW, Pettingell WH, Kovacs DM. 2005. Presenilin/γ-secretase-mediated cleavage of the voltage-gated sodium channel β2-subunit regulates cell adhesion and migration. *J Biol Chem* **280:** 23251–23261.

Kim DY, Carey BW, Wang H, Ingano LAM, Binshtok AM, Wertz MH, Pettingell WH, He P, Lee VM-Y, Woolf CJ, et al. 2007. BACE1 regulates voltage-gated sodium channels and neuronal activity. *Nat Cell Biol* **9:** 755–764.

Kim DY, Gersbacher MT, Inquimbert P, Kovacs DM. 2011. Reduced sodium channel Na_v1.1 levels in *BACE1*-null mice. *J Biol Chem* **286:** 8106–8116.

Ko H, Gelb BD. 2014. Concise review: Drug discovery in the age of the induced pluripotent stem cell. *Stem Cells Transl Med* **3:** 396–402.

Krafte DS. 1990. Inactivation of cloned Na channels expressed in *Xenopus* oocytes. *J Gen Physiol* **96:** 689–706.

Kraner SD, Tanaka JC, Barchi RL. 1985. Purification and functional reconstitution of the voltage-sensitive sodium channel from rabbit T-tubular membranes. *J Biol Chem* **260:** 6341–6347.

Kwan P, Arzimanoglou A, Berg AT, Brodie MJ, Hauser WA, Mathern G, Moshé SL, Perucca E, Wiebe S, French J. 2010. Definition of drug resistant epilepsy: Consensus proposal by the Ad Hoc Task Force of the ILAE Commission on Therapeutic Strategies. *Epilepsia* **51:** 1069–1077.

Kwan P, Schachter SC, Brodie MJ. 2011. Drug-resistant epilepsy. *N Engl J Med* **365:** 919–926.

Kwong K, Carr MJ. 2015. Voltage-gated sodium channels. *Curr Opin Pharmacol* **22:** 131–139.

Lancaster MA, Renner M, Martin C-A, Wenzel D, Bicknell LS, Hurles ME, Homfray T, Penninger JM, Jackson AP, Knoblich JA. 2013. Cerebral organoids model human brain development and microcephaly. *Nature* **501:** 373–379.

Laniado ME, Lalani EN, Fraser SP, Grimes JA, Bhangal G, Djamgoz MB, Abel PD. 1997. Expression and functional analysis of voltage-activated Na⁺ channels in human prostate cancer cell lines and their contribution to invasion in vitro. *Am J Pathol* **150:** 1213–1221.

Larramendi L, Lorente De No R, Vidal F. 1956. Restoration of sodium-deficient frog nerve fibres by an isotonic solution of guanidinium chloride. *Nature* 178: 316–317.

Li D, Qiu Z, Shao Y, Chen Y, Guan Y, Liu M, Li Y, Gao N, Wang L, Lu X, et al. 2013. Heritable gene targeting in the mouse and rat using a CRISPR-Cas system. *Nat Biotechnol* 31: 681–683.

Lin X, O'Malley H, Chen C, Auerbach D, Foster M, Shekhar A, Zhang M, Coetzee W, Jalife J, Fishman GI, et al. 2015. *Scn1b* deletion leads to increased tetrodotoxin-sensitive sodium current, altered intracellular calcium homeostasis and arrhythmias in murine hearts. *J Physiol* 593: 1389–1407.

Liu Y, Lopez-Santiago LF, Yuan Y, Jones JM, Zhang H, O'Malley HA, Patino GA, O'Brien JE, Rusconi R, Gupta A, et al. 2013. Dravet syndrome patient-derived neurons suggest a novel epilepsy mechanism. *Ann Neurol* 74: 128–139.

Lombet A, Lazdunski M. 1984. Characterization, solubilization, affinity labeling and purification of the cardiac Na$^+$ channel using *Tityus* toxin γ. *Eur J Biochem* 141: 651–660.

Lopez-Santiago LF, Pertin M, Morisod X, Chen C, Hong S, Wiley J, Decosterd I, Isom LL. 2006. Sodium channel β2 subunits regulate tetrodotoxin-sensitive sodium channels in small dorsal root ganglion neurons and modulate the response to pain. *J Neurosci* 26: 7984–7994.

Lopez-Santiago LF, Meadows LS, Ernst SJ, Chen C, Malhotra JD, McEwen DP, Speelman A, Noebels JL, Maier SKG, Lopatin AN, et al. 2007. Sodium channel *Scn1b* null mice exhibit prolonged QT and RR intervals. *J Mol Cell Cardiol* 43: 636–647.

Lopez-Santiago LF, Brackenbury WJ, Chen C, Isom LL. 2011. Na$^+$ channel *Scn1b* gene regulates dorsal root ganglion nociceptor excitability in vivo. *J Biol Chem* 286: 22913–22923.

Lorente De No R, Vidal F, Larramendi L. 1957. Restoration of sodium-deficient frog nerve fibres by onium ions. *Nature* 179: 737–738.

Maier SKG, Westenbroek RE, McCormick KA, Curtis R, Scheuer T, Catterall WA. 2004. Distinct subcellular localization of different sodium channel α and β subunits in single ventricular myocytes from mouse heart. *Circulation* 109: 1421–1427.

Malhotra JD, Kazen-Gillespie K, Hortsch M, Isom LL. 2000. Sodium channel β subunits mediate homophilic cell adhesion and recruit ankyrin to points of cell–cell contact. *J Biol Chem* 275: 11383–11388.

Malhotra JD, Koopmann MC, Kazen-Gillespie KA, Fettman N, Hortsch M, Isom LL. 2002. Structural requirements for interaction of sodium channel β1 subunits with ankyrin. *J Biol Chem* 277: 26681–26688.

Malhotra JD, Thyagarajan V, Chen C, Isom LL. 2004. Tyrosine-phosphorylated and nonphosphorylated sodium channel β1 subunits are differentially localized in cardiac myocytes. *J Biol Chem* 279: 40748–40754.

Marini C, Scheffer IE, Nabbout R, Suls A, De Jonghe P, Zara F, Guerrini R. 2011. The genetics of Dravet syndrome. *Epilepsia* 52: 24–29.

Massey CA, Sowers LP, Dlouhy BJ, Richerson GB. 2014. Mechanisms of sudden unexpected death in epilepsy: The pathway to prevention. *Nat Rev Neurol* 10: 271–282.

McCormick KA, Isom LL, Ragsdale D, Smith D, Scheuer T, Catterall WA. 1998. Molecular determinants of Na$^+$ channel function in the extracellular domain of the β1 subunit. *J Biol Chem* 273: 3954–3962.

McCusker EC, Bagnéris C, Naylor CE, Cole AR, D'Avanzo N, Nichols CG, Wallace BA. 2012. Structure of a bacterial voltage-gated sodium channel pore reveals mechanisms of opening and closing. *Nat Commun* 3: 1102.

McEwen DP, Isom LL. 2004. Heterophilic interactions of sodium channel β1 subunits with axonal and glial cell adhesion molecules. *J Biol Chem* 279: 52744–52752.

McEwen DP, Chen C, Meadows LS, Lopez-Santiago LF, Isom LL. 2009. The voltage-gated Na$^+$ channel β3 subunit does not mediate trans homophilic cell adhesion or associate with the cell adhesion molecule contactin. *Neurosci Lett* 462: 272–275.

Meng H, Xu H-Q, Yu L, Lin G-W, He N, Su T, Shi Y-W, Li B, Wang J, Liu X-R, et al. 2015. The *SCN1A* mutation database: Updating information and analysis of the relationships among genotype, functional alteration, and phenotype. *Hum Mutat* 36: 573–580.

Miller JA, Agnew WS, Levinson SR. 1983. Principal glycopeptide of the tetrodotoxin/saxitoxin binding protein from *Electrophorus electricus*: Isolation and partial chemical and physical characterization. *Biochemistry* 22: 462–470.

Miller AR, Hawkins NA, McCollom CE, Kearney JA. 2014. Mapping genetic modifiers of survival in a mouse model of Dravet syndrome. *Genes Brain Behav* 13: 163–172.

Mistry AM, Thompson CH, Miller AR, Vanoye CG, George AL, Kearney JA. 2014. Strain- and age-dependent hippocampal neuron sodium currents correlate with epilepsy severity in Dravet syndrome mice. *Neurobiol Dis* 65: 1–11.

Morgan K, Stevens EB, Shah B, Cox PJ, Dixon AK, Lee K, Pinnock RD, Hughes J, Richardson PJ, Mizuguchi K, et al. 2000. β3: An additional auxiliary subunit of the voltage-sensitive sodium channel that modulates channel gating with distinct kinetics. *Proc Natl Acad Sci* 97: 2308–2313.

Namadurai S, Balasuriya D, Rajappa R, Wiemhöfer M, Stott K, Klingauf J, Edwardson JM, Chirgadze DY, Jackson AP. 2014. Crystal structure and molecular imaging of the Na$_v$ channel β3 subunit indicates a trimeric assembly. *J Biol Chem* 289: 10797–10811.

Nelson M, Millican-Slater R, Forrest LC, Brackenbury WJ. 2014. The sodium channel β1 subunit mediates outgrowth of neurite-like processes on breast cancer cells and promotes tumour growth and metastasis. *Int J Cancer* 135: 1–14.

Noda M, Shimizu S, Tanabe T, Takai T, Kayano T, Ikeda T, Takahashi H, Nakayama H, Kanaoka Y, Minamino N, et al. 1984. Primary structure of Electrophorus electricus sodium channel deduced from cDNA sequence. *Nature* 312: 121–127.

Noda M, Suzuki H, Numa S, Stühmer W. 1989. A single point mutation confers tetrodotoxin and saxitoxin insensitivity on the sodium channel II. *FEBS Lett* 259: 213–216.

Ogiwara I, Nakayama T, Yamagata T, Ohtani H, Mazaki E, Tsuchiya S, Inoue Y, Yamakawa K. 2012. A homozygous mutation of voltage-gated sodium channel β$_I$ gene *SCN1B* in a patient with Dravet syndrome. *Epilepsia* 53: e200–e203.

O'Malley HA, Isom LL. 2015. Sodium channel β subunits: Emerging targets in channelopathies. *Annu Rev Physiol* 77: 481–504.

Onganer PU, Djamgoz MBA. 2005. Small-cell lung cancer (human): Potentiation of endocytic membrane activity by voltage-gated Na$^+$ channel expression in vitro. *J Membr Biol* 75: 67–75.

Parent JM, Anderson SA. 2015. Reprogramming patient-derived cells to study the epilepsies. *Nat Neurosci* 18: 360–366.

Patel F, Brackenbury WJ. 2015. Dual roles of voltage-gated sodium channels in development and cancer. *Int J Dev Biol* 59: 357–366.

Patino GA, Isom LL. 2010. Electrophysiology and beyond: Multiple roles of Na$^+$ channel β subunits in development and disease. *Neurosci Lett* 486: 53–59.

Patino GA, Claes LRF, Lopez-Santiago LF, Slat EA, Dondeti RSR, Chen C, O'Malley HA, Gray CBB, Miyazaki H, Nukina N, et al. 2009. A functional null mutation of *SCN1B* in a patient with Dravet syndrome. *J Neurosci* 29: 10764–10778.

Patino GA, Brackenbury WJ, Bao Y, Lopez-Santiago LF, O'Malley HA, Chen C, Calhoun JD, Lafrenière RG, Cossette P, Rouleau GA, et al. 2011. Voltage-gated Na$^+$ channel β1B: A secreted cell adhesion molecule involved in human epilepsy. *J Neurosci* 31: 14577–14591.

Payandeh J, Scheuer T, Zheng N, Catterall WA. 2011. The crystal structure of a voltage-gated sodium channel. *Nature* 475: 353–358.

Raman IM, Bean BP. 1997. Resurgent sodium current and action potential formation in dissociated cerebellar Purkinje neurons. *J Neurosci* 17: 4517–4526.

Ratcliffe CF, Qu Y, McCormick KA, Tibbs VC, Dixon JE, Scheuer T, Catterall WA. 2000. A sodium channel signaling complex: Modulation by associated receptor protein tyrosine phosphatase β. *Nat Neurosci* 3: 437–444.

Ratcliffe CF, Westenbroek RE, Curtis R, Catterall WA. 2001. Sodium channel β1 and β3 subunits associate with neurofascin through their extracellular immunoglobulin-like domain. *J Cell Biol* 154: 427–434.

Recio-Pinto E, Thornhill WB, Duch DS, Levinson SR, Urban BW. 1990. Neuraminidase treatment modifies the function of electroplax sodium channels in planar lipid bilayers. *Neuron* 5: 675–684.

Riuró H, Beltran-Alvarez P, Tarradas A, Selga E, Campuzano O, Vergés M, Pagans S, Iglesias A, Brugada J, Brugada P, et al. 2013. A missense mutation in the sodium channel β2 subunit reveals *SCN2B* as a new candidate gene for Brugada syndrome. *Hum Mutat* 34: 961–966.

Cite this introduction as *Cold Spring Harb Perspect Boil* doi:10.1101/cshperspect.a029264

Roberts RH, Barchi RL. 1987. The voltage-sensitive sodium channel from rabbit skeletal muscle. Chemical characterization of subunits. *J Biol Chem* **262**: 2298–2303.

Roger S, Rollin J, Barascu A, Besson P, Raynal P-I, Iochmann S, Lei M, Bougnoux P, Gruel Y, Le Guennec J-Y. 2007. Voltage-gated sodium channels potentiate the invasive capacities of human non-small-cell lung cancer cell lines. *Int J Biochem Cell Biol* **39**: 774–786.

Rohl CA, Boeckman FA, Baker C, Scheuer T, Catterall WA, Klevit RE. 1999. Solution structure of the sodium channel inactivation gate. *Biochemistry* **38**: 855–861.

Rosker C, Lohberger B, Hofer D, Steinecker B, Quasthoff S, Schreibmayer W. 2007. The TTX metabolite 4, 9-anhydro-TTX is a highly specific blocker of the Na$_v$1. 6 voltage-dependent sodium channel. *Am J Physiol Cell Physiol* **293**: 783–789.

Schmidt JW, Catterall WA. 1986. Biosynthesis and processing of the α subunit of the voltage-sensitive sodium channel in rat brain neurons. *Cell* **46**: 437–444.

Schmidt J, Rossie S, Catterall WA. 1985. A large intracellular pool of inactive Na channel α subunits in developing rat brain. *Proc Natl Acad Sci* **82**: 4847–4851.

Schrey M, Codina C, Kraft R, Beetz C, Kal R, Stefan W, Patt S. 2002. Molecular characterization of voltage-gated sodium channels in human gliomas. **13**: 20–25.

Shi X, Yasumoto S, Kurahashi H, Nakagawa E, Fukasawa T, Uchiya S, Hirose S. 2012. Clinical spectrum of *SCN2A* mutations. *Brain Dev* **34**: 541–545.

Siva N. 2015. UK gears up to decode 100,000 genomes from NHS patients. *Lancet* **385**: 103–104.

Steinlein OK. 2014. Mechanisms underlying epilepsies associated with sodium channel mutations. *Prog Brain Res* **213**: 97–111.

Stevens M, Peigneur S, Tytgat J. 2011. Neurotoxins and their binding areas on voltage-gated sodium channels. *Front Pharmacol* **2**: 71.

Stocker PJ, Bennett ES. 2006. Differential sialylation modulates voltage-gated Na⁺ channel gating throughout the developing myocardium. *J Gen Physiol* **127**: 253–265.

Takahashi K, Tanabe K, Ohnuki M, Narita M, Ichisaka T, Tomoda K, Yamanaka S. 2007. Induction of pluripotent stem cells from adult human fibroblasts by defined factors. *Cell* **131**: 861–872.

Tasaki I, Singer I. 1966. Membrane macromolecules and nerve excitability: A physico-chemical interpretation of excitation in squid giant axons. *Ann NY Acad Sci* **137**: 792–806.

Tasaki I, Singer I, Watanabe A. 1965. Excitation of internally perfused squid giant axons in sodium-free media. *Proc Natl Acad Sci* **54**: 763–769.

Tasaki I, Singer I, Watanabe A. 1966. Excitation of squid giant axons in sodium-free external media. *Am J Physiol* **211**: 746–754.

Terlau H, Heinemann SH, Stühmer W, Pusch M, Conti F, Imoto K, Numa S. 1991. Mapping the site of block by tetrodotoxin and saxitoxin of sodium channel II. *FEBS Lett* **293**: 93–96.

Theile JW, Cummins TR. 2011. Recent developments regarding voltage-gated sodium channel blockers for the treatment of inherited and acquired neuropathic pain syndromes. *Front Pharmacol* **2**: 54.

Tyrrell L, Renganathan M, Dib-Hajj SD, Waxman SG. 2001. Glycosylation alters steady-state inactivation of sodium channel Na$_v$1.9/NaN in dorsal root ganglion neurons and is developmentally regulated. *J Neurosci* **21**: 9629–9637.

Uebachs M, Opitz T, Royeck M, Dickhof G, Horstmann M-T, Isom LL, Beck H. 2010. Efficacy loss of the anticonvulsant carbamazepine in mice lacking sodium channel β subunits via paradoxical effects on persistent sodium currents. *J Neurosci* **30**: 8489–8501.

Vargas E, Yarov-Yarovoy V, Khalili-Araghi F, Catterall WA, Klein ML, Tarek M, Lindahl E, Schulten K, Perozo E, Bezanilla F, et al. 2012. An emerging consensus on voltage-dependent gating from computational modeling and molecular dynamics simulations. *J Gen Physiol* **140**: 587–594.

Vassilev PM, Scheuer T, Catterall WA. 1988. Identification of an intracellular peptide segment involved in sodium channel inactivation. *Science* **241**: 1658–1661.

Vassilev P, Scheuer T, Catterall WA. 1989. Inhibition of inactivation of single sodium channels by a site-directed antibody. *Proc Natl Acad Sci* **86**: 8147–8151.

Wagnon JL, Meisler MH. 2015. Recurrent and non-recurrent mutations of *SCN8A* in epileptic encephalopathy. *Front Neurol* **6**: 1–7.

Wallace RH, Wang DW, Singh R, Scheffer IE, George AL, Phillips HA, Saar K, Reis A, Johnson EW, Sutherland GR, et al. 1998. Febrile seizures and generalized epilepsy associated with a mutation in the Na⁺-channel α1 subunit gene *SCN1B*. *Nat Genet* **19**: 366–370.

West JW, Patton DE, Scheuer T, Wang Y, Goldin AL, Catterall WA. 1992. A cluster of hydrophobic amino acid residues required for fast Na⁺-channel inactivation. *Proc Natl Acad Sci* **89**: 10910–10914.

Wong H-K, Sakurai T, Oyama F, Kaneko K, Wada K, Miyazaki H, Kurosawa M, De Strooper B, Saftig P, Nukina N. 2005. β subunits of voltage-gated sodium channels are novel substrates of β-site amyloid precursor protein-cleaving enzyme (BACE1) and γ-secretase. *J Biol Chem* **280**: 23009–23017.

Xiao ZC, Ragsdale DS, Malhotra JD, Mattei LN, Braun PE, Schachner M, Isom LL. 1999. Tenascin-R is a functional modulator of sodium channel β subunits. *J Biol Chem* **274**: 26511–26517.

Xiao Y, Blumenthal K, Cummins TR, Sciences B. 2014. Gating pore currents demonstrate selective and specific modulation of individual sodium channel voltage sensors by biological toxins. *Mol Pharmacol* **46202**: 159–167.

Yang M, Kozminski DJ, Wold LA, Modak R, Calhoun JD, Isom LL, Brackenbury WJ. 2012. Therapeutic potential for phenytoin: Targeting Na$_v$1.5 sodium channels to reduce migration and invasion in metastatic breast cancer. *Breast Cancer Res Treat* **134**: 603–615.

Yang D, Xu J, Zhu T, Fan J, Lai L, Zhang J, Chen YE. 2014a. Effective gene targeting in rabbits using RNA-guided Cas9 nucleases. *J Mol Cell Biol* **6**: 97–99.

Yang H, Wang H, Jaenisch R. 2014b. Generating genetically modified mice using CRISPR/Cas-mediated genome engineering. *Nat Protoc* **9**: 1956–1968.

Yereddi NR, Cusdin FS, Namadurai S, Packman LC, Monie TP, Slavny P, Clare JJ, Powell AJ, Jackson AP. 2013. The immunoglobulin domain of the sodium channel β3 subunit contains a surface-localized disulfide bond that is required for homophilic binding. *FASEB J* **27**: 568–580.

Yu FH, Westenbroek RE, Silos-Santiago I, McCormick KA, Lawson D, Ge P, Ferriera H, Lilly J, DiStefano PS, Catterall WA, et al. 2003. Sodium channel β4, a new disulfide-linked auxiliary subunit with similarity to β2. *J Neurosci* **23**: 7577–7585.

Yu FH, Mantegazza M, Westenbroek RE, Robbins CA, Kalume F, Burton KA, Spain WJ, McKnight GS, Scheuer T, Catterall WA. 2006. Reduced sodium current in GABAergic interneurons in a mouse model of severe myoclonic epilepsy in infancy. *Nat Neurosci* **9**: 1142–1149.

Zhang Y, Hartmann HA, Satin J. 1999. Glycosylation influences voltage-dependent gating of cardiac and skeletal muscle sodium channels. *J Membr Biol* **171**: 195–207.

Zhang X, Ren W, DeCaen P, Yan C, Tao X, Tang L, Wang J, Hasegawa K, Kumasaka T, He J, et al. 2012. Crystal structure of an orthologue of the NaChBac voltage-gated sodium channel. *Nature* **486**: 130–134.

Zhang M-M, Wilson MJ, Azam L, Gajewiak J, Rivier JE, Bulaj G, Olivera BM, Yoshikami D. 2013. Co-expression of Na$_v$β subunits alters the kinetics of inhibition of voltage-gated sodium channels by pore-blocking μ-conotoxins. *Br J Pharmacol* **168**: 1597–1610.

Zhang Y, Kecskés A, Copmans D, Langlois M, Crawford AD, Ceulemans B, Lagae L, de Witte PAM, Esguerra CV. 2015. Pharmacological characterization of an antisense knockdown zebrafish model of Dravet syndrome: Inhibition of epileptic seizures by the serotonin agonist fenfluramine. *PLoS ONE* **10**: e0125898.

Measuring Ca²⁺-Dependent Modulation of Voltage-Gated Ca²⁺ Channels in HEK-293T Cells

Jessica R. Thomas[1,2] and Amy Lee[1]

[1]*Departments of Molecular Physiology and Biophysics, Otolaryngology—Head and Neck Surgery, and Neurology, University of Iowa, Iowa City, Iowa 52242;* [2]*Interdisciplinary Graduate Program in Neuroscience, University of Iowa, Iowa City, Iowa 52242*

Voltage-gated Ca²⁺ (Ca_v) channels regulate a variety of biological processes, such as muscle contraction, gene expression, and neurotransmitter release. Ca_v channels are subject to diverse forms of regulation, including those involving the Ca²⁺ ions that permeate the pore. High voltage-activated Ca_v channels undergo Ca²⁺-dependent inactivation (CDI) and facilitation (CDF), which can regulate processes such as cardiac rhythm and synaptic plasticity. CDI and CDF differ slightly between Ca_v1 (L-type) and Ca_v2 (P/Q-, N-, and R-type) channels. Human embryonic kidney cells transformed with SV40 large T-antigen (HEK-293T) are advantageous for studying CDI and CDF of a particular type of Ca_v channel. HEK-293T cells do not express endogenous Ca_v channels, but Ca_v channels can be expressed exogenously at high levels in these cells by transient transfection. This protocol describes how to characterize and analyze Ca²⁺-dependent modulation of recombinant Ca_v channels in HEK-293T cells.

MATERIALS

It is essential that you consult the appropriate Material Safety Data Sheets and your institution's Environmental Health and Safety Office for proper handling of equipment and hazardous material used in this protocol.

RECIPES: Please see the end of this protocol for recipes indicated by <R>. Additional recipes can be found online at http://cshprotocols.cshlp.org/site/recipes.

Reagents

cDNA encoding enhanced green fluorescent protein (pEGFP)
  pEGFP fluorescence is used to identify transfected cells.

cDNAs encoding the voltage-gated Ca²⁺ (Ca_v) channel components of interest (i.e., $Ca_v\alpha_2\delta$, $Ca_v\beta$, $Ca_v\alpha_1$ subunits)
  These cDNAs can be purchased from Addgene.

Dulbecco's modified Eagle medium, high-glucose (DMEM; Thermo Fisher Scientific, 11965)
Extracellular recording solution <R>
Fetal bovine serum (FBS; Atlanta Biologicals, S11150)
FuGENE 6 transfection reagent (Promega, E2691)
HEK-293T/17 cells (ATCC, CRL-11268)
Intracellular recording solution <R>
Sylgard (optional; see Step 9)
Versene (Thermo Fisher Scientific, 15040066)

Equipment

Capillaries, borosilicate glass, 4-inch, 1.5-mm OD, 1.12-mm ID (WPI, TW150-4)

Cell-culture dish, 35 mm × 10 mm (Corning)

Data acquisition and analysis software

Microforge

Microcentrifuge tubes, sterile, 1.5-mL

Micropipette puller

Micropipettor, P1000

Patch-clamp electrophysiology setup including inverted fluorescence microscope, micromanipulators, amplifier, and perfusion setup

Tissue culture hood

Tissue culture incubator, humidified, 5% CO_2, 37°C

METHOD

Transient Transfection

1. Plate HEK-293T cells in 35-mm culture dishes. Maintain cells in 2 mL of culture medium (DMEM supplemented with 10% FBS) in a tissue culture incubator at 37°C with a humidified atmosphere and 5% CO_2.

   Transfect cells when they are ~70% confluent.

2. Approximately 30 min before transfection, aspirate culture medium. Replace with fresh medium.

3. Combine 6 µL FuGENE 6 transfection reagent with 100 µL of FBS-free DMEM in a sterile 1.5-mL microcentrifuge tube. Mix gently.

4. To the same tube, add 1.5 µg $Ca_v\alpha_1$ cDNA, 0.5 µg $Ca_v\beta$ cDNA, 0.5 µg $Ca_v\alpha_2\delta$ cDNA, and 50 ng pEGFP cDNA. Mix gently.

5. Incubate for 15 min at room temperature.

6. Add transfection mixture dropwise to cells, covering the entire dish.

7. Incubate for 24–72 h at 37°C. Replace spent culture medium with fresh culture medium every 24 h.

8. Isolate cells for electrophysiological recordings:

   i. Aspirate culture medium from transfected cells. Add 1 mL of Versene to dish. Incubate for 1 min.

   Do not use trypsin–EDTA to isolate cells; it can proteolyze channels.

   ii. Aspirate the Versene. Gently triturate the cells in 1 mL of culture medium by pipetting up and down 10 times using a P1000 micropipettor.

   iii. Plate cells at low density by adding 1–2 drops of the cell suspension to culture dishes containing 2 mL of culture medium.

   Allow cells to rest for at least 2 h before recording at 37°C.

Whole-Cell Voltage-Clamp Recording

9. Using a micropipette puller, produce electrodes with a resistance of 4–6 megaohms (MΩ) in the extracellular recording solution. Polish the tips with a microforge.

   Electrodes can be coated with Sylgard to reduce pipette capacitance, although this is generally unnecessary.

10. Remove culture medium from dish. Replace with 1 mL of Ca^{2+}-containing extracellular recording solution.

Cite this protocol as *Cold Spring Harb Protoc*; doi:10.1101/pdb.prot087213

11. Identify a GFP-positive cell with a suitable spherical/boxlike morphology.

    Avoid recording from oval- and triangular-shaped cells. These can yield distorted currents because of an inability to properly clamp the membrane voltage.

12. Fill electrode with intracellular recording solution. Begin electrophysiological recordings.

    The intracellular recording solution contains adenosine triphosphate, which can undergo autohydrolysis under some conditions. Although working solutions can be maintained at room temperature, maintain the intracellular recording solution on ice for prolonged experiments.

13. Obtain a gigaohm (GΩ) seal. Change the holding voltage (V_h) to −80 mV.

14. Apply negative pressure to rupture the membrane to gain whole-cell access.

15. Before running voltage protocols, reduce capacitance and series resistance electronically to 60%–70%.

Characterization of Ca^{2+}-Dependent Inactivation (CDI)

16. Evoke a Ca^{2+} current (I_{Ca}) using a 1-sec depolarizing test pulse from V_h (−80 mV) to voltages ranging from −20 mV to +20 mV (Fig. 1A). Measure inactivation as the residual current amplitude at the end of the pulse normalized to the peak current amplitude (I_{res}/I_{peak}).

    Plotting I$_{res}$/I$_{peak}$ against test voltages produces a U-shaped curve, with maximal inactivation at the test voltage evoking maximal inward I$_{Ca}$ (Fig. 1).

17. Measure voltage-dependent inactivation (VDI):

    i. Use a gravity-driven or pressurized perfusion system to exchange the Ca^{2+}-containing extracellular recording solution for a Ba^{2+}-containing solution.

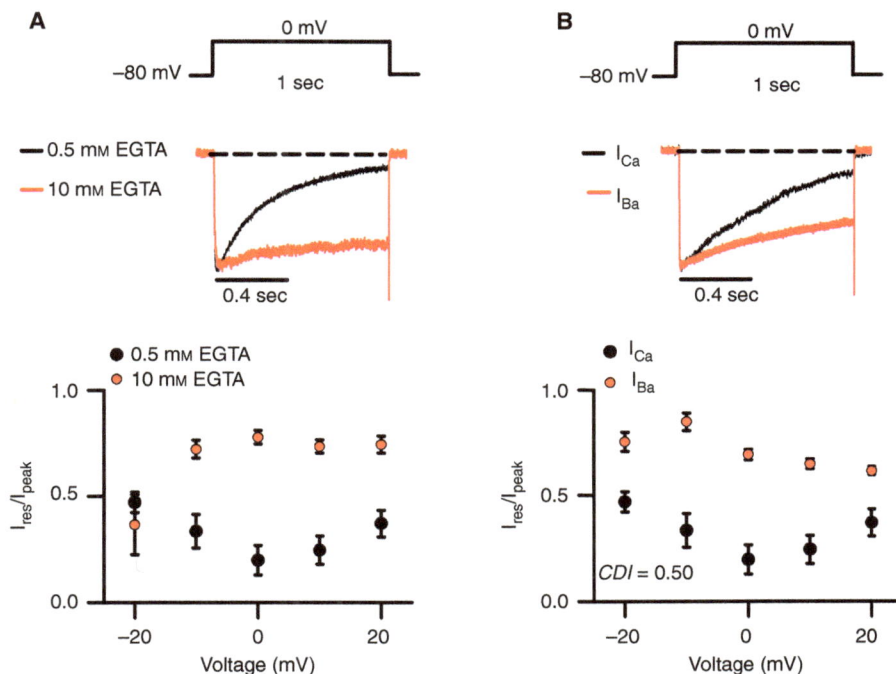

FIGURE 1. CDI of Ca$_v$2.2 channels. (A,B) (upper) Voltage protocols and representative current traces. Currents were leak-subtracted using the P/4 method. (lower) Plots of the ratios (I_{res}/I_{peak}) of the residual current amplitude at the end of the pulse (I_{res}) normalized to the peak current amplitude (I_{peak}) versus the test voltage. (A) Intracellular recording solution contained 0.5 or 10 mM EGTA. (B) Intracellular recording solution contained 0.5 mM EGTA; extracellular solution contained 10 mM Ca^{2+} (I_{Ca}) or Ba^{2+} (I_{Ba}). Currents and averaged data for I_{Ca} ($n = 10$) and I_{Ba} ($n = 8$) are from different cells. CDI was calculated as the difference between the I_{res}/I_{peak} of I_{Ba} and I_{Ca}, where I_{res}/I_{peak} for $I_{Ba} = 0.69 \pm 0.02$ and $I_{Ca} = 0.19 \pm 0.07$ for a test pulse to 0 mV.

Unlike Ca^{2+}, Ba^{2+} binds poorly to calmodulin, an essential mediator of CDI. Using Ba^{2+} as the permeant ion, Ca_v channels undergo inactivation that is voltage- rather than Ca^{2+}-dependent.

ii. Repeat Step 16 using the Ba^{2+}-containing extracellular recording solution.

The difference between I_{res}/I_{peak} for I_{Ca} and I_{Ba} is a convenient metric for CDI (Fig. 1B). Alternatively, because the differences in CDI and VDI are robust, population averages of cells recorded in Ca^{2+}- or Ba^{2+}-containing solution are often used.

Characterization of Ca^{2+}-Dependent Facilitation (CDF) by $Ca_v2.1$ Channels

18. Fill electrode an intracellular recording solution containing a high concentration of EGTA (e.g., 10 mM).

The elevated EGTA concentration minimizes CDI.

19. To measure CDF (Fig. 2A):
    i. Give a test pulse from V_h (−80 mV) to 0 mV (P1).

    ii. One second later, give a 50-msec prepulse to a voltage within the desired experimental range.

    iii. Five milliseconds later, give a test pulse from V_h (−80 mV) to 0 mV (P2).

20. Repeat Step 19 using Ba^{2+}-containing extracellular recording solution as the charge carrier (Fig. 2B) and reducing the voltage of the test pulse for P1 and P2 (Steps 19.i and 19.iii, respectively) to −10 mV.

Ba^{2+} causes a negative shift in the voltage-dependence of activation as a result of surface charge screening effects that must be compensated for in the channel modulation protocols.

21. Calculate facilitation (F) by plotting the ratio of P2 to P1 against the prepulse voltage.

If P2/P1 is >1, then the current is facilitated. A convenient metric for CDF is the difference between facilitation of I_{Ca} and I_{Ba} at the prepulse voltage eliciting maximal facilitation of I_{Ca} (Fig. 2B).

DISCUSSION

Voltage-gated Ca_v channels mediate Ca^{2+} signals that regulate cellular excitability, muscle contraction, gene expression, and hormone/neurotransmitter release. Ca_v channels are multisubunit complexes

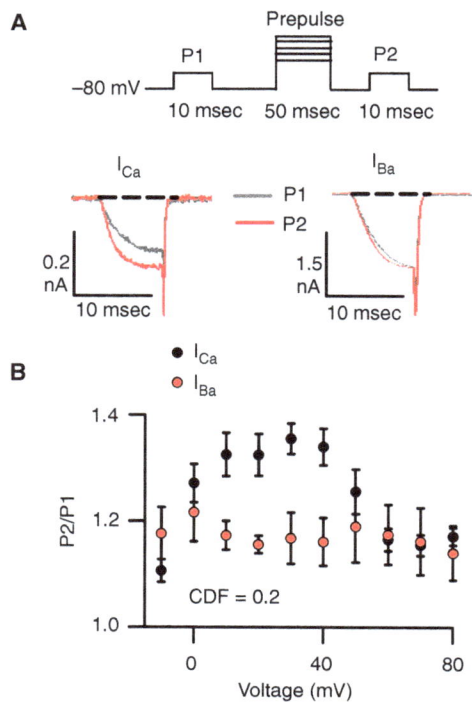

FIGURE 2. CDF of $Ca_v2.1$ channels. Currents were recorded using 10 mM Ca^{2+} or Ba^{2+} in the extracellular solution and 10 mM EGTA in the intracellular solution; measurements and averaged data for I_{Ca} and I_{Ba} are from different cells. Currents were leak-subtracted using the P/4 method. (A). Voltage protocol (*upper*) and representative currents (*lower*) evoked by 10-msec test pulses of 0 mV (for I_{Ca}) and −10 mV (for I_{Ba}) before (gray) and after (red) a 50-msec prepulse to +30 mV. (B) The ratio of P2/P1 current amplitudes plotted against prepulse voltage. CDF = the difference between P2/P1 for I_{Ca} and I_{Ba}, where P2/P1 for I_{Ca} = 1.4 ± 0.03 ($n = 4$) and for $I_{Ba} = 1.2 \pm 0.05$ ($n = 4$) for a +30-mV prepulse.

that consist of a pore-forming α_1 subunit and auxiliary subunits Ca$_V\beta$ and Ca$_V\alpha_2\delta$ (Simms and Zamponi 2014). The α_1 subunit mediates Ca^{2+} entry; the auxiliary subunits regulate trafficking and other properties of the channel. Ten genes encode distinct α_1 subunits that display distinct pharmacological and biophysical properties (Ca$_V$1.1–4, Ca$_V$2.1–3, and Ca$_V$3.1–3). Mutations in these genes are associated with disorders such as epilepsy, migraine, deafness, and congenital stationary night blindness (Pietrobon 2010; Striessnig et al. 2010).

The Ca$_V$1 and Ca$_V$2 α_1 subunits are constitutively associated with calmodulin (CaM), which is essential for CDI and CDF. The amino- and carboxy-terminal lobes of CaM each contain two EF-hand Ca^{2+} binding domains. Increases in global and local Ca^{2+} entry through Ca$_V$ channels are detected by the carboxy- and amino-terminal lobes of CaM, which play distinct roles in the CDI of Ca$_V$1 and Ca$_V$2 channels, respectively. The CDI of Ca$_V$1 channels is mediated by the carboxy-terminal lobe of CaM and is insensitive to strong intracellular Ca^{2+} buffering. In contrast, the CDI of Ca$_V$2 channels is mediated by the amino-terminal lobe of CaM and is inhibited by strong intracellular Ca^{2+} buffering. Interestingly, CDF of Ca$_V$2.1 channels is mediated by the carboxy-terminal lobe of CaM, and is spared by high concentrations of intracellular EGTA. The complex regulation of Ca$_V$ channels by CaM has been described in recent reviews (Christel and Lee 2012; Ben-Johny and Yue 2014).

Factors other than Ca^{2+} can influence inactivation of Ca$_V$1 and Ca$_V$2 channels. For example, inactivation of Ca$_V$ channels can be driven by VDI, as seen by using the Ba^{2+}-containing extracellular recording solution. "Fast" VDI occurs from the resting (closed) state (Patil et al. 1998), whereas "slow" VDI occurs not only from the fast-inactivated state but also the open state (Sokolov et al. 2000). Both types of VDI are influenced by Ca$_V\beta$ subunits. Some Ca$_V\beta$ subunits (such as Ca$_V\beta_{1b}$) promote VDI, whereas the membrane-associated subunit Ca$_V\beta_{2a}$ diminishes VDI (Buraei and Yang 2010). Because CDI can be masked by strong VDI, the use of Ca$_V$ channels containing Ca$_V\beta_{2a}$ is a common approach for studying CDI (Lee et al. 2000; DeMaria et al. 2001).

The α_1 subunit of Ca$_V$ channels is subject to alternative splicing, which increases the functional diversity of Ca$_V$ channels (Lipscombe et al. 2013). Such splicing events can either heighten or dampen Ca^{2+}-dependent regulation of Ca$_V$ channels. For example, inclusion of the alternatively spliced exon 42A of Ca$_V$1.3 channels enhances CDI as a result of truncation of a distal carboxy-terminal autoregulatory domain (Singh et al. 2008). Also, alternative splicing of exon 37 in Ca$_V$2.1 alters CDF (Soong et al. 2002; Chaudhuri et al. 2005). These examples illustrate the capability of splicing events to serve as molecular switches for CDI and CDF of Ca$_V$ channels.

Other regulators of the CDI and CDF of Ca$_V$ channels include a family of Ca^{2+}-binding proteins (CaBPs) related to CaM (Haeseleer et al. 2000). Unlike CaM, CaBPs contain at least one nonfunctional EF-hand, and are almost exclusively expressed in neurons. CaBPs can compete with and/or allosterically modulate CaM interactions with Ca$_V$1 channels, inhibiting CDI. For Ca$_V$2.1 channels, one CaBP family member (CaBP1) strongly inhibits CDI, VDI, and voltage-dependent activation. The regulation of Ca$_V$ channels by CaBPs has been reviewed recently (Christel and Lee 2012; Lee et al. 2014).

RECIPES

Extracellular Recording Solution

Reagent	Concentration
CaCl$_2$ or BaCl$_2$	10 mM
MgCl$_2$	1 mM
Tris	150 mM

Prepare Ca^{2+}- and Ba^{2+}-containing variants of the solution to measure Ca^{2+}-dependent inactivation and voltage-dependent inactivation, respectively, of voltage-gated Ca^{2+} channels. Adjust pH to 7.3 using methanesulfonic acid. (The osmolarity should be ~290–310.) Filter and store at 4°C.

Intracellular Recording Solution

Reagent	Concentration
EGTA	0.5 or 10 mM
HEPES	10 mM
MgCl$_2$	2 mM
Mg-ATP	2 mM
N-methyl-D-glucamine	140 mM

Use solutions containing 0.5 mM EGTA to record Ca^{2+}-dependent inactivation (CDI) in Ca$_v$2 channels. Use solutions containing 0.5 or 10 mM EGTA to record the CDI of Ca$_v$2 or Ca$_v$1 channels, respectively, and 10 mM EGTA to record the CDF of Ca$_v$2.1 channels (high concentrations of EGTA inhibit the CDI of Ca$_v$2 channels). Adjust pH to 7.3 using methanesulfonic acid. (The osmolarity should be ~290–310.) Filter-sterilize. Store aliquots at −20°C.

ACKNOWLEDGMENTS

This work was supported by grants from the National Institutes of Health (DC009433, NS084190, NS045549) and a Carver Research Program of Excellence Award.

REFERENCES

Ben-Johny M, Yue DT. 2014. Calmodulin regulation (calmodulation) of voltage-gated calcium channels. *J Gen Physiol* **143**: 679–692.

Buraei Z, Yang J. 2010. The β subunit of voltage-gated Ca^{2+} channels. *Physiol Rev* **90**: 1461–1506.

Chaudhuri D, Alseikhan BA, Chang SY, Soong TW, Yue DT. 2005. Developmental activation of calmodulin-dependent facilitation of cerebellar P-type Ca^{2+} current. *J Neurosci* **25**: 8282–8294.

Christel C, Lee A. 2012. Ca^{2+}-dependent modulation of voltage-gated Ca^{2+} channels. *Biochim Biophys Acta* **1820**: 1243–1252.

DeMaria CD, Soong T, Alseikhan BA, Alvania RS, Yue DT. 2001. Calmodulin bifurcates the local Ca^{2+} signal that modulates P/Q-type Ca^{2+} channels. *Nature* **411**: 484–489.

Haeseleer F, Sokal I, Verlinde CL, Erdjument-Bromage H, Tempst P, Pronin AN, Benovic JL, Fariss RN, Palczewski K. 2000. Five members of a novel Ca^{2+}-binding protein (CABP) subfamily with similarity to calmodulin. *J Biol Chem* **275**: 1247–1260.

Lee A, Scheuer T, Catterall WA. 2000. Ca^{2+}/calmodulin-dependent facilitation and inactivation of P/Q-type Ca^{2+} channels. *J Neurosci* **20**: 6830–6838.

Lee A, Fakler B, Kaczmarek LK, Isom LL. 2014. More than a pore: Ion channel signaling complexes. *J Neurosci* **34**: 15159–15169.

Lipscombe D, Allen SE, Toro CP. 2013. Control of neuronal voltage-gated calcium ion channels from RNA to protein. *Trends Neurosci* **36**: 598–609.

Patil PG, Brody DL, Yue DT. 1998. Preferential closed-state inactivation of neuronal calcium channels. *Neuron* **20**: 1027–1038.

Pietrobon D. 2010. Ca$_v$2.1 channelopathies. *Pflügers Arch* **460**: 375–393.

Simms BA, Zamponi GW. 2014. Neuronal voltage-gated calcium channels: Structure, function, and dysfunction. *Neuron* **82**: 24–45.

Singh A, Gebhart M, Fritsch R, Sinnegger-Brauns MJ, Poggiani C, Hoda JC, Engel J, Romanin C, Striessnig J, Koschak A. 2008. Modulation of voltage- and Ca^{2+}-dependent gating of Ca$_v$1.3 L-type calcium channels by alternative splicing of a C-terminal regulatory domain. *J Biol Chem* **283**: 20733–20744.

Sokolov S, Weiss RG, Timin EN, Hering S. 2000. Modulation of slow inactivation in class A Ca^{2+} channels by β-subunits. *J Physiol* **527**: 445–454.

Soong TW, DeMaria CD, Alvania RS, Zweifel LS, Liang MC, Mittman S, Agnew WS, Yue DT. 2002. Systematic identification of splice variants in human P/Q-type channel α$_1$2.1 subunits: Implications for current density and Ca^{2+}-dependent inactivation. *J Neurosci* **22**: 10142–10152.

Striessnig J, Bolz HJ, Koschak A. 2010. Channelopathies in Ca$_v$1.1, Ca$_v$1.3, and Ca$_v$1.4 voltage-gated L-type Ca^{2+} channels. *Pflügers Arch* **460**: 361–374.

Cite this protocol as *Cold Spring Harb Protoc*; doi:10.1101/pdb.prot087213

Strategies for Investigating G-Protein Modulation of Voltage-Gated Ca^{2+} Channels

Van B. Lu and Stephen R. Ikeda[1]

Section on Transmitter Signaling, Laboratory of Molecular Physiology, National Institute on Alcohol Abuse and Alcoholism, National Institutes of Health, Bethesda, Maryland 20892-9411

G-protein-coupled receptor modulation of voltage-gated ion channels is a common means of fine-tuning the response of channels to changes in membrane potential. Such modulation impacts physiological processes such as synaptic transmission, and hence therapeutic strategies often directly or indirectly target these pathways. As an exemplar of channel modulation, we examine strategies for investigating G-protein modulation of Ca$_V$2.2 or N-type voltage-gated Ca^{2+} channels. We focus on biochemical and genetic tools for defining the molecular mechanisms underlying the various forms of Ca$_V$2.2 channel modulation initiated following ligand binding to G-protein-coupled receptors.

INTRODUCTION

The primary factor determining the probability of a voltage-gated ion channel being open or closed is the transmembrane potential. An additional layer of regulation adds subtle alterations to the open probability versus membrane potential relationship—a process termed modulation. Modulation is present for nearly all voltage-gated ion channels, takes place over milliseconds to hours, is initiated by a wide range of stimuli originating from both extra- and intracellular sources, and has a large impact on physiological processes such as synaptic transmission. To illustrate experimental approaches for exploring ion channel modulation, we focus on a canonical signaling pathway linking G-protein-coupled receptors (GPCRs) to Ca$_V$2.2 (N-type) Ca^{2+} channels in mammalian neurons. For details of a method to examine G-protein modulation of N-type voltage-gated Ca^{2+} channels (VGCCs), see Protocol 1: G-Protein Modulation of Voltage-Gated Ca^{2+} Channels from Isolated Adult Rat Superior Cervical Ganglion Neurons (Lu and Ikeda 2016).

GPCRs are seven-transmembrane-spanning proteins that are coupled to heterotrimeric G-proteins consisting of Gα, Gβ, and Gγ subunits, with the latter two subunits serving as a functional monomer termed G$\beta\gamma$ (Oldham and Hamm 2008). Following activation by ligand binding (e.g., a neurotransmitter), Gα exchanges the bound guanine nucleotide GDP for GTP, which triggers the dissociation of the G$\beta\gamma$ subunit (Fig. 1A). Both Gα-GTP and G$\beta\gamma$ interact with downstream effectors to bring about cellular changes. The pathway can be simple (e.g., direct binding of G$\beta\gamma$ to ion channels) or complex (e.g., initiating a cascade of enzymatic processes resulting in post-translational protein modifications such as phosphorylation). The G-protein cycle is terminated when GTP on Gα is hydrolyzed to GDP, resulting in the increased affinity for G$\beta\gamma$ and consequent reforming of the G$\alpha\beta\gamma$ heterotrimer. Accessory proteins, such as regulators of G-protein signaling (RGS proteins), are

---

[1]Correspondence: sikeda@mail.nih.gov

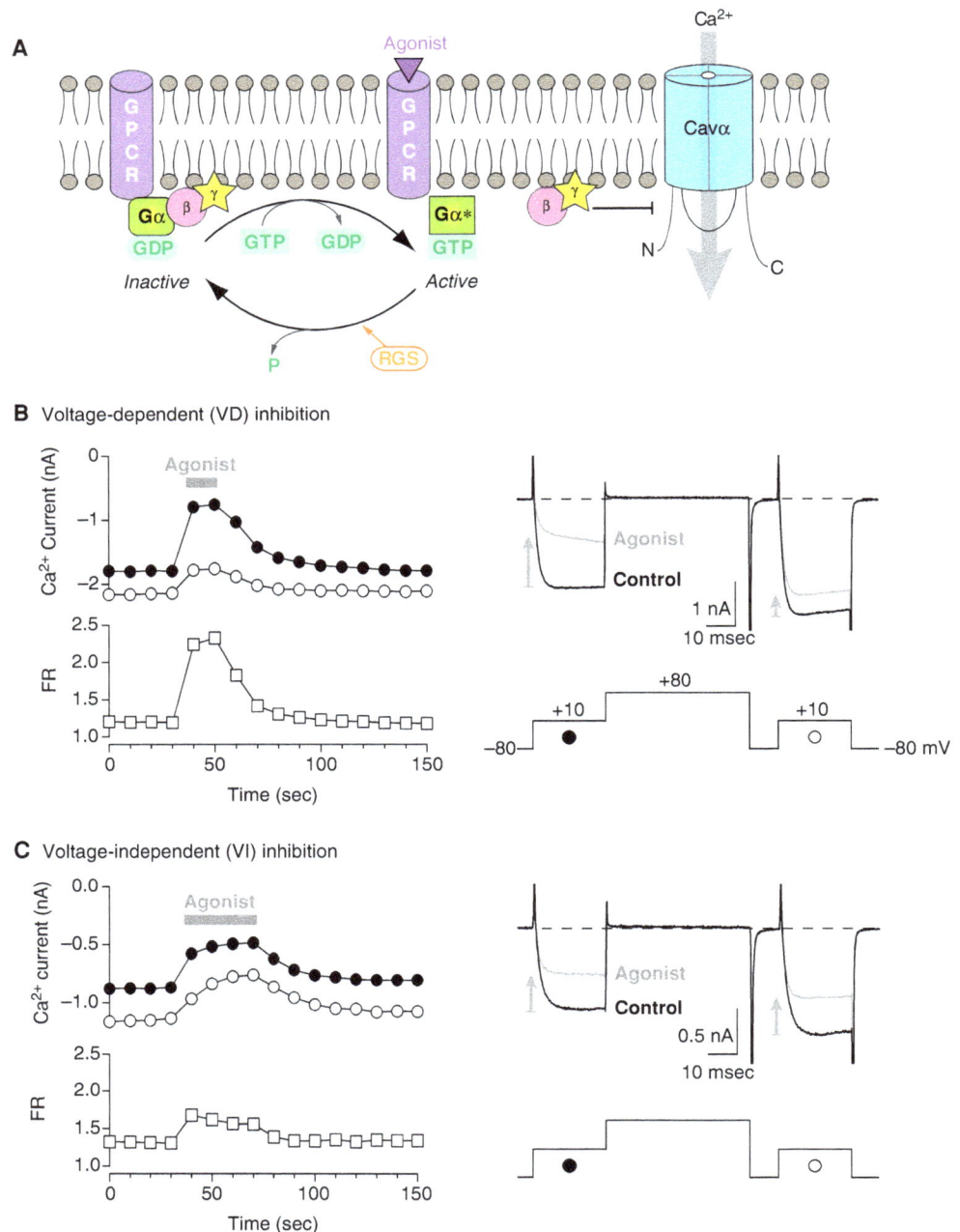

FIGURE 1. Mechanisms of G-protein modulation of VGCCs. (*A*) Schematic diagram of the heterotrimeric G-protein cycle. Heterotrimeric G-proteins, consisting of Gα and Gβγ subunits, are bound together in an inactive state (*left*). On agonist binding and activation of G-protein-coupled receptors (GPCRs), the Gα subunit exchanges bound GDP with GTP and dissociates from Gβγ (*middle*). Active GTP-bound Gα (indicated by asterisk) can engage downstream signaling cascades, and free Gβγ subunits can interact directly with ion channels such as VGCCs (e.g., $Ca_V\alpha$ [*right*]). Intrinsic Gα subunit hydrolysis of GTP to GDP restores the heterotrimeric G-protein complex and can be accelerated by accessory proteins such as regulators of G-protein signaling (RGS proteins). (*B,C*) Examples of voltage-dependent (*B*) and voltage-independent (*C*) inhibition of VGCCs. (*Left*) Plots of Ca^{2+} current amplitude over time during agonist application, indicated by a solid gray bar. Ca^{2+} currents were recorded using the triple-pulse protocol, shown at the *bottom right* of (*B*) and (*C*), which includes two 25-msec test pulses to +10 mV from a holding potential of −80 mV, separated by a 50-msec conditioning pulse to +80 mV. Filled black circles represent Ca^{2+} current amplitude during the first test pulse or prepulse; open circles represent Ca^{2+} current amplitude during the second test pulse or postpulse. The facilitation ratio (FR) was calculated as the postpulse/prepulse Ca^{2+} current amplitude. (*Right*) exemplar recordings of Ca^{2+} current before (black) and during (gray) agonist application. Gray arrows indicate the magnitude of Ca^{2+} current inhibition induced by agonist.

capable of accelerating the intrinsic GTPase activity of Gα and thereby altering the magnitude and time course of responses (Ross and Wilkie 2000).

VGCCs of the $Ca_V2.x$ family are inhibited by many types of GPCRs whether natively or heterologously expressed (Hille 1994; Ikeda and Dunlap 1999; Elmslie 2003; Dolphin 2003; Tedford and Zamponi 2006). Two canonical pathways have been studied over the past 30 yr. The first, termed voltage-dependent (VD) (Bean 1989; Ikeda 1991), is readily identified during voltage-clamp experiments that use a triple-pulse voltage protocol (Elmslie et al. 1990) by two signature characteristics: (1) VD slowing of current activation and (2) relief of inhibition following a large depolarizing conditioning pulse (Fig. 1B, right). The latter is parameterized by the FR, defined as the postpulse/prepulse Ca^{2+} current amplitude. Hence, a large increase in FR during agonist application is diagnostic of VD modulation (Fig. 1B, left). Mechanistically, VD inhibition arises from direct interaction of the Gβγ subunit with the α-subunit of the Ca^{2+} channel (Ikeda 1996; Herlitze et al. 1996). The second, termed voltage-independent (VI), displays neither of these characteristics. Thus, the inhibited current approximates a scaled-down version of the control current (Fig. 1C, right) and the FR changes little during GPCR activation (Fig. 1C, left). Although VI inhibition arises from a variety of mechanisms, the best-studied VI pathway arises from depletion of plasma membrane PIP_2 following activation of PLCβ by $Gα_q$-GTP (Gamper et al. 2004; Delmas et al. 2005). Variables such as the composition of the Ca^{2+} channel (e.g., the presence of a specific Ca^{2+} channel β-subunit) can alter the characteristics of VD and VI modulation, thus complicating this dichotomy (Mitra-Ganguli et al. 2009; Heneghan et al. 2009; Keum et al. 2014). However, the simplistic model put forward suffices as a basis for discussing tools available to investigate mechanisms of ion channel modulation by GPCRs.

Once a response is identified (e.g., channel modulation following agonist application), a typical strategy for probing G-protein participation involves establishing the following: (1) participation of a heterotrimeric G-protein; (2) the subfamily of Gα heterotrimer (e.g., $G_{i/o}$, $G_{q/11}$, G_s, $G_{12/13}$) involved; and (3) whether Gα-GTP, Gβγ, or both mediate the effect. We will discuss some of the available reagents and tools (Table 1) loosely in this order proceeding from simpler-to-use reagents to more complex genetic manipulations.

DETERMINING G-PROTEIN INVOLVEMENT

The earliest G-protein reagents used to exploit the intracellular dialysis provided by the whole-cell patch-clamp technique were modifications of GTP and GDP (Holz et al. 1986; Lewis et al. 1986; Dolphin and Scott 1987). Guanosine 5′-O-[γ-thio]triphosphate (GTP-γ-S) and guanosine 5′-[β,γ-imido]triphosphate (Gpp(NH)p) are hydrolysis-resistant analogs of GTP that can substitute for GTP when included in the patch pipette. Depending on the cell type (and perhaps other factors), two different responses are seen when GTP-γ-S or Gpp(NH)p is substituted for GTP in the patch pipette solution. In one scenario, responses are normal before and during agonist application but fail to return to baseline following agonist washout, as they would with a GTP-containing pipette solution (Lewis et al. 1986). In the second scenario, modulation occurs spontaneously (in the absence of agonist) during dialysis with GTP-γ-S or Gpp(NH)p (Ikeda 1996), and, when fully developed, addition of agonist produces no further effect (Ikeda and Schofield 1989). High basal GDP–GTP turnover appears to dictate the latter effect, although the precise mechanisms are unknown. Aluminum fluoride (AlF_4^-) produces effects similar to GTP-γ-S or Gpp(NH)p but is little used in contemporary electrophysiology experiments (Elmslie et al. 1992). A caveat for using F^- as an intracellular anion in patch experiments is that trace levels of contaminating Al^{3+} can potentially confound experiments via AlF_4^--mediated effects. Guanosine 5′-[β-thio] diphosphate (GDP-β-S) is a hydrolysis-resistant analog of GDP that, when substituted for GTP in the pipette, blocks the activation of G-proteins by receptors (Eckstein et al. 1979). In general, high concentrations (e.g., 2 mM) are used and prolonged dialysis times (e.g., 5 min) are optimal for blocking of G-protein-mediated responses (Chen et al. 2005). It should be noted that GTP/GDP analogs are rather blunt instruments, and although useful, the ubiquity of GTP-binding proteins (e.g., the RAS family of small GTPases) puts limits on interpretation.

TABLE 1. Tools to study G-protein modulation of ion channels

Name (abbreviations or alternative names)	Mechanism of action	Inhibits/activates	Additional notes
Gα$_s$			
Cholera toxin (CTX)	ADP-ribosylates Arg-201 residue of Gα$_s$ proteins, locks Gα$_s$ in GTP-bound state, activates downstream effectors (e.g., adenylyl cyclase)	Activates initially but later down-regulates Gα$_s$	A/α subunit = catalytic domain
Gα$_{i/o}$			
Pertussis toxin (PTX)	ADP-ribosylates Cys residue (fourth position from the carboxyl terminus) of Gα$_{i/o}$ proteins, locks Gα$_{i/o}$ in GDP-bound state, uncouples G-protein-coupled receptors (GPCRs) from Gα$_{i/o}$ proteins	Inhibits	Slow acting (rate-limiting step: entry into cell by endocytosis) S1 subunit of toxin = catalytic domain
N-ethylmaleimide (NEM)	Alkylates cysteine residues	Inhibits	pH-sensitive (pH 6.5–7.5) Nonspecific/targets all sulfhydryl groups
Gα$_{q/11}$			
Phospholipase C β3 subunit carboxyl terminus (PLCβ3-ct)	Binds activated Gα$_{q/11}$ but lacks phospholipase enzymatic activity, may also display intrinsic Gα$_q$ GTPase-activating protein (GAP) activity	Inhibits	
Regulator of G-protein signaling 2 (RGS2)	RGS2 selectively interacts with Gα$_{q/11}$ and overexpression blocks signaling mediated by Gα$_{q/11}$ by accelerating intrinsic GTPase activity of the Gα subunit	Inhibits	
Gβγ			
G-protein-coupled receptor kinase 2 or 3 carboxyl terminus (MAS-GRK2-ct, MAS-GRK3-ct)	Devoid of kinase activity but retains ability to bind Gβγ	Inhibits	Myristic acid attachment signal (MAS) on the amino terminus increases potency of Gβγ buffering
Transducin (Gα$_t$)	Exists primarily in the GDP-bound state with high affinity for Gβγ	Inhibits	Overexpression of different Gα subtypes can also buffer Gβγ, but transducin more advantageous because interactions with neurotransmitter receptors should be minimal
Gallein (chemical name: 3′,4′,5′,6′-tetrahydroxyspiro [isobenzofuran-1(3H),9′-(9H)xanthen]-3-one)	Small molecule binds a protein–protein interface of Gβ (the Gα switch II binding region on Gβ), disrupts interaction with downstream Gβγ effectors (e.g., PI3-kinase γ, Rac1)	Inhibits	Other small-molecule analogs (M119)

 Cite this introduction as *Cold Spring Harb Protoc*; doi:10.1101/pdb.top087072

Tool	Description	Effect	Comments
Synthetic peptides: SIRKALNILGYPDYD (SIRK peptide)	Binds Gβγ subunits and blocks interaction of Gβγ with downstream effectors (e.g., PI3-kinase, PLCβ2)	Inhibits[a]	Other similar peptide: SIGKAFKILGYPDYD or SIGK peptide. No effect of SIRK peptide on Gβγ-mediated inhibition of N-type Ca^{2+} channels in SCG neurons or Gβγ-dependent inhibition of adenylyl cyclase in vitro
G-proteins (all)			
Guanosine 5'-[-thio]triphosphate tetralithium salt (GTP-γ-S)	Nonhydrolyzable GTP analog, prevents hydrolysis of GTP bound to Gα leading to constitutive activity (prevents inactivation of Gα subunits)	Activates	Other GTP analogs: guanosine 5'-[β γ-imino]triphosphate or Gpp[NH]p
Guanosine 5'-O-(2-thio)diphosphate (GDP-β-S)	Nonhydrolyzable GDP analog	Inhibits	
AlF_4^-	Mimics the γ-phosphate of GTP bound to Gα protein and inhibits GTPase activity	Activates	
Antibodies against Gα subunits	Recognizes epitope specific for Gα protein	May inhibit	Very large protein (~150 kD); limited intracellular diffusion of large molecules through the patch pipette
Genetic tools Dominant-negative (DN) Gα Group of Gα mutants including mutations within: 1) G1 box 2) G3 box/Switch II 3) G4 box/NKXD sequence	Decrease G-protein signaling by various mechanisms: -sequester Gβγ subunits -sequester activated GPCR by heterotrimeric complex -sequester activated GPCR by nucleotide-free Gα (e.g., Gαo D273N/Q205L)	Inhibits	Not a complete knockout of Gα protein, mixed population of DN-Gα and endogenous Gα
DREADD designed GPCRs	Custom designed for activation by nonendogenous ligand	Activates	Well suited for in vivo studies
Knockout animals	Genetically modified animals where Gα gene is disrupted or removed	Inhibits	Low survival of knockouts, compensatory up-regulation of other Gα proteins
Short interfering RNA (siRNA)	Synthetic double-stranded RNA (dsRNA) designed to specifically target a complementary mRNA sequence for degradation, therefore reducing mRNA that undergoes translation	Inhibits	Concerns with use: –May induce immune reaction because of similarity to viral dsRNA –Off-target effects –Incomplete knockdown of protein –Compensatory up-regulation of other Gα proteins
Optogenetic tools Rhodopsin	Light activation of Gi/o proteins	Activates	May require exogenous 11-cis-retinal
Opto-X, rhodopsin/GPCR chimera	Light activation of Gq or Gs proteins	Activates	Trafficking issues

[a]Cell-permeant versions of SIRK (mSIRK or tatSIRK) in intact cells "activated" ERK/MAP kinase pathways in a Gβγ-dependent manner.

G-PROTEIN FAMILY IDENTIFICATION

ADP-ribosylating bacterial exotoxins (Krueger and Barbieri 1995) are often the first approach to defining the G-protein family involved in a response. *Bordetella pertussis* toxin (PTX) contains an enzymatic subunit S1 or A protomer (Ribeiro-Neto et al. 1985) that ADP-ribosylates a cysteine residue in the carboxyl terminus of $G_{i/o}$-family proteins ($G\alpha_{i1-3}$, $G\alpha_o$, transducin, but not $G\alpha_z$). The reaction uncouples the G-protein heterotrimer from the GPCR and thus ablates modulation dependent on this family of G-proteins (Holz et al. 1986; Lewis et al. 1986). PTX does not, however, allow discrimination between $G\alpha$- or $G\beta\gamma$-mediated modulation. Typically, PTX holotoxin is added to culture media for overnight (i.e., >12 h) incubation at 37°C as the rate-limiting step is the temperature-dependent S1-subunit internalization via cell surface glycoprotein binding. These conditions are problematic for short-lived preparations such as brain slices or dissociated central nervous system (CNS) neurons. Shorter incubations (e.g., 4 h) at higher PTX concentrations or inclusion of PTX in the patch pipette have been tried but with varying success. If the latter is attempted, a preparation containing only the S1 subunit (i.e., not the holotoxin) should be used. *N*-ethylmaleimide (NEM), a sulfhydryl alkylating reagent, has been used when acute PTX-like effects were desired (Nakajima et al. 1990; Shapiro et al. 1994). However, NEM is inherently nonspecific and conditions for use are so fastidious that the reagent has fallen out of favor. It should be noted that U73122, a drug used to block phospholipase C (PLC), likely acts similarly to NEM and thus use of this agent as a diagnostic for PLC involvement is not recommended (Horowitz et al. 2005; Won et al. 2009; Klein et al. 2011). The PTX S1-subunit can also be heterologously expressed in mammalian cells (Ikeda et al. 1999) or from a conditional knock-in mouse model (Regard et al. 2007).

Analogous to PTX, an ADP-ribosylating toxin from *Vibrio cholera* (CTX) specifically modifies an interior arginine in $G\alpha_s$ or $G\alpha_{olf}$. However, unlike PTX, the modification constitutively activates the $G\alpha$ subunit, leading to stimulation of adenylyl cyclase and subsequent elevations in cAMP (Ribeiro-Neto et al. 1985). Following stimulation, the $G\alpha$ subunit is rapidly degraded, leading to a disruption of the modulatory pathways (Chang and Bourne 1989). For pathways independent of cAMP, CTX has proven to be a useful tool for identifying the G-protein family involved (Zhu and Ikeda 1994). However, if cAMP modifies the response under study, the interpretation is less clear as small amounts of activated $G\alpha_{s/olf}$ can produce prolonged elevations of cAMP.

A toxin from *Pasteurella multocida* (PMT) has been proposed for probing $G\alpha_{q/11}$-dependent pathways (Lei et al. 2001). However, PMT appears to affect numerous G-protein families and thus interpretation is compromised (Surguy et al. 2014). A cyclic depsipeptide isolated from *Chromobacterium*, YM254890, appears to be a selective inhibitor of $G\alpha_{q/11}$ and is potentially useful for identifying pathways dependent on this G-protein subfamily (Takasaki et al. 2004; Veale et al. 2007; Nishimura et al. 2010). At this time, the availability of YM254890 or a related compound, UBO-QIC (Inamdar et al. 2015), appears limited, and only a few studies using the compound have been published.

Antibodies targeted to the carboxyl terminus of specific $G\alpha$ subunits and delivered to the intracellular compartment via diffusion from patch pipettes (Wilk-Blaszczak et al. 1994) or microinjected into cells (McFadzean et al. 1989; Zhu and Ikeda 1994) have been used to identify the participation of $G\alpha$ subfamilies involved in channel modulation. The diffusion of a 150-kDa protein from even low-resistance patch pipettes is predicted to be slow (Pusch and Neher 1988), and a comprehensive study identifying antibodies that possess functional blocking activity and recognize epitopes in living cells is not currently available.

IDENTIFYING $G\alpha$- AND $G\beta\gamma$-DEPENDENT PATHWAYS

The realization that $G\beta\gamma$ mediates VD modulation of $Ca_V2.x$ Ca^{2+} channels was accomplished by heterologous expression of $G\beta\gamma$ in cell lines and neurons (Herlitze et al. 1996; Ikeda 1996). This procedure disrupted the resting 1:1 stoichiometry of $G\alpha:G\beta\gamma$, resulting in basal (absence of agonist) Ca^{2+} currents that mimicked the VD modulated state. An inducible version of this strategy using rapamycin-mediated heterodimerization to target a modified $G\beta\gamma$ to the plasma membrane has been reported (Putyrski

Cite this introduction as *Cold Spring Harb Protoc*; doi:10.1101/pdb.top087072

and Schultz 2011). Unlike Gβγ, overexpression of Gα did not produce the VD phenotype but did occlude modulation when agonist was applied, presumably because excess Gα-GDP binds with Gβγ released from the heterotrimer during GPCR activation (Ikeda 1996; Jeong and Ikeda 1999).

Molecules that specifically interact with free Gβγ allow evaluation of their role in channel modulation. Overexpression of Gα subunits, although effective at impairing modulation, suffers from interpretative issues. For example, overexpressed Gα might "steal" Gβγ from native heterotrimers to form entities that do not couple to the GPCR under investigation. Hence, Gβγ-binding modules derived from downstream effectors provide an alternative as affinity for Gβγ is likely to be less than for Gα-GDP. One such module is the carboxy-terminus pleckstrin homology (PH) domain from GRK2 or GRK3 (Koch et al. 1994; Touhara et al. 1994; Daaka et al. 1997). Expression of a membrane-targeted version of this module has been shown to block Gβγ effects while leaving Gα-GTP effects unaltered (Kammermeier and Ikeda 1999; Kammermeier et al. 2000). Moreover, studies using resonance energy transfer techniques (FRET or BRET) provide evidence that Gβγ reversibly interacts with the module during agonist application (Hollins et al. 2009). A downside to this tool is the requirement for genetic overexpression rather than acute application. To this end, peptides and small molecules have been designed to interact with the surface of Gβγ that is buried in the heterotrimer and thus exposed on subunit dissociation. SIRK peptide (SIRKALNILGYPDYD) and gallein ($3',4',5',6'$-tetrahydroxyspiro[isobenzofuran-1($3H$),9'-($9H$)xanthen]-3-one) are examples of such molecules (Davis et al. 2005; Bonacci et al. 2006). Unfortunately, these molecules may interact with residues on Gβ that are not critical for VD Ca^{2+} channel modulation (Lin and Smrcka 2011). Further testing is required to validate these tools for ion channel modulation studies.

Molecules that interact specifically with $Gα_{q/11}$-GTP while leaving the Gβγ untouched have revealed that both entities participate in $Ca_V2.2$ modulation. RGS2 or the carboxyl terminus of PLCβ1 (PLCβ1-ct), when overexpressed, binds to $Gα_{q/11}$-GTP following GPCR activation and inhibits responses mediated by the Gα subunit (Kammermeier and Ikeda 1999). For example, activation of $Gα_{q/11}$-linked receptors in sympathetic neurons produced primarily VI $Ca_V2.2$ Ca^{2+}-channel inhibition. However, in neurons expressing RGS2 or PLCβ1-ct, the response was primarily VD, indicating that Gβγ derived from $Gα_{q/11}$ activation is capable of producing VD inhibition (Kammermeier et al. 2000). Unfortunately, cognate tools for investigating potential actions of Gα-GTP of the $G_{i/o}$ and G_s family have not been validated in ion channel systems.

DOMINANT-NEGATIVE MUTANTS AND KNOCKDOWN TECHNOLOGIES

Dominant-negative mutants are used extensively in the small GTPase protein field, but well-characterized Gα dominant-negative mutants are lacking. Gα subunits engineered to alter the nucleotide binding specificity from GTP to XTP (e.g., $Gα_o$ D273N/Q205L) have been put forward as G-protein dominant-negative mutants (Yu and Simon 1998). Specificity across G-protein families has not been systematically tested and thus caution is advised when interpreting results from experiments using these constructs.

As with many proteins, genetic knockdown of G-protein subunits has provided insight into specificity of signaling pathways. However, whether using knockout animals (constitutive or inducible) or transient technologies (e.g., antisense, siRNA), it should be noted that heterotrimeric G-proteins display redundancy and resistance to perturbations in stoichiometry (i.e., compensation). A rigorous and comprehensive G-protein-subunit RNAi silencing study performed in HeLa cells showed the resiliency of the system to perturbation (Krumins and Gilman 2006). Thus, caution is required when interpreting studies performed before documentation of this phenomenon.

DESIGNER AND OPTICALLY ACTIVATED GPCRs

The final two categories of tools are useful for probing the functional consequences of ion channel modulation in vivo rather than dissecting the pathways at a cellular and molecular level. Designer receptors exclusively activated by designer drugs (DREADDs) are modified native GPCRs engineered to bind a synthetic ligand with no endogenous counterpart (Lee et al. 2014). DREADDs coupling to specific G-protein signaling pathways (e.g., $G_{i/o}$ vs $G_{q/11}$) combined with modern gene targeting technology allow investigation, in vivo, of how specific pathways impact channel function and the integrative impact of modulation.

Although most GPCRs use peptides or small organic molecules as endogenous agonists, light-sensing organs use photons to activate a covalently bound endogenous ligand (retinal) bound to an opsin, a prototypical GPCR. In mammals, opsins are coupled to transducins, members of the $G\alpha_{i/o}$ family. The native expression of the opsins and transducins is restricted almost exclusively to light-sensing cells such as rods and cones. Outside of the visual system, heterologously expressed rhodopsin coupled to native $G\alpha_{i/o}$ leads to $Ca_V2.x$ Ca^{2+} channel modulation similar to that produced by traditional GPCRs (Li et al. 2005; Gutierrez et al. 2011; Masseck et al. 2014). Although native mammalian opsins couple exclusively to the $G\alpha_{i/o}$ family, chimeric constructs, termed optoXRs, extend the coupling to the $G\alpha_{q/11}$ and $G\alpha_s$ family (Airan et al. 2009). In vivo expression of rhodopsin or optoXRs in mouse CNS produces changes in behavior following illumination with an appropriate wavelength of light. The spatial and temporal resolution provided by such tools should facilitate our understanding of how channel modulation impacts higher-order physiological processes.

REFERENCES

Airan RD, Thompson KR, Fenno LE, Bernstein H, Deisseroth K. 2009. Temporally precise in vivo control of intracellular signalling. *Nature* 458: 1025–1029.

Bean BP. 1989. Neurotransmitter inhibition of neuronal calcium currents by changes in channel voltage dependence. *Nature* 340: 153–156.

Bonacci TM, Mathews JL, Yuan C, Lehmann DM, Malik S, Wu D, Font JL, Bidlack JM, Smrcka AV. 2006. Differential targeting of Gβγ-subunit signaling with small molecules. *Science* 312: 443–446.

Chang FH, Bourne HR. 1989. Cholera toxin induces cAMP-independent degradation of Gs. *J Biol Chem* 264: 5352–5357.

Chen H, Puhl HL, Niu SL, Mitchell DC, Ikeda SR. 2005. Expression of Rem2, an RGK family small GTPase, reduces N-type calcium current without affecting channel surface density. *J Neurosci* 25: 9762–9772.

Daaka Y, Pitcher JA, Richardson M, Stoffel RH, Robishaw JD, Lefkowitz RJ. 1997. Receptor and Gβγ isoform-specific interactions with G protein–coupled receptor kinases. *Proc Natl Acad Sci* 94: 2180–2185.

Davis TL, Bonacci TM, Sprang SR, Smrcka AV. 2005. Structural and molecular characterization of a preferred protein interaction surface on G protein βγ subunits. *Biochemistry* 44: 10593–10604.

Delmas P, Coste B, Gamper N, Shapiro MS. 2005. Phosphoinositide lipid second messengers: New paradigms for calcium channel modulation. *Neuron* 47: 179–182.

Dolphin AC. 2003. G protein modulation of voltage-gated calcium channels. *Pharmacol Rev* 55: 607–627.

Dolphin AC, Scott RH. 1987. Calcium channel currents and their inhibition by (–)-baclofen in rat sensory neurones: Modulation by guanine nucleotides. *J Physiol (Lond)* 386: 1–17.

Eckstein F, Cassel D, Levkovitz H, Lowe M, Selinger Z. 1979. Guanosine 5′-O-(2-thiodiphosphate). An inhibitor of adenylate cyclase stimulation by guanine nucleotides and fluoride ions. *J Biol Chem* 254: 9829–9834.

Elmslie KS. 2003. Neurotransmitter modulation of neuronal calcium channels. *J Bioenerg Biomembr* 35: 477–489.

Elmslie KS, Zhou W, Jones SW. 1990. LHRH and GTP-γ-S modify calcium current activation in bullfrog sympathetic neurons. *Neuron* 5: 75–80.

Elmslie KS, Kammermeier PJ, Jones SW. 1992. Calcium current modulation in frog sympathetic neurones: L-current is relatively insensitive to neurotransmitters. *J Physiol (Lond)* 456: 107–123.

Gamper N, Reznikov V, Yamada Y, Yang J, Shapiro MS. 2004. Phosphatidylinositol 4,5-bisphosphate signals underlie receptor-specific Gq/11-mediated modulation of N-type Ca^{2+} channels. *J Neurosci* 24: 10980–10992.

Gutierrez DV, Mark MD, Masseck O, Maejima T, Kuckelsberg D, Hyde RA, Krause M, Kruse W, Herlitze S. 2011. Optogenetic control of motor coordination by Gi/o protein–coupled vertebrate rhodopsin in cerebellar Purkinje cells. *J Biol Chem* 286: 25848–25858.

Heneghan JF, Mitra-Ganguli T, Stanish LF, Liu L, Zhao R, Rittenhouse AR. 2009. The Ca^{2+} channel β subunit determines whether stimulation of Gq-coupled receptors enhances or inhibits N current. *J Gen Physiol* 134: 369–384.

Herlitze S, Garcia DE, Mackie K, Hille B, Scheuer T, Catterall WA. 1996. Modulation of Ca^{2+} channels by G-protein βγ subunits. *Nature* 380: 258–262.

Hille B. 1994. Modulation of ion-channel function by G-protein-coupled receptors. *Trends Neurosci* 17: 531–536.

Hollins B, Kuravi S, Digby GJ, Lambert NA. 2009. The C-terminus of GRK3 indicates rapid dissociation of G protein heterotrimers. *Cell Signal* 21: 1015–1021.

Holz GG, Rane SG, Dunlap K. 1986. GTP-binding proteins mediate transmitter inhibition of voltage-dependent calcium channels. *Nature* 319: 670–672.

Horowitz LF, Hirdes W, Suh B-C, Hilgemann DW, Mackie K, Hille B. 2005. Phospholipase C in living cells: Activation, inhibition, Ca^{2+} requirement, and regulation of M current. *J Gen Physiol* 126: 243–262.

Ikeda SR. 1991. Double-pulse calcium channel current facilitation in adult rat sympathetic neurones. *J Physiol (Lond)* 439: 181–214.

Ikeda SR. 1996. Voltage-dependent modulation of N-type calcium channels by G-protein βγ subunits. *Nature* 380: 255–258.

Ikeda SR, Dunlap K. 1999. Voltage-dependent modulation of N-type calcium channels: Role of G protein subunits. *Adv Second Messenger Phosphoprotein Res* 33: 131–151.

Ikeda SR, Schofield GG. 1989. Somatostatin blocks a calcium current in rat sympathetic ganglion neurones. *J Physiol (Lond)* 409: 221–240.

Ikeda SR, Jeong SW, Kammermeier PJ, Ruiz-Velasco V, King MM. 1999. Heterologous expression of a green fluorescent protein-pertussis toxin S1 subunit fusion construct disrupts calcium channel modulation in rat superior cervical ganglion neurons. *Neurosci Lett* 271: 163–166.

Inamdar V, Patel A, Manne BK, Dangelmaier C, Kunapuli SP. 2015. Characterization of UBO-QIC as a $G\alpha_q$ inhibitor in platelets. *Platelets* 26: 771–778.

Jeong S-W, Ikeda SR. 1999. Sequestration of G-protein βγ subunits by different G-protein α subunits blocks voltage-dependent modulation of Ca^{2+} channels in rat sympathetic neurons. *J Neurosci* 19: 4755–4761.

Kammermeier PJ, Ikeda SR. 1999. Expression of RGS2 alters the coupling of metabotropic glutamate receptor 1a to M-type K^+ and N-type Ca^{2+} channels. *Neuron* 22: 819–829.

Kammermeier PJ, Ruiz-Velasco V, Ikeda SR. 2000. A voltage-independent calcium current inhibitory pathway activated by muscarinic agonists in rat sympathetic neurons requires both $G\alpha_{q/11}$ and $G\beta\gamma$. *J Neurosci* **20**: 5623–5629.

Keum D, Baek C, Kim DI, Kweon HJ, Suh BC. 2014. Voltage-dependent regulation of $Ca_V2.2$ channels by G_q-coupled receptor is facilitated by membrane-localized β subunit. *J Gen Physiol* **46**: 891.

Klein RR, Bourdon DM, Costales CL, Wagner CD, White WL, Williams JD, Hicks SN, Sondek J, Thakker DR. 2011. Direct activation of human phospholipase C by its well known inhibitor U73122. *J Biol Chem* **286**: 12407–12416.

Koch WJ, Hawes BE, Inglese J, Luttrell LM, Lefkowitz RJ. 1994. Cellular expression of the carboxyl terminus of a G protein-coupled receptor kinase attenuates $G_{\beta\gamma}$-mediated signaling. *J Biol Chem* **269**: 6193–6197.

Krueger KM, Barbieri JT. 1995. The family of bacterial ADP-ribosylating exotoxins. *Clin Microbiol Rev* **8**: 34–47.

Krumins AM, Gilman AG. 2006. Targeted knockdown of G protein subunits selectively prevents receptor-mediated modulation of effectors and reveals complex changes in non-targeted signaling proteins. *J Biol Chem* **281**: 10250–10262.

Lee H-M, Giguère PM, Roth BL. 2014. DREADDs: Novel tools for drug discovery and development. *Drug Discov Today* **19**: 469–473.

Lei Q, Talley EM, Bayliss DA. 2001. Receptor-mediated inhibition of G protein–coupled inwardly rectifying potassium channels involves $G\alpha_q$ family subunits, phospholipase C, and a readily diffusible messenger. *J Biol Chem* **276**: 16720–16730.

Lewis DL, Weight FF, Luini A. 1986. A guanine nucleotide-binding protein mediates the inhibition of voltage-dependent calcium current by somatostatin in a pituitary cell line. *Proc Natl Acad Sci* **83**: 9035–9039.

Li X, Gutierrez DV, Hanson MG, Han J, Mark MD, Chiel H, Hegemann P, Landmesser LT, Herlitze S. 2005. Fast noninvasive activation and inhibition of neural and network activity by vertebrate rhodopsin and green algae channelrhodopsin. *Proc Natl Acad Sci* **102**: 17816–17821.

Lin Y, Smrcka AV. 2011. Understanding molecular recognition by G protein $\beta\gamma$ subunits on the path to pharmacological targeting. *Mol Pharmacol* **80**: 551–557.

Lu VB, Ikeda SR. 2016. G-protein modulation of voltage-gated Ca^{2+} channels from isolated adult rat superior cervical ganglion neurons. *Cold Spring Harb Protoc* doi: 10.1101/pdb.prot.091223.

Masseck OA, Spoida K, Dalkara D, Maejima T, Rubelowski JM, Wallhorn L, Deneris ES, Herlitze S. 2014. Vertebrate cone opsins enable sustained and highly sensitive rapid control of $G_{i/o}$ signaling in anxiety circuitry. *Neuron* **81**: 1263–1273.

McFadzean I, Mullaney I, Brown DA, Milligan G. 1989. Antibodies to the GTP binding protein, G_o, antagonize noradrenaline-induced calcium current inhibition in NG108-15 hybrid cells. *Neuron* **3**: 177–182.

Mitra-Ganguli T, Vitko I, Perez-Reyes E, Rittenhouse AR. 2009. Orientation of palmitoylated $Ca_V\beta2a$ relative to $Ca_V2.2$ is critical for slow pathway modulation of N-type Ca^{2+} current by tachykinin receptor activation. *J Gen Physiol* **134**: 385–396.

Nakajima T, Irisawa H, Giles W. 1990. *N*-ethylmaleimide uncouples muscarinic receptors from acetylcholine-sensitive potassium channels in bullfrog atrium. *J Gen Physiol* **96**: 887–903.

Nishimura A, Kitano K, Takasaki J, Taniguchi M, Mizuno N, Tago K, Hakoshima T, Itoh H. 2010. Structural basis for the specific inhibition of heterotrimeric G_q protein by a small molecule. *Proc Natl Acad Sci* **107**: 13666–13671.

Oldham WM, Hamm HE. 2008. Heterotrimeric G protein activation by G-protein-coupled receptors. *Nat Rev Mol Cell Biol* **9**: 60–71.

Pusch M, Neher E. 1988. Rates of diffusional exchange between small cells and a measuring patch pipette. *Pflugers Arch* **411**: 204–211.

Putyrski M, Schultz C. 2011. Switching heterotrimeric G protein subunits with a chemical dimerizer. *Chem Biol* **18**: 1126–1133.

Regard JB, Kataoka H, Cano DA, Camerer E, Yin L, Zheng Y-W, Scanlan TS, Hebrok M, Coughlin SR. 2007. Probing cell type-specific functions of G_i in vivo identifies GPCR regulators of insulin secretion. *J Clin Invest* **117**: 4034–4043.

Ribeiro-Neto FA, Mattera R, Hildebrandt JD, Codina J, Field JB, Birnbaumer L, Sekura RD. 1985. ADP-ribosylation of membrane components by pertussis and cholera toxin. *Methods Enzymol* **109**: 566–572.

Ross EM, Wilkie TM. 2000. GTPase-activating proteins for heterotrimeric G proteins: Regulators of G protein signaling (RGS) and RGS-like proteins. *Annu Rev Biochem* **69**: 795–827.

Shapiro MS, Wollmuth LP, Hille B. 1994. Modulation of Ca^{2+} channels by PTX-sensitive G-proteins is blocked by *N*-ethylmaleimide in rat sympathetic neurons. *J Neurosci* **14**: 7109–7116.

Surguy SM, Duricki DA, Reilly JM, Lax AJ, Robbins J. 2014. The actions of *Pasteurella multocida* toxin on neuronal cells. *Neuropharmacol* **77**: 9–18.

Takasaki J, Saito T, Taniguchi M, Kawasaki T, Moritani Y, Hayashi K, Kobori M. 2004. A novel $G\alpha_{q/11}$-selective inhibitor. *J Biol Chem* **279**: 47438–47445.

Tedford HW, Zamponi GW. 2006. Direct G protein modulation of Ca_V2 calcium channels. *Pharmacol Rev* **58**: 837–862.

Touhara K, Inglese J, Pitcher JA, Shaw G, Lefkowitz RJ. 1994. Binding of G protein $\beta\gamma$-subunits to pleckstrin homology domains. *J Biol Chem* **269**: 10217–10220.

Veale EL, Kennard LE, Sutton GL, MacKenzie G, Sandu C, Mathie A. 2007. $G\alpha_q$-mediated regulation of TASK3 two-pore domain potassium channels: The role of protein kinase C. *Mol Pharmacol* **71**: 1666–1675.

Wilk-Blaszczak MA, Singer WD, Gutowski S, Sternweis PC, Belardetti F. 1994. The G protein G_{13} mediates inhibition of voltage-dependent calcium current by bradykinin. *Neuron* **13**: 1215–1224.

Won Y-J, Puhl HL, Ikeda SR. 2009. Molecular reconstruction of mGluR5a-mediated endocannabinoid signaling cascade in single rat sympathetic neurons. *J Neurosci* **29**: 13603–13612.

Yu B, Simon MI. 1998. Interaction of the xanthine nucleotide binding $G_o\alpha$ mutant with G protein-coupled receptors. *J Biol Chem* **273**: 30183–30188.

Zhu Y, Ikeda SR. 1994. VIP inhibits N-type Ca^{2+} channels of sympathetic neurons via a pertussis toxin-insensitive but cholera toxin-sensitive pathway. *Neuron* **13**: 657–669.

Protocol 1

G-Protein Modulation of Voltage-Gated Ca²⁺ Channels from Isolated Adult Rat Superior Cervical Ganglion Neurons

Van B. Lu and Stephen R. Ikeda[1]

Section on Transmitter Signaling, Laboratory of Molecular Physiology, National Institute on Alcohol Abuse and Alcoholism, National Institutes of Health, Bethesda, Maryland 20892-9411

Sympathetic neurons isolated from adult rat superior cervical ganglia (SCG) are a well-established model to study G-protein modulation of voltage-gated Ca^{2+} channels (VGCCs). SCG neurons can be easily dissociated and are amendable to heterologous expression of genes, including genetic tools to study G-protein signaling pathways, within a time frame to maintain good spatial voltage-clamp control of membrane potential during electrophysiological recordings (8–36 h postdissociation). This protocol focuses on examining G-protein modulation of VGCCs; however, the procedures and experimental setup for acute application of agonists can be applied to study modulation of other ion channels (e.g., M-current, G-protein-coupled inwardly rectifying K^+ channels). We also discuss some common sources of artifacts that can arise during acute drug application onto dissociated neurons, which can mislead interpretation of results.

MATERIALS

It is essential that you consult the appropriate Material Safety Data Sheets and your institution's Environmental Health and Safety Office for proper handling of equipment and hazardous materials used in this protocol.

RECIPES: Please see the end of this protocol for recipes indicated by <R>. Additional recipes can be found online at http://cshprotocols.cshlp.org/site/recipes.

Reagents

Adult rat (250–350 g)
Agarose (2%) in 150 mM NaCl (for construction of ground bridge; see Step 14)
Culture medium for superior cervical ganglia (SCG) neurons <R>
Dissection medium (e.g., Hanks balanced salt solution [HBSS])
Dissociation medium for superior cervical ganglia (SCG) neurons <R>
Drug solution(s) of interest
Ethanol (70%)
External recording solution for voltage-gated Ca^{2+} channels <R>
Internal recording solution for voltage-gated Ca^{2+} channels <R>

---

[1]Correspondence: sikeda@mail.nih.gov

Poly-L-lysine solution in 0.1 M borate buffer <R>

Coat 35-mm tissue culture dishes with this solution for at least 4 h in a laminar flow tissue culture hood. Aspirate excess solution and rinse dishes with filter-sterilized (0.2 µm) deionized water. Store coated dishes in a dry place for up to 6 mo at room temperature. These dishes will be used in Step 8.

Equipment

Beaker (250-mL)

Dissecting microscope with fiber-optic illumination

Dissecting trays (60-mm and 150-mm Petri dishes filled one-half-full with Sylgard 184)

Drug applicator system (including drug applicator tip, drug applicator manifold, drug reservoirs, drug reservoir platform/holder)

Figure 1B shows the design for a custom-built, multichannel drug application device. Commercial drug applicator devices are also available.

Electrophysiology recording rig (including inverted microscope, amplifier, analog-to-digital converter, oscilloscope, micromanipulator, computer, data acquisition program)

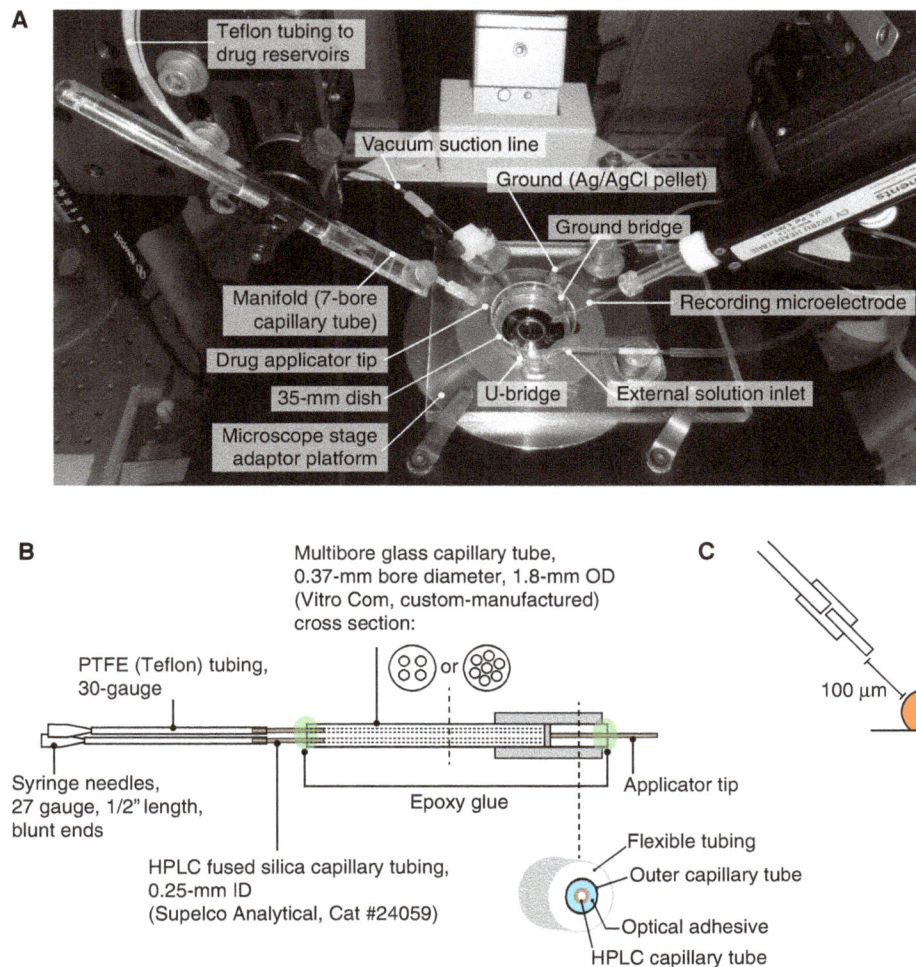

FIGURE 1. Suggested design for a manually operated, gravity-fed perfusion system and drug applicator device. (*A*) Design layout of inverted microscope stage for performing electrophysiological recordings on cells plated in 35-mm dishes. (*B*) Design for a custom-built, multichannel drug application device. There are commercial drug applicator devices available as well, but descriptions of the individual parts to build a custom device are listed in the figure. (*C*) Schematic of the placement of drug applicator device relative to a plated SCG neuron being patched. PTFE, polytetrafluoroethylene; HPLC, high-performance liquid chromatography.

External solution perfusion system (including vacuum suction line, external solution inlet line with a flow regulator, external solution reservoir and holder, perfusion chamber, ground [e.g., Ag/AgCl pellet] and U-bridges)

A 35-mm dish (Step 11) serves as the perfusion chamber. See Figure 1A for suggested setup of the perfusion system.

Forceps, large and fine-tip (e.g., Dumont #5 forceps)
Glass microelectrodes, fabricated from a multistep micropipette puller
Guillotine
Laminar flow tissue culture hood
Shaking water bath (37°C)
Source of 95% O_2/5% CO_2 (see Step 4)
Spring scissors (e.g., Fine Science Tools #15006-09)
Surgical scissors
Swinging bucket centrifuge capable of low centrifuge speeds (~60g; e.g., Eppendorf #5804R)
Tissue culture dishes (35-mm)
Tissue culture flask (25-cm^2) with plug seal cap (e.g., Corning #430168)
Tissue culture incubator (37°C, humidified atmosphere of 5% CO_2 in air)

METHOD

Dissection and Dissociation of SCG Neurons

The following is a brief description of the procedures to isolate adult rat SCG neurons for study. Previous reviews on the procedure have been published (Ikeda 1997, 2004; Ikeda and Jeong 2004) and should be referred to for further details.

1. Anesthetize rat following approved methods from the animal welfare guidelines of your country and institution. Decapitate rat with a laboratory guillotine. Place the head in a 250-ml beaker half-full of chilled dissection medium (e.g., Hanks balanced salt solution [HBSS]) to rinse away blood.

2. Pin the head down on a 150-mm dissecting tray and view using a dissecting microscope with fiber-optic illumination. Remove both superior cervical ganglia, located on the medial side of the carotid artery bifurcation (i.e., where the common carotid artery branches into the internal and external carotid arteries), using a 60-mm dissecting tray containing dissection medium, forceps, and surgical scissors.

   Keep dissection area moist by repeatedly applying dissection medium with a transfer pipette.

3. Place ganglia in a 35-mm tissue culture dish containing chilled dissection medium and remove connective tissue sheath surrounding ganglia using fine-tip forceps and spring scissors. Make several transverse cuts (perpendicular to the long axis) to allow access to digestion enzymes in dissociation medium in the next step.

4. Transfer ganglia to a 25-cm^2 flask containing filtered dissociation medium (see recipe) using a fire-polished Pasteur pipette. Flush flask with 95% O_2/5% CO_2 for at least 30 sec and then tightly seal the flask cap.

5. Incubate the flask in a shaking water bath (~110 rpm/min) for 1 h at 37°C.

6. Disperse neurons by vigorously shaking flask for 10 sec. Continue the rest of the protocol in a laminar flow tissue culture hood when possible.

7. Add 5 mL of preheated culture medium. Transfer contents to a 15-mL polypropylene centrifuge tube. Centrifuge at 60g in a swinging bucket centrifuge for 6 min at room temperature. Remove as much solution as possible, resuspend pellet in 10 mL fresh culture medium, and repeat centrifugation.

8. Carefully remove the supernatant and resuspend pellet in 1.2 mL of culture medium. Triturate gently with a 1-mL aerosol-resistant (filtered) pipette tip (15×–20×). Plate cells on six poly-L-

lysine-coated 35-mm culture dishes. Slowly add 2 mL of culture medium to each dish. Incubate dishes in a tissue culture incubator (37°C, humidified atmosphere of 5% CO_2 in air) for at least 3–4 h, to allow neurons to adhere firmly to substrate, before beginning experiments.

Recording of Whole-Cell Ca^{2+} Currents and Acute Application of G-Protein-Coupled Receptor (GPCR) Agonists on SCG Neurons

9. Prime external solution inlet line. Set flow to ~1 mL/min with flow regulator. Place inlet line into well on stage adaptor platform filled with external recording solution. Close external solution inlet line.

10. Prime each line of drug applicator system and the drug applicator tip.

    One line must be filled with external recording solution; this avoids flow-induced artifacts and accelerates washout (see Discussion).

    It is critical that ALL lines are filled with solution. If a line is not being used for drug application, fill it with external recording solution.

11. Place a 35-mm dish with plated SCG neurons on the microscope stage (see Fig. 1A for suggested setup).

12. Place vacuum suction line into dish. Adjust the height of the vacuum suction line to avoid overflow but not so low that it dries out the dish.

13. Connect the well on the stage adaptor platform filled with external recording solution (shown below the 35-mm dish in Fig. 1A) to the 35-mm dish with a U-bridge filled with external recording solution. Open the external recording solution inlet line. Completely exchange culture medium in dish with external recording solution (10–15 min).

14. Insert ground (e.g., Ag/AgCl pellet) into a well on the stage adaptor platform filled with internal recording solution. Electrically couple the ground with the 35-mm dish using a ground bridge (U-bridge filled with 2% agarose in 150 mM NaCl).

15. Select a cell of interest in the dish. Position drug applicator tip 100 μm away from cell of interest (Fig. 1C).

16. Fill the glass microelectrode with internal recording solution (not overfilling pipette, just sufficient to contact the chlorided region of the silver wire). Patch the cell in voltage clamp mode as follows.

    i. Bring the glass microelectrode into the dish. Apply constant positive pressure. Zero the voltage offset between the glass microelectrode and the bath solution. Apply a 5-mV test pulse.

    ii. Lower the microelectrode gently onto the cell of interest. Release positive pressure. Form a seal between the glass microelectrode and the cell membrane by applying gentle suction to the tubing connecting the glass microelectrode to the pipette holder. Apply a hyperpolarizing pulse to aid in formation of a gigaohm seal (−80 mV).

    iii. Neutralize the pipette capacitative transient currents.

        Fast capacitance transients should be gone.

    iv. Obtain whole-cell configuration by rupturing the membrane patch with gentle suction through tubing on the microelectrode holder.

        Whole-cell configuration is evident as an increase in the magnitude and time course of capacitative transients and a drop in access resistance.

    v. Record capacitative transient currents by applying a 10-msec +10-mV step.

    vi. Compensate whole-cell capacitance and series resistance (≥80%) using controls on the amplifier.

    vii. Allow sufficient time for cell dialysis with intracellular recording solution and for current levels to stabilize (few minutes). Once currents are stable, proceed with experiment.

17. Start flow of external recording solution from drug applicator.

18. Start recording protocols (e.g., triple-pulse protocol).

19. To switch solutions, start flow from a new line in the drug applicator and then close the line with external recording solution.

20. For washout of applied agonist/drug, perform the reverse of the previous step (start flow of external recording solution and then close line with drug solution).

21. At the end of the experiment, clean all lines by running 70% ethanol, followed by distilled water, through all lines.

DISCUSSION

Heterologous Expression of Genetic Tools to Study G-Protein Modulation of Ca^{2+} Channels

Many genetic tools are available to study G-protein signaling pathways (see Table 1 in Introduction: Strategies for Investigating G-Protein Modulation of Voltage-Gated Ca^{2+} Channels [Lu and Ikeda 2016]). Utilization of these tools in a neuronal context, complete with endogenous G-proteins and ion channels, has provided a physiologically relevant background for interpretation of results. However, primary cultures of neurons are resistant to cDNA transfection methods typically used in clonal cell lines such as calcium phosphate, electroporation, and lipofection. Consequently, alternative transfection approaches have been used to heterologously express genes of interest in neurons. A few of these methods are listed below with references to reviews detailing the technique.

1. Direct microinjection of cDNA into the nucleus (Ikeda 1997, 2004; Lu et al. 2009)

2. Direct microinjection of mRNA into the cytoplasm (Ikeda 1997)

3. mRNA transfection (Williams et al. 2010)

High levels of heterologous expression can be achieved by nuclear microinjection of cDNA constructs. In addition, consistent ratios of expression for multiple constructs are typically observed, which is essential in rebalancing the stoichiometry of the heterotrimeric G-protein complex during reconstitution experiments. However, the nuclear injection procedure is laborious, expensive, and typically yields a low success rate (10%–30%). In contrast, mRNA transfection protocols are becoming increasing popular because of their ease of use and higher transfection efficiency compared with nuclear microinjections. However, expression levels are lower and transient compared with those from cDNA transfection methods. The limitations of mRNA transfection may be overcome by transfecting more mRNA or synthesizing mRNA with modified ribonucleotides.

Common Sources of Flow-Induced Artifacts

Stagnant Recording Chamber Solution

Replacing the culture medium used during incubation of cells with external recording solution is important for control of ionic gradients and to pharmacologically isolate currents during electrophysiological recordings. However, continuous flow of fresh recording solution is highly recommended, especially during recordings from secretory cells, such as neurons and chromaffin cells, which can release factors or mediators that may influence drug responses of nearby cells. Also, continuous flow of fresh external solution will assist in washout of applied drugs, thus allowing experiments on multiple cells within the same dish.

Chemical Vehicle Used to Dissolve Drugs

Organic solvents can aid in dissolving water-insoluble drugs into solution, but can also produce their own effects on ion channels (Lewohl et al. 1999; Kobayashi et al. 1999), or even react with the components of the drug applicator system (McDonald et al. 2008). Even the counter-anions of

Cite this protocol as *Cold Spring Harb Protoc*; doi:10.1101/pdb.prot091223

drug salts may drastically change the pH of solution to influence cell health or may even produce direct effects on ion channels or GPCRs (Brown et al. 2003). Reducing the solvent concentration used (e.g., DMSO ≤0.1%) and including vehicle controls in experiments are essential in detecting these issues.

Physical Effect of OutFlow of Drug Applicator

The close proximity of the drug perfusion device to the recorded cells (~100 μm) is important for achieving fast and accurate local drug concentrations. However, the mechanical pressure from the flow of solution out of the drug perfusion device can also produce a slight potentiation of current amplitude (5%–10%). We recommend a constant flow of external solution at the start of the experiment and in between multiple drug applications, hence the need for one line to be filled with external recording solution, to negate this effect.

Blocked Flow from Drug Perfusion Device

A partial or complete block in solution flow out of the drug perfusion device may not be immediately apparent during experiments, especially when investigating novel agonists. Below are just a few indicators of blocked solution flow from drug perfusion devices.

- No response or small responses (5%–10% reduction in I_{Ca}), even from positive controls. Keep in mind that the removal of solution flow from an external solution line to a line with no solution flow will remove the potentiation caused by the mechanical pressure of solution flow.
- Abnormal drug responses (responses that start after drug flow is turned off)
- Highly variable responses, both in speed and magnitude of responses, when using different perfusion lines
- Carryover of responses, especially at the beginning of drug application

The most common source of blocked solution flow is air bubbles, which provide a space that traps solution from other lines or completely occludes flow. Air bubbles arise more often when using bovine serum albumin (BSA) as a carrier for lipid-soluble drugs (e.g., endocannabinoids). Ideally, keep the concentration of BSA below 0.5 mg/mL. Air bubbles can arise at any part of the drug applicator system (in any of the tubing lines, in the manifold, in between the manifold and applicator tip, etc.). Careful attention should be paid while priming the device before use, and solution flow from the drug perfusion device should be checked regularly during the course of an experiment. Also consider using alternating lines to ensure there is no bias and no carryover of drug caused by insufficient washing of the drug applicator.

Considerations for Electrophysiological Recording Solutions

In general, electrophysiological recording solutions are designed to isolate specific currents, control ionic gradients, and maintain good cell health for long-lasting, high-quality electrophysiological recordings. The recording solutions provided in this protocol may vary from other solutions used to study voltage-gated Ca^{2+} channels. For instance, barium ions (Ba^{2+}) may replace calcium ions (Ca^{2+}) as the charge carrier. Ba^{2+} is better suited for studies requiring block of Ca^{2+}-dependent responses, but keep in mind that the permeability of Ba^{2+} through voltage-gated Ca^{2+} channels is slightly different than that of Ca^{2+} and results in altered responses (i.e., increased magnitude of current and shift in the peak of the IV curve), so voltage protocols should be adjusted accordingly.

Each chemical component of electrophysiological recording solutions should be carefully chosen for a particular purpose and have minimal off-target effects. However, this is not always possible. For instance, TEA$^+$ used in recording solutions is an effective K$^+$ channel blocker but is also a weak muscarinic antagonist. As mentioned previously, even the counterion of salts used to make solutions can produce unanticipated responses (Brown et al. 2003). Furthermore, nonphysiological permeation of ions can arise given the nonphysiological nature of the solutions being used (e.g., Na^+ flowing through voltage-gated Ca^{2+} channels in the absence of Ca^{2+}). It is not possible to predict all off-target effects, but being vigilant and aware that effects observed may arise from the chemical components used to isolate specific currents will help in the detection of such effects and prevent misinterpretation of results.

RECIPES

Culture Medium for SCG Neurons

Reagent	Final concentration
Minimum essential medium with Earle's salts and L-glutamine (MEM; e.g., Life Technologies 11095)	90%
Fetal bovine serum	10%
Penicillin	500 U/mL
Streptomycin	500 μg/mL

Prepare medium under sterile conditions. Divide into 50-mL aliquots. Store at 4°C for up to 1 mo.

Dissociation Medium for SCG Neurons

Reagent	Final concentration
Earle's balanced salt solution (EBSS)	100%
HEPES	10 mM
D-glucose	20 mM
Collagenase	1.3 mg/mL
Trypsin	0.60 mg/mL
DNase I	0.05 mg/mL

Prepare EBSS in deionized water. Supplement EBSS with HEPES and D-glucose at the concentrations indicated. Adjust the pH to 7.4 and filter the solution by using a 0.2-μm-pore membrane filter. Divide into 5-mL aliquots and store for up to 3 mo at 4°C. Add enzymes to the supplemented EBSS immediately before the start of dissociation. (The concentrations of collagenase and trypsin must be empirically optimized with each new lot of enzymes.) Filter dissociation medium with enzymes by using a 0.22-μm syringe filter (Millex-GV; Millipore #SLGV033RB).

External Recording Solution for Voltage-Gated Ca^{2+} Channels

Reagent	Final concentration
Methanesulfonic acid (CH_3SO_3H)	140 mM
Tetraethylammonium hydroxide (TEA-OH)	145 mM
HEPES	10 mM
D-glucose	15 mM
$CaCl_2 \cdot 2H_2O$	10 mM
Tetrodotoxin (TTX)	0.0003 mM

Prepare solution containing all but TTX in deionized water. Adjust pH to 7.4 with TEA-OH. (Osmolality should range between 310 and 325 mmol/kg.) Filter using a 0.2-μm-pore membrane filter. Store up to 3 mo at room temperature (20°C–22°C) or up to 6 mo at 4°C.
Add TTX before the start of recordings. Be aware that TTX is susceptible to breakdown under basic (high-pH) conditions.

Cite this protocol as *Cold Spring Harb Protoc*; doi:10.1101/pdb.prot091223

Internal Recording Solution for Voltage-Gated Ca^{2+} Channels

Reagent	Final concentration
HCl	20 mM
N-methyl-D-glucamine (NMG)	120 mM
Tetraethylammonium hydroxide (TEA-OH)	20 mM
EGTA	11 mM
HEPES	10 mM
Sucrose	10 mM
CaCl$_2$	1 mM
Tris-creatine phosphate	14 mM
Mg-ATP	4 mM
Na$_2$-GTP	0.3 mM

Prepare in deionized water. Adjust pH to ~8.5 with methanesulfonic acid (CH$_3$SO$_3$H) before adding Tris-creatine phosphate, Mg-ATP, and Na$_2$-GTP. Adjust final pH to 7.2 with methanesulfonic acid (CH$_3$SO$_3$H). (Osmolality should range between 290 and 305 mmol/kg.) Filter internal recording solution using a 0.2-μm-pore membrane filter. Divide into 5-mL aliquots and store up to 6 mo at −20°C.

Poly-L-Lysine Solution in 0.1 M Borate Buffer

Reagent	Final concentration
Poly-L-lysine	0.1 mg/mL
Borate buffer (pH 8.5)	0.1 M

Prepare poly-L-lysine in borate buffer (1.24 g boric acid, 1.90 g sodium tetraborate, pH 8.5, in 500 mL deionized water). Filter the solution by using a 0.2-μm-pore membrane filter. Store filtered solution up to 1 mo at 4°C.

ACKNOWLEDGMENTS

This work was supported by the Intramural Research Program of the National Institutes of Health (National Institute on Alcohol Abuse and Alcoholism).

REFERENCES

Brown AJ, Goldsworthy SM, Barnes AA, Eilert MM, Tcheang L, Daniels D, Muir AI, Wigglesworth MJ, Kinghorn I, Fraser NJ, et al. 2003. The orphan G protein–coupled receptors GPR41 and GPR43 are activated by propionate and other short chain carboxylic acids. *J Biol Chem* **278:** 11312–11319.

Ikeda SR. 1997. Heterologous expression of receptors and signaling proteins in adult mammalian sympathetic neurons by microinjection. *Methods Mol Biol* **83:** 191–202.

Ikeda SR. 2004. Expression of G-protein signaling components in adult mammalian neurons by microinjection. *Methods Mol Biol* **259:** 167–181.

Ikeda SR, Jeong S-W. 2004. Use of RGS-insensitive Gα subunits to study endogenous RGS protein action on G-protein modulation of N-type calcium channels in sympathetic neurons. *Methods Enzymol* **389:** 170–189.

Kobayashi T, Ikeda K, Kojima H, Niki H, Yano R, Yoshioka T, Kumanishi T. 1999. Ethanol opens G-protein-activated inwardly rectifying K$^+$ channels. *Nat Neurosci* **2:** 1091–1097.

Lewohl JM, Wilson WR, Mayfield RD, Brozowski SJ, Morrisett RA, Harris RA. 1999. G-protein-coupled inwardly rectifying potassium channels are targets of alcohol action. *Nat Neurosci* **2:** 1084–1090.

Lu VB, Ikeda SR. 2016. Strategies for investigating G-protein modulation of voltage-gated Ca^{2+} channels. *Cold Spring Harb Protoc* doi: 10.1101/pdb.top087072.

Lu VB, Williams DJ, Won Y-J, Ikeda SR. 2009. Intranuclear microinjection of DNA into dissociated adult mammalian neurons. *J Vis Exp* 1–4.

McDonald GR, Hudson AL, Dunn SMJ, You H, Baker GB, Whittal RM, Martin JW, Jha A, Edmondson DE, Holt A. 2008. Bioactive contaminants leach from disposable laboratory plasticware. *Science* **322:** 917.

Williams DJ, Puhl HL, Ikeda SR. 2010. A simple, highly efficient method for heterologous expression in mammalian primary neurons using cationic lipid-mediated mRNA transfection. *Front Neurosci* **4:** 181.

CHAPTER 6

Hyperpolarization-Activated Cyclic Nucleotide-Gated Channel Currents in Neurons

Mala M. Shah[1]

Department of Pharmacology, UCL School of Pharmacy, University College London, London WC1N 1AX, United Kingdom

Hyperpolarization-activated cyclic nucleotide-gated (HCN) channels are voltage-gated ion channels that activate at potentials more negative than −50 mV and are predominantly permeable to Na^+ and K^+ ions. Four HCN subunits (HCN1–4) have been cloned. These subunits have distinct expression patterns and biophysical properties. In addition, cyclic nucleotides as well as multiple intracellular substances including various kinases and phosphatases modulate the expression and function of the subunits. Hence, the characteristics of the current, I_h, are likely to vary among neuronal subtypes. In many neuronal subtypes, I_h is present postsynaptically, where it plays a critical role in setting the resting membrane potential and the membrane resistance. By influencing these intrinsic properties, I_h will affect synaptic potential shapes and summation and thereby affect neuronal excitability. Additionally, I_h can have an effect on resonance properties and intrinsic neuronal oscillations. In some neurons, I_h may also be present presynaptically in axons and synaptic terminals, where it modulates neuronal transmitter release. Hence the effects of I_h on neuronal excitability are complex. It is, however, necessary to fully understand these as I_h has a significant impact on physiological conditions such as learning as well as pathophysiological states such as epilepsy.

HCN CHANNELS

Hyperpolarization-activated cyclic nucleotide-gated (HCN) channels are voltage-gated ion channels (Robinson and Siegelbaum 2003; Biel et al. 2009; Shah 2014). Each channel is composed of four α subunits, with each subunit having six transmembrane segments. The voltage sensor is located in the fourth transmembrane segment (S4 segment) of each subunit (Männikkö et al. 2002). HCN channels open at potentials more negative than −50 mV (i.e., with hyperpolarization) and are thus active at rest in many neurons (Robinson and Siegelbaum 2003; Biel et al. 2009; Shah 2014). They are permeable to both Na^+ and K^+ (Ludwig et al. 1998; Santoro et al. 1998). At the normal resting membrane potential (RMP) of neurons (−60 to −70 mV), though, the HCN channel current (I_h) is predominantly formed by a net inward Na^+ current. Hence, I_h is depolarizing at rest (Robinson and Siegelbaum 2003; Biel et al. 2009; Shah 2014).

Four HCN channel subunits, HCN1–4, have been cloned so far (Ludwig et al. 1998; Santoro et al. 1998). These subunits can assemble in homomers or heteromers. The biophysical properties of the HCN1–4 homomeric channels differ significantly (Wainger et al. 2001). Thus, HCN1 homomeric channels activate relatively rapidly compared with HCN2, HCN3, and HCN4 homomeric channels. This variation in the subunit biophysical properties together with the distinct expression profile of the subunits is likely to impart diverse biophysical characteristics of I_h in particular neurons. To determine

[1]Correspondence: mala.shah@ucl.ac.uk

Cite this introduction as *Cold Spring Harb Protoc*; doi:10.1101/pdb.top087346

the biophysical properties, it is essential to record the current. The accompanying protocol provides methods for recording I_h under whole-cell voltage-clamp conditions, in cell-attached mode, or with outside-out patches (see Protocol 1: Recording Hyperpolarization-Activated Cyclic Nucleotide-Gated Channel Currents (I_h) in Neurons [Shah 2016]).

Multiple proteins affect the expression and biophysical properties of HCN subunits. The activation curves of HCN1–4 current are shifted to the right by cyclic nucleotides, though the extent of the shift varies depending on the subunit composition (Ludwig et al. 1998; Santoro et al. 1998; Wainger et al. 2001; Robinson and Siegelbaum 2003; Biel et al. 2009). HCN subunit expression and channel function are also regulated by kinases, phosphatases, and auxiliary subunits such as tetratricopeptide repeat (TPR)-containing Rab8b-interacting protein (TRIP8b) (Biel et al. 2009; Shah 2014). This dynamic modulation of HCN channels differs within neurons as well as within individual neuronal subcellular compartments (axons, dendrites, and somata). Consequently, the current density and kinetics will vary considerably within neurons and among neuronal subtypes.

HCN CURRENTS

The differential I_h characteristics will have a significant impact on neuronal function. I_h acts as a high-pass filter, opposing any slow changes in membrane potential. Membrane hyperpolarization slowly activates I_h. The resulting inward current then causes the membrane potential to depolarize, producing a "sag" (Fig. 1). In contrast, membrane depolarization triggers slow deactivation of the HCN channels that are open at rest. Consequently, the membrane potential hyperpolarizes (Fig. 1). Hence, I_h significantly affects the neuronal input resistance (defined as the change in voltage induced by a given current injection). If I_h is open at rest, it will affect the membrane resistance too. By affecting the cell impedance, I_h regulates neuronal excitability in multiple ways including affecting subthreshold resonance properties, intrinsic oscillations, and synaptic integration (Robinson and Siegelbaum 2003; Biel et al. 2009; Shah 2014).

How I_h modulates neuronal excitability will depend on the subcellular and neuronal expression level of HCN subunits. For example, some hippocampal and cortical projection neurons (including pyramidal cells) express relatively low levels of HCN1 and HCN2 subunits at the soma (Lörincz et al. 2002; Notomi and Shigemoto 2004). Inhibition of these somatic HCN channels results in RMP hyperpolarization but no significant change in neuronal firing (Magee 1998; Shah et al. 2004; Shah 2014). In contrast, HCN subunits are located at a much higher density in pyramidal cell dendrites (Lörincz et al. 2002; Notomi and Shigemoto 2004). Interestingly, inhibiting dendritic I_h in pyramidal neurons causes a significant increase in the number of action potentials elicited by depolarization, despite the RMP being significantly hyperpolarized (Shah et al. 2004). This is at least in part because the decrease in I_h results in a greater enhancement of dendritic than somatic input resistance (Robinson and Siegelbaum 2003; Biel et al. 2009; Shah 2014). In addition, the RMP hyperpolarization caused by reduced I_h alters the activity of other dendritic ion channels such as T-type Ca^{2+} channels (Tsay et al. 2007), which may contribute to the augmented dendritic input resistance and excitability in the absence of I_h. The amplified dendritic input resistance together with the possible alterations in

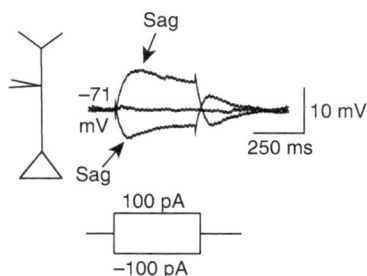

FIGURE 1. Example of "sag" recorded from an entorhinal cortical layer III pyramidal cell dendrite when short current pulses are injected. The sag in the hyperpolarizing direction is caused by activation of I_h whereas that in the depolarizing direction is caused by deactivation of I_h.

Cite this introduction as *Cold Spring Harb Protoc*; doi:10.1101/pdb.top087346

other dendritic ion channel properties is likely to boost synaptic potential amplitudes and slow the decay of synaptic potentials, leading to greater synaptic potential summation in pyramidal cell dendrites compared with somata (Magee 2000; Shah 2014). This is at least one of the reasons for the increased long-term potentiation at distal hippocampal CA1 pyramidal dendrites in HCN1 null mice compared with wild types (Nolan et al. 2004).

I_h also exists in subsets of inhibitory and excitatory synaptic terminals, where it regulates neurotransmitter release (Shah 2014). Given that I_h is subject to activity-dependent regulation (Robinson and Siegelbaum 2003; Biel et al. 2009; Shah 2014), the contribution of I_h toward neuronal network excitability is likely to be complex. I_h has been suggested to affect theta rhythm power in the hippocampus (Nolan et al. 2004) and contribute to learning and memory (Nolan et al. 2004) as well as to spatial navigation (Giocomo et al. 2011; Hussaini et al. 2011), though the cellular mechanisms for this phenomenon remain to be fully evaluated. Moreover, alterations in I_h have been suggested to occur during many disorders such as epilepsy and neuropathic pain (Robinson and Siegelbaum 2003; Biel et al. 2009; Shah 2014). Hence, fully understanding how I_h affects cellular function is likely to be critical for determining how it contributes to physiological states such as learning as well as to pathophysiological conditions. The accompanying protocol may be useful in such investigations (see Protocol 1: Recording Hyperpolarization-Activated Cyclic Nucleotide-Gated Channel Currents (I_h) in Neurons [Shah 2016]).

ACKNOWLEDGMENTS

Work in the laboratory is supported by the European Research Council and the Biotechnology and Biological Sciences Research Council UK.

REFERENCES

Biel M, Wahl-Schott C, Michalakis S, Zong X. 2009. Hyperpolarization-activated cation channels: From genes to function. *Physiol Rev* **89**: 847–885.

Giocomo LM, Hussaini SA, Zheng F, Kandel ER, Moser MB, Moser EI. 2011. Grid cells use HCN1 channels for spatial scaling. *Cell* **147**: 1159–1170.

Hussaini SA, Kempadoo KA, Thuault SJ, Siegelbaum SA, Kandel ER. 2011. Increased size and stability of CA1 and CA3 place fields in HCN1 knockout mice. *Neuron* **72**: 643–653.

Lörincz A, Notomi T, Tamás G, Shigemoto R, Nusser Z. 2002. Polarized and compartment-dependent distribution of HCN1 in pyramidal cell dendrites. *Nat Neurosci* **5**: 1185–1193.

Ludwig A, Zong X, Jeglitsch M, Hofmann F, Biel M. 1998. A family of hyperpolarization-activated mammalian cation channels. *Nature* **393**: 587–591.

Magee JC. 1998. Dendritic hyperpolarization-activated currents modify the integrative properties of hippocampal CA1 pyramidal neurons. *J Neurosci* **18**: 7613–7624.

Magee JC. 2000. Dendritic integration of excitatory synaptic input. *Nat Rev* **1**: 181–190.

Männikkö R, Elinder F, Larsson HP. 2002. Voltage-sensing mechanism is conserved among ion channels gated by opposite voltages. *Nature* **419**: 837–841.

Nolan MF, Malleret G, Dudman JT, Buhl DL, Santoro B, Gibbs E, Vronskaya S, Buzsáki G, Siegelbaum SA, Kandel ER, et al. 2004. A behavioral role for dendritic integration: HCN1 channels constrain spatial memory and

plasticity at inputs to distal dendrites of CA1 pyramidal neurons. *Cell* **119**: 719–732.

Notomi T, Shigemoto R. 2004. Immunohistochemical localization of I_h channel subunits, HCN1-4, in the rat brain. *J Comp Neurol* **471**: 241–276.

Robinson RB, Siegelbaum SA. 2003. Hyperpolarization-activated cation currents: From molecules to physiological function. *Annu Rev Physiol* **65**: 453–480.

Santoro B, Liu DT, Yao H, Bartsch D, Kandel ER, Siegelbaum SA, Tibbs GR. 1998. Identification of a gene encoding a hyperpolarization-activated pacemaker channel of brain. *Cell* **93**: 717–729.

Shah MM. 2014. Cortical HCN channels: Function, trafficking and plasticity. *J Physiol* **592**: 2711–2719.

Shah MM. 2016. Recording hyperpolarization-activated cyclic nucleotide-gated channel currents (I_h) in neurons. *Cold Spring Harb Protoc* doi: 10.1101/pdb.prot091462.

Shah MM, Anderson AE, Leung V, Lin X, Johnston D. 2004. Seizure-induced plasticity of h channels in entorhinal cortical layer III pyramidal neurons. *Neuron* **44**: 495–508.

Tsay D, Dudman JT, Siegelbaum SA. 2007. HCN1 channels constrain synaptically evoked Ca^{2+} spikes in distal dendrites of CA1 pyramidal neurons. *Neuron* **56**: 1076–1089.

Wainger BJ, DeGennaro M, Santoro B, Siegelbaum SA, Tibbs GR. 2001. Molecular mechanism of cAMP modulation of HCN pacemaker channels. *Nature* **411**: 805–810.

Protocol 1

Recording Hyperpolarization-Activated Cyclic Nucleotide-Gated Channel Currents (I_h) in Neurons

Mala M. Shah[1]

Department of Pharmacology, UCL School of Pharmacy, University College London, London, WC1N 1AX, United Kingdom

Hyperpolarization-activated cyclic nucleotide–gated (HCN) channels are voltage-gated ion channels that play a crucial role in many physiological processes such as memory formation and spatial navigation. Alterations in expression and function of HCN channels have also been associated with multiple disorders including epilepsy, neuropathic pain, and anxiety/depression. Interestingly, neuronal HCN currents (I_h) have diverse biophysical properties in different neurons. This is likely to be in part caused by the heterogeneity of the HCN subunits expressed in neurons. This variation in biophysical characteristics is likely to influence how I_h affects neuronal activity. Thus, it is important to record I_h directly from individual neurons. This protocol describes voltage-clamp methods that can be used to record neuronal I_h under whole-cell voltage-clamp conditions, in cell-attached mode, or with outside-out patches. The information obtained using this approach can be used in combination with other techniques such as computational modeling to determine the significance of I_h for neuronal function.

MATERIALS

It is essential that you consult the appropriate Material Safety Data Sheets and your institution's Environmental Health and Safety Office for proper handling of equipment and hazardous material used in this protocol.

RECIPES: Please see the end of this protocol for recipes indicated by <R>. Additional recipes can be found online at http://cshprotocols.cshlp.org/site/recipes.

Reagents

Biological sample
  This can be a slice preparation or cultured neurons.

External recording solution for I_h recording, freshly prepared <R>
Internal pipette solution for I_h recording in cell-attached mode <R>
Internal pipette solution for I_h recording in whole-cell voltage-clamp mode or outside-out patch mode <R>
Pharmacological inhibitor (ZD7288 [10 µM] or Cs^+ [1–2 mM])

Equipment

Amplifier (e.g., Axopatch 200B or MultiClamp 700B [Molecular Devices])
Analog–digital converter

---

[1]Correspondence: mala.shah@ucl.ac.uk

Computer with acquisition software (e.g., pCLAMP [Molecular Devices])

Micromanipulator

Microscope

The microscope should have sufficiently good optics (objectives with at least 40× magnification and a numerical aperture of 1.0) to visualize individual cultured neurons and neurons present in a slice preparation. To visualize neurons present in a slice preparation using a microscope, it is preferably to use a microscope with differential infrared optics coupled to a good camera (e.g., Rolera Bolt CMOS [QImaging]) and monitor (for details of equipment specification, see Davie et al. 2006; Shah 2013).

Pipette holder

Pipette puller and glass

It is preferable to use thick-walled borosilicate glass for these recordings.

Temperature controller, e.g., Model CL-100 Bipolar Temperature controller (Harvard Apparatus Ltd) (optional; see Step 1)

Tubing (to provide pressure to pipette) and stopcock or switchable valve

METHOD

I_h can be recorded under whole-cell voltage-clamp conditions, in cell-attached mode, or with outside-out patches (see Step 8).

1. Place the biological sample under a microscope and perfuse the sample with the external solution.

   I_h is highly sensitive to temperature, with Q10 values ranging from 4 to 6 (Robinson and Siegelbaum 2003). Thus, while I_h can be recorded at room temperature, the biophysical characteristics are likely to be different from those recorded at near-physiological temperatures. To obtain recordings at near-physiological temperature, the external solution can be heated with a temperature controller such as the Model CL-100 Bipolar Temperature controller (Harvard Apparatus Ltd).

2. Identify the soma or dendrite of a healthy neuron present in culture or in a slice preparation.

3. Pull patch-pipettes.

4. Fill a micropipette with the appropriate internal solution for the recording method being used. Insert the pipette into a holder that is securely attached to the amplifier headstage and micromanipulator.

   The micropipette resistance should be between 3 and 6 $M\Omega$.

   The internal solution should be filtered to remove any debris. We typically attach a Nalgene filter (0.2-µm pore size, 4-mm diameter, cellulose acetate membrane) onto the end of the syringe containing the internal solution to filter it as it is being inserted into the patch pipette.

5. Add positive pressure to the pipette via tubing attached to the pipette holder. Secure the pressure within the pipette using a stopcock or a switchable valve.

6. Using the micromanipulator, lower the pipette on top of the visually identified soma or dendrite. Zero any offset recorded with the pipette before touching the cell with it.

   If the cell is healthy, a dimple will form on the cell membrane.

7. Release the pressure and form a gigaohm seal.

   Gentle suction may need to be applied to form a good gigaohm seal.

   For cell-attached recording (Steps 8.iii–8.iv), the voltage-clamp protocol described below can be applied to record HCN channel currents or I_h.

   For whole-cell recording (Steps 8.i–8.ii), further gentle suction will be required to be rupture the membrane. This should be done while holding the cell near the normal resting membrane potential (RMP) (i.e., −70 mV). Once a whole-cell recording has been established, the pipette can be gently pulled away to obtain an outside-out patch.

8. Perform recording using whole-cell voltage-clamp or outside-out patch recording (Steps 8.i–8.ii), or in cell-attached mode (Steps 8.iii–8.iv).

Whole-Cell Voltage-Clamp and Outside-Out Patch Recordings

i. For whole-cell voltage-clamp recordings, carry out appropriate series resistance compensation (at least 60%–70%) before commencing the recordings.

ii. For both whole-cell voltage-clamp and outside-out patch recordings, apply the protocol shown in Figure 1A.

> Briefly, the voltage potential should be stepped up to −40 mV and 2-sec hyperpolarizing square steps applied in 10-mV increments to −120 mV or beyond to fully activate the current. The voltage is stepped between successive increments to a potential such as −60 mV to record the tail Ih current (Fig. 1A). Once the recorded current is stable, the pharmacological inhibitor ZD7288 (10 μm) or Cs$^+$ (1–2 mm) should be applied for 15 min and the protocol repeated in the presence of this compound. The currents in the presence of the inhibitor should be subtracted from those under control conditions to obtain I$_h$

Cell-Attached Recordings

iii. Ensure that the background noise levels are between 1 and 5 pA.

iv. Use the protocol described in Steps 8.i–8.ii to elicit I_h (Fig. 1B).

> Because the internal side of the cell membrane is not accessible to the pipette, the internal RMP must be initially estimated. This then should be subtracted from the required potentials. For example, if the RMP is known to be on average −70 mV, the cell-attached potential will be +25 mV to achieve a membrane potential of −45 mV. In addition, because under cell-attached conditions, it is difficult to inhibit the recorded current with externally applied pharmacological inhibitors, it is advisable to measure the leak current by applying a +10 mV step from a holding potential of −45 mV. This leak current can then be subtracted from the current acquired using the protocol in Figure 1 to record I$_h$. At the end of the cell-attached recording, the patch can be ruptured to obtain the RMP. This can then be used to determine the membrane potential values at which I$_h$ was recorded. For example, if on rupturing the patch, the RMP was −68 mV, then a potential of +25 mV would be equivalent to holding the cell at −43 mV.

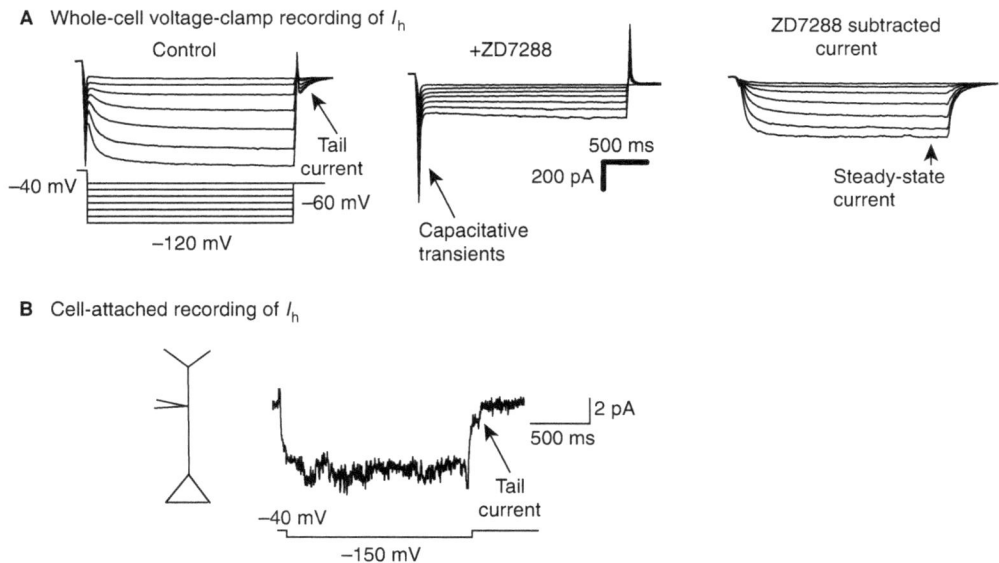

FIGURE 1. (A) Example showing a whole-cell voltage-clamp recording of current from a hippocampal pyramidal neuron somata present in a rat brain slice under control conditions and in the presence of ZD7288 (15 μM) when a series of 2-sec hyperpolarizing steps between −40 and −120 mV were applied as shown. The ZD7288-subtracted current, I_h, is shown on the *right*. (B) Example of a cell-attached recording of I_h recorded from an entorhinal cortical layer III pyramidal neuron present in a rat brain slice when a step from −40 to −150 mV was applied (adapted from Shah et al. 2004). The trace was leak subtracted. The leak current was obtained by applying a step from −40 to −20 mV.

Cite this protocol as *Cold Spring Harb Protoc*; doi:10.1101/pdb.prot091462

9. Filter the obtained currents at 1–2 kHz and acquire using appropriate software.

10. (Optional) Determine liquid junction potentials.

    Liquid junction potentials are a potential source of error for whole-cell voltage-clamp or outside-out patch recordings and can be measured by determining the difference between the offset of an intact micropipette and that of a micropipette with a broken tip. Other methods for determining the liquid junction potential are described in Neher (1992).

11. Calculate current amplitudes.

    From the acquired records, the tail current amplitudes elicited at the various voltages can be expressed as a fraction of the maximal tail current amplitude (Magee 1998; Poolos et al. 2002; Shah et al. 2004). The fractional current amplitudes can then be plotted against the voltage potential from which they were evoked and fitted using a Boltzmann function.

12. To estimate the reversal potential, plot the current amplitudes at a given voltage against the voltage (Magee 1998).

    The best line fit can be extrapolated to the x-axis to obtain an approximate value for the reversal potential.

13. Obtain the activation and deactivation kinetics by fitting the steady-state currents and tail currents, respectively, with exponential functions as described by Magee (1998).

DISCUSSION

The voltage-clamp protocols described above are useful for obtaining the biophysical characteristics of I_h. While the presence of a "sag" obtained under current-clamp conditions (see Fig. 1 in Introduction: Hyperpolarization-Activated Cyclic Nucleotide-Gated Channel Currents in Neurons [Shah 2016]) is indicative that the neuron expresses HCN channels, this will not provide an insight into the biophysical characteristics of I_h. The I_h biophysical characteristics this information can then be used in computational models to determine the effect of I_h on neuronal and network activity.

Whole-cell voltage-clamp conditions have been used by many laboratories to record I_h from neuronal somata. There are, though, several limitations of the method including rundown of the current, space-clamp and series resistance issues. A solution to reducing series resistance errors is to measure the applied voltage using a second patch-pipette from the cell. This information can then be used to estimate membrane potential errors and to construct a more accurate activation curve.

The voltage errors associated with the whole-cell voltage-clamp method can be avoided using the cell-attached protocol and outside-out patch method. With the outside-out patch method, the intracellular milieu is disturbed and so the biophysical properties may not be representative of the native current. This is avoided by using the cell-attached method (Williams and Wozny 2011). There are, though, several drawbacks of the cell-attached method, too. Errors associated with the transmembrane voltage changes caused by activation of voltage-gated ion channels in the patch in contact with the tip of the micropipette may still occur (Williams and Wozny 2011). An additional limitation is that the HCN channel density may not be uniform in all parts of the cell membrane. Thus, the current recorded from a single patch may not fully represent the density of the current in the particular subcellular compartment of the neuron. Therefore, a large number of recordings from a given cell might be necessary to accurately determine the amplitude and kinetics of the current. Despite their shortcomings, the methods described can be useful for estimating I_h densities and characteristics within a given neuron.

ACKNOWLEDGMENTS

Work in the laboratory is supported by the European Research Council and the Biotechnology and Biological Sciences Research Council UK.

RECIPES

External Recording Solution for I_h Recording

Reagents	Final concentration
NaCl	125 mM
KCl	2.5 mM
NaH_2PO_4	1.25 mM
$NaHCO_3$	25 mM
$CaCl_2$	2 mM
$MgCl_2$	2 mM
Glucose	10 mM
Tetrodotoxin	0.001 mM
$CdCl_2$	0.1 mM
Tetraethylammonium chloride (TEA-Cl)	10 mM
4-aminopyridine (4-AP)	1 mM
$BaCl_2$	1 mM

Bubble with 95% O_2/5% CO_2. (After bubbling, the solution will have a pH of 7.2; osmolarity will be ~320 mOsm. Ideally, this solution should be made up fresh on the day of the experiment and kept bubbling throughout the experiment.)

Internal Pipette Solution for I_h Recording in Cell-Attached Mode

Reagents	Final concentration
KCl	120 mM
Tetraethylammonium chloride (TEA-Cl)	20 mM
4-aminopyridine (4-AP)	5 mM
$BaCl_2$	1 mM
HEPES	10 mM
$MgCl_2$	1 mM
$CaCl_2$	2 mM
Tetrodotoxin	0.001 mM
$NiCl_2$	0.1 mM

Adjust the pH of the solution to 7.3. (The osmolarity will be 300 mOsm.) Store at 4°C for up to 1 mo.

Internal Pipette Solution for I_h Recording in Whole-Cell Voltage-Clamp Mode or Outside-Out Patch Mode

Reagents	Final concentration
$KMeSO_4$	120 mM
KCl	20 mM
4-(2-hydroxyethyl)-1-piperazineethanesulfonic acid (HEPES)	10 mM
$MgCl_2$	2 mM
Ethylene glycol-bis(2-aminoethylether)-N,N,N', N'-tetraacetic acid (EGTA)	0.2 mM
Na_2ATP	4 mM
Tris–GTP	0.3 mM
Tris–phosphocreatine	14 mM

Adjust the pH of the solution to 7.3. (The osmolarity will be 300 mOsm.) Aliquot and store at −20°C for up to 3 mo. Keep the solution on ice during use.

Cite this protocol as *Cold Spring Harb Protoc*; doi:10.1101/pdb.prot091462

REFERENCES

Davie JT, Kole MH, Letzkus JJ, Rancz EA, Spruston N, Stuart GJ, Hausser M. 2006. Dendritic patch-clamp recording. *Nat Protoc* **1**: 1235–1247.

Magee JC. 1998. Dendritic hyperpolarization-activated currents modify the integrative properties of hippocampal CA1 pyramidal neurons. *J Neurosci* **18**: 7613–7624.

Neher E. 1992. Correction for liquid junction potentials in patch clamp experiments. *Methods Enzymol* **207**: 123–131.

Poolos NP, Migliore M, Johnston D. 2002. Pharmacological upregulation of h-channels reduces the excitability of pyramidal neuron dendrites. *Nat Neurosci* **5**: 767–774.

Robinson RB, Siegelbaum SA. 2003. Hyperpolarization-activated cation currents: From molecules to physiological function. *Annu Rev Physiol* **65**: 453–480.

Shah MM. 2013. Recording dendritic ion channel properties and function from cortical neurons. *Methods Mol Biol* **998**: 303–309.

Shah MM. 2016. Hyperpolarization-activated cyclic nucleotide-gated channel currents in neurons. *Cold Spring Harb Protoc* doi: 10.1101/pdb.top087346.

Shah MM, Anderson AE, Leung V, Lin X, Johnston D. 2004. Seizure-induced plasticity of h channels in entorhinal cortical layer III pyramidal neurons. *Neuron* **44**: 495–508.

Williams SR, Wozny C. 2011. Errors in the measurement of voltage-activated ion channels in cell-attached patch-clamp recordings. *Nat Commun* **2**: 242.

Single-Channel Recording of Ligand-Gated Ion Channels

Andrew J.R. Plested[1]

*Leibniz-Institut für Molekulare Pharmakologie (FMP), 13125 Berlin, Germany; NeuroCure,
Charité-Universitätsmedizin Berlin, 10117 Berlin, Germany*

Single-channel recordings reveal the microscopic properties of individual ligand-gated ion channels. Such recordings contain much more information than measurements of ensemble behavior and can yield structural and functional information about the receptors that participate in fast synaptic transmission in the brain. With a little care, a standard patch-clamp electrophysiology setup can be adapted for single-channel recording in a matter of hours. Thenceforth, it is a realistic aim to record single-molecule activity with microsecond resolution from arbitrary cell types, including cell lines and neurons.

INTRODUCTION

Single-channel recordings are a label-free method to obtain microscopic kinetic information from the ion channels that participate in synaptic transmission in neurons (Hamill et al. 1981). The time course of synaptic currents depends on the kinetics of ligand-gated channels (Wyllie et al. 1998), and single-channel conductance determines charge transfer during a synaptic potential. In the age of super-resolution imaging, we can track individual ion channels as they diffuse in and out of synapses (Opazo et al. 2010). So, in general, microscopic properties of ion channels are highly relevant for understanding basal signaling in neurons and plastic changes in synaptic strength both short and long in duration. Furthermore, because they can yield rich kinetic information, revealing events from microseconds to seconds, single-channel patch-clamp recordings can be used to assess conformational changes (Burzomato et al. 2004) or the number of subunits (Rosenmund et al. 1998) in an ion channel complex.

DESIGN OF SINGLE-CHANNEL RECORDING EXPERIMENTS

First, the optimal recording configuration must be chosen. The three main options are cell-attached (with ligand inside the pipette), inside-out, and outside-out (Fig. 1). Intracellular ligand-gated channels (e.g., some transient receptor potential [TRP] channels) can be patched in analogous ways, with organelles obtained by fractionation (Kirichok et al. 2004) or by slicing cells open with a patch pipette (Dong et al. 2008). Further, biochemically purified channels can be reconstituted into membranes for bilayer recordings to verify their activity (Baranovic et al. 2013; Dellisanti et al. 2013). Some ligand-gated channels (e.g., glycine receptors [Fucile et al. 2000]) undergo substantial changes in activity following patch excision. Such receptors are best studied in the cell-attached configuration, allowing records to be collected for tens of minutes (Burzomato et al. 2004; Plested et al. 2007; Kussius and

---

[1]Correspondence: Plested@fmp-berlin.de

Cite this introduction as *Cold Spring Harb Protoc*; doi:10.1101/pdb.top087239

A

B

Cell-attached recording of rat glycine α1, −80 mV

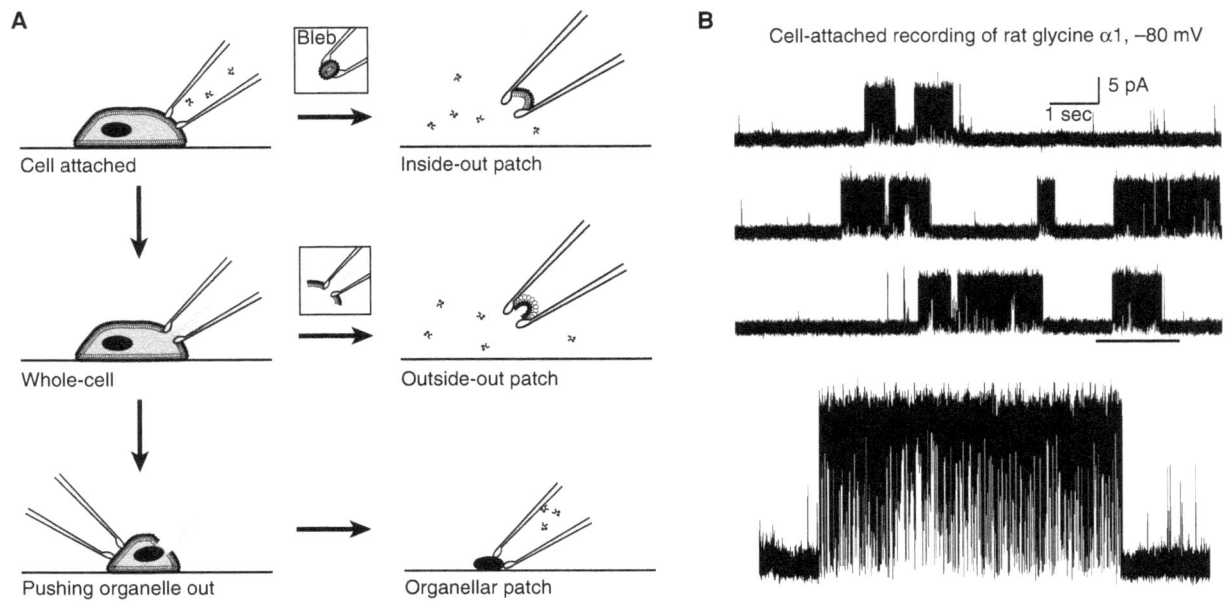

FIGURE 1. Recording configurations in mammalian cells. (*A*) Excising an inside-out patch from the cell-attached configuration often results in a bleb, from which no recording can be made. The outer surface can be removed by brief exposure of the patch to air. Outside-out patches form within the pipette from the resealing of the patch membrane after a smooth and assertive withdrawal of the patch pipette from the whole-cell configuration. For an organellar patch, the plasma membrane is sliced and the organelle of interest (preferably labeled; here shown as a black oval) is pushed out through the incision. Agonist molecules (black) are shown in the pipette or bath solution, as appropriate. The outer leaflet of the plasma membrane is shown in black to aid orientation. (*B*) A typical recording (30 sec) is shown for the rat glycine α1 receptor in cell-attached mode. Chloride-conducting openings are upward at negative membrane potential (pipette potential was approximately +70 mV in this recording). The bar under the trace indicates the magnified section (shown at *bottom*) illustrating a cluster with high open probability.

Popescu 2009; Purohit and Auerbach 2009). The ligand is included in the pipette and cannot be exchanged.

The noise in excised patches tends to be higher, but patch excision brings the opportunity to do both equilibrium and nonequilibrium experiments (Rosenmund et al. 1998), which can provide specific information about channel kinetics. It is also essential with a channel of unknown properties to start in a configuration where the patch can be perfused, to establish the dependence of activity on relevant agonists or drug compounds. For most receptors, this process means recording in outside-out configuration. The same logic applies to channels with intracellular ligand-binding sites (such as cyclic nucleotide-gated [CNG] channels), for which excision into the inside-out configuration is required. A precisely timed perfusion system is helpful for isolating specific activity induced by, say, an agonist from random noise, other channels, and breakdown. For glutamate receptors, we use a fast perfusion system with resolution of better than a millisecond (Carbone and Plested 2012), even though we generally apply glutamate for seconds at a time. Recording episodically with timed perfusion allows episodes that show spurious activity outside of the agonist pulse to be discarded.

For nonequilibrium experiments, it is generally necessary to obtain patches with one channel (Rosenmund et al. 1998), which is achieved only with very low expression. Channel activity may also run down during the recording, yielding a section with only one channel. To obtain useful equilibrium recordings, it is essential that the channel density in the membrane be neither too high nor too low. The aim is to isolate stretches of recording where we are certain only one channel is active. In this context, the ideal level of expression depends on the activity of the particular channel. If the channels have substantial open probability, it is likely that only patches with a few channels will be useful, otherwise activity will be overlapping. Normally, the random nature of expression will lead to some patches being "blanks"—that is, no activity at all is observed. If the channel in question desensitizes

Cite this introduction as *Cold Spring Harb Protoc*; doi:10.1101/pdb.top087239

substantially, it is not a problem to have multiple channels in the patch, because there is a good chance that for some of the record, only one of the channels will be active, and the rest will be desensitized.

A further critical consideration is what voltage to record at. We want the highest potential at which patches can survive and give a decent conductance. Most ligand-gated channels, including the nicotinic receptors and glycine receptors (Dionne and Stevens 1975; Gill et al. 2006), however, show some weak voltage dependence in their activity. Thus, the recording voltage cannot be chosen at random. Recordings at potentials up to 100 mV can last tens of minutes in good conditions. More extreme potentials tend to dramatically shorten patch lifetime, probably because of voltage-dependent "creeping" deformation (Suchyna et al. 2009).

PREPARING THE SETUP

The luxury of building a single-channel setup from scratch is less common than adapting an existing rig. Most equipment is standard for any patch-clamp electrophysiology, but some attention should be paid to the choice of manipulator, to ensure mechanically stable, low-noise recordings. Hydraulic manipulators are naturally electrically silent but do seem to suffer from drift more than their motorized counterparts. Mechanical drift is as problematic for cell-attached recording as it is for whole-cell patch clamp. In severe cases, patch excision will remove this constraint. Scientifica manipulators have admirable stability and apparently no noise. Stepper-motor-driven manipulators can also work well, but most designs require that current flows to keep the manipulator still. In the popular Sutter manipulators, the stepper holding current can be adjusted (reduced) by using a secret menu at the expense of slightly more drift (contact Sutter for details). In our hands, Luigs & Neumann manipulators, although having incredible mechanical stability, are not optimized for low electrical noise. These manipulators, and some others, can be switched off after a recording begins, without mechanical disturbance.

REDUCING NOISE

Electrical noise in the patch-clamp record has numerous sources, both those intrinsic to the amplifier and those extrinsic (i.e., from the environment) (Benndorf 1995). Noise adds in quadrature: In practice, the largest source of noise swamps the rest. Thus, after removing a major source of noise, all previously ineffective manipulations should be repeated. The practical process is illustrated in Table 1. Recently, inexpensive digital oscilloscopes became available that can generate noise spectra on the fly. This feature can be useful for detecting major sources of noise.

The first assumption should be that every electrical device is a major source of noise until proven otherwise. Typical troublemakers are the often inexpensive power supplies provided for cameras and illuminators. Video monitors, mobile phones, refrigerators, and centrifuges are further familiar culprits. Unfortunately, some microscopes were not manufactured with low electrical noise in mind. In our hands, the IX series from Olympus can be grounded entirely from one point at the back corner of the movable stage.

Solution lines that are not needed should be removed—especially those that run outside the Faraday cage. In any case, the best noise conditions are obtained if the patch is held near the top of the bath, so that the immersion of the pipette is minimized. For cell-attached recordings, perfusion can be removed altogether, as long as bathing solution is exchanged regularly by hand. Perfusion of excised patches brings noise. The major source is microphonic pickup of acoustic noise by the pipette as solution is sucked from the bath. We avoid this problem by removing solution from the bath only between episodes and otherwise stopping the suction using a pinch valve under computer control. The slow increase in the level of the bath alters capacitance of the pipette, and thus brings noise pickup, but

TABLE 1. Hypothetical noise troubleshooting

Noise source	Amplifier limit[a]	Manip	Lamp	Camera	Earth loop	Dirty holder	Total	Fraction in round (%)
Round #1								
Setup	50	50	100	300	400	800	951	
Switch off manipulator	50		100	300	400	800	950	100
Insulate lamp cable	50	50		300	400	800	946	99
Turn off camera	50	50	100		400	800	903	95
Rewire	50	50	100	300		800	863	91
Clean the holder	50	50	100	300	400		515	54
Conclusion: Removing stray salt and dirt from the pipette holder reduces noise by ~50%.								
Round #2								
Setup	50	50	100	300	400		515	
Switch off manipulator	50		100	300	400		512	100
Insulate lamp cable	50	50		300	400		505	98
Turn off camera	50	50	100		400		418	81
Rewire	50	50	100	300			324	63
Conclusion: Removing an earth loop, caused by too many grounding cables, reduces noise by ~40%.								
Round #3								
Setup	50	50	100	300			324	
Switch off manipulator	50		100	300			320	99
Insulate lamp cable	50	50		300			308	95
Turn off camera	50	50	100				122	38
Conclusion: Turning off the camera after finding the cell reduces noise by 60%.								
Round #4								
Setup	50	50	100				122	
Switch off manipulator	50		100				112	91
Insulate lamp cable	50	50					71	58
Conclusion: Insulating the lamp cable from the tabletop reduces noise by 40%.								

Typical noise contributions (in fA RMS) in the 5-kHz band are indicated. Four successive rounds of searching for the major source of noise return a major reduction from a new source each time. For example, turning off the camera has no effect in round #1 but has a major effect in round #3. Sources of noise in each row (σ_i, corresponding to each component) add in quadrature to give the total:

$$\sigma_{total} = \sqrt{\sigma_1^2 + \sigma_2^2 + \cdots}.$$

The boxed value is the lowest for each round.
Manip, micromanipulator.
[a]Amplifier limit cannot be changed or removed.

this effect is preferable to continuous suction noise. If required, the suction line from the bath can be internally grounded with a long length of silver wire.

A key principle is to avoid ground loops by limiting the number of grounding connections and by testing all of them for electrical integrity. Major optimizations can be done in a few hours simply by cleaning out the setup. Start with a minimal setup and only add back essential items. To achieve very low noise levels is extremely challenging but a little careful work at this stage can produce major dividends. Once the setup is adjusted, it will not necessarily be trivial to switch between different recording configurations. Maddeningly, counterintuitive modifications (such as disconnecting unique earth leads) may improve noise.

For excluding noise whose source cannot be throttled, structural shielding of the prep with wire mesh or metallic fabric (e.g., Shieldex Bremen) is a useful adjunct to the Faraday cage. Metallic fabric can act as a very effective full curtain at the front of a cage as well.

Once the noise levels are minimized at the setup, the main source of noise will become the seal itself—tight, multi-gigaohm seals are indispensable for low-noise recording.

DATA ANALYSIS

A common question is: How much data do I need? Conductance measurements require relatively few isolated openings. For fitting of kinetics, it is necessary to collect 5000–10,000 or more events in each

Cite this introduction as *Cold Spring Harb Protoc*; doi:10.1101/pdb.top087239

recording. However, what can be collected depends on channel activity. Ideally, measuring events with a mean lifetime of 10 msec, there will be 100 events per second, which means recording for a couple of minutes will yield enough. This goal is not trivially achieved.

It is crucial to examine the digitized record by eye to confirm that the recording is stable. Glitches, other channels, or noise and drift will play havoc with even the simplest measurements of conductance. Baseline drifts can be corrected by hand or, in some software packages, automatically. But no automated procedure can assure a clean record without checking.

To measure kinetic information seriously, the digitized record must be idealized into a list of events, with their durations, amplitudes, and other properties. The fastest events that can be seen depend on the relation between the single-channel conductance and the background noise, because more noise means a more closed (and thus slower) filter.

Several packages are available to evaluate putative kinetic models against idealized data, but this approach is beyond this scope of this introduction. To make meaningful inferences at this stage, it is essential to reliably identify stretches of the record where only one channel is active. There may be a degree of iteration required to identify the right recording conditions to get meaningful information. In low-activity conditions, we expect bursts (groups of openings) to be well spread out. However, the degree to which bursts should be separated to be well defined can be surprisingly large—about 100-fold longer than the mean duration of the next shortest shut interval (Lape et al. 2012).

APPLICATIONS

Most single-channel studies are done in human embryonic kidney (HEK) 293 cells or from patches excised from *Xenopus* oocytes. (For methods to record from HEK cells, see Protocol 1: Single-Channel Recording of Glycine Receptors in Human Embryonic Kidney (HEK) Cells [Plested and Baranovic

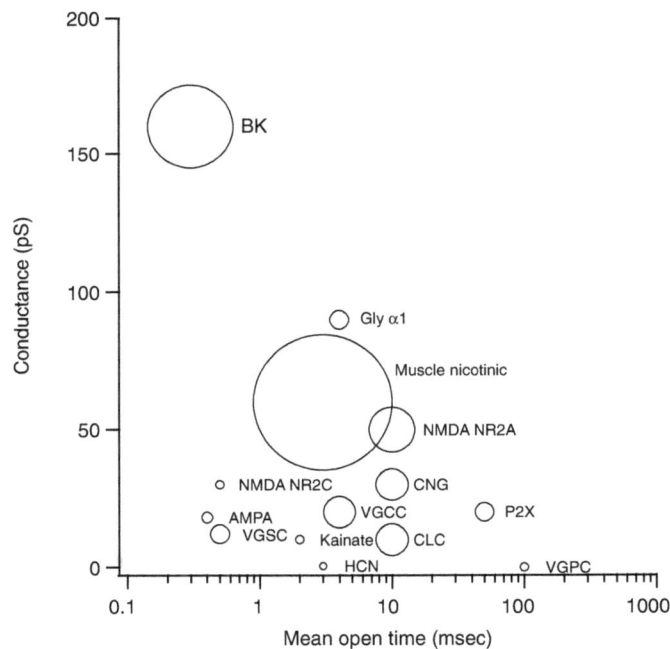

FIGURE 2. The single-channel recording landscape. The larger the conductance, or the slower the kinetics, the easier it is to obtain recordings, because the signal-to-noise ratio is larger at a given filter setting. The size of each circle corresponds to the approximate number of single-channel papers for the channel in question in the PubMed database (http://www.ncbi.nlm.nih.gov/pubmed/) in 2014. The range is from ~5 (HCN) to ~600 (muscle nicotinic). Abbreviations: VGSC, voltage-gated sodium channel; VGPC, voltage-gated proton channel; VGCC, voltage-gated calcium channel; HCN, hyperpolarization-gated cation channel; CNG, cyclic nucleotide-gated channel.

2016].) Recording single channels from neurons and other native cells has somewhat fallen out of vogue, mainly because so many different channel variants are expressed at somatic membranes, and there are few ways to control expression. However, there is great value in this approach if channels can be "fingerprinted" from their conductance or kinetics. In recent years, there have been considerable advances in the recording of intracellular channels or channels in specialized cells such as sperm. These approaches are often key for answering the question of whether a given protein is a channel or a transporter. When choosing a topic of study, the simple relation plotted in Figure 2 shows what is likely to be achievable. This figure shows the single-channel research landscape—how many papers were found in PubMed for each type of channel plotted against conductance and open time. It is notable that publishing on small or fast channels is comparatively much more difficult, or less popular, than large-conductance channels such as the BK potassium channels. But the lesser-studied channels remain important for physiology and represent understudied targets that may still yield fascinating insights.

ACKNOWLEDGMENTS

I thank Jelena Baranovic for comments and suggestions. Work in my laboratory was funded by the Deutsche Forschungsgemeinschaft (DFG) (PL619/1 and EXC-257) and the European Research Council (ERC) ("Gluactive").

REFERENCES

Baranovic J, Ramanujan CS, Kasai N, Midgett CR, Madden DR, Torimitsu K, Ryan JF. 2013. Reconstitution of homomeric GluA2$_{flop}$ receptors in supported lipid membranes: Functional and structural properties. *J Biol Chem* **288**: 8647–8657.

Benndorf K. 1995. Low-noise recording. In *Single-channel recording* (ed. Sakmann B, Neher E), pp. 483–587. Plenum, New York.

Burzomato V, Beato M, Groot-Kormelink PJ, Colquhoun D, Sivilotti LG. 2004. Single-channel behavior of heteromeric α1β glycine receptors: An attempt to detect a conformational change before the channel opens. *J Neurosci* **24**: 10924–10940.

Carbone AL, Plested AJ. 2012. Coupled control of desensitization and gating by the ligand binding domain of glutamate receptors. *Neuron* **74**: 845–857.

Dellisanti CD, Ghosh B, Hanson SM, Raspanti JM, Grant VA, Diarra GM, Schuh AM, Satyshur K, Klug CS, Czajkowski C. 2013. Site-directed spin labeling reveals pentameric ligand-gated ion channel gating motions. *PLoS Biol* **11**: e1001714.

Dionne VE, Stevens CF. 1975. Voltage dependence of agonist effectiveness at the frog neuromuscular junction: Resolution of a paradox. *J Physiol* **251**: 245–270.

Dong XP, Cheng X, Mills E, Delling M, Wang F, Kurz T, Xu H. 2008. The type IV mucolipidosis-associated protein TRPML1 is an endolysosomal iron release channel. *Nature* **455**: 992–996.

Fucile S, De Saint Jan D, de Carvalho LP, Bregestovski P. 2000. Fast potentiation of glycine receptor channels of intracellular calcium in neurons and transfected cells. *Neuron* **28**: 571–583.

Gill SB, Veruki ML, Hartveit E. 2006. Functional properties of spontaneous IPSCs and glycine receptors in rod amacrine (AII) cells in the rat retina. *J Physiol* **575**: 739–759.

Hamill OP, Marty A, Neher E, Sakmann B, Sigworth FJ. 1981. Improved patch-clamp techniques for high-resolution current recording from cells and cell-free membrane patches. *Pflugers Arch* **391**: 85–100.

Kirichok Y, Krapivinsky G, Clapham DE. 2004. The mitochondrial calcium uniporter is a highly selective ion channel. *Nature* **427**: 360–364.

Kussius CL, Popescu GK. 2009. Kinetic basis of partial agonism at NMDA receptors. *Nat Neurosci* **12**: 1114–1120.

Lape R, Plested AJ, Moroni M, Colquhoun D, Sivilotti LG. 2012. The α1K276E startle disease mutation reveals multiple intermediate states in the gating of glycine receptors. *J Neurosci* **32**: 1336–1352.

Opazo P, Labrecque S, Tigaret CM, Frouin A, Wiseman PW, De Koninck P, Choquet D. 2010. CaMKII triggers the diffusional trapping of surface AMPARs through phosphorylation of Stargazin. *Neuron* **67**: 239–252.

Plested AJR, Baranovic J. 2016. Single-channel recording of glycine receptors in human embryonic kidney (HEK) cells. *Cold Spring Harb Protoc* doi: 10.1101/pdb.prot091652.

Plested AJ, Groot-Kormelink PJ, Colquhoun D, Sivilotti LG. 2007. Single-channel study of the spasmodic mutation α1A52S in recombinant rat glycine receptors. *J Physiol* **581**: 51–73.

Purohit P, Auerbach A. 2009. Unliganded gating of acetylcholine receptor channels. *Proc Natl Acad Sci* **106**: 115–120.

Rosenmund C, Stern-Bach Y, Stevens CF. 1998. The tetrameric structure of a glutamate receptor channel. *Science* **280**: 1596–1599.

Suchyna TM, Markin VS, Sachs F. 2009. Biophysics and structure of the patch and the gigaseal. *Biophys J* **97**: 738–747.

Wyllie DJ, Béhé P, Colquhoun D. 1998. Single-channel activations and concentration jumps: Comparison of recombinant NR1a/NR2A and NR1a/NR2D NMDA receptors. *J Physiol* **510**: 1–18.

Cite this introduction as *Cold Spring Harb Protoc*; doi:10.1101/pdb.top087239

Single-Channel Recording of Glycine Receptors in Human Embryonic Kidney (HEK) Cells

Andrew J.R. Plested[1] and Jelena Baranovic[1]

Leibniz-Institut für Molekulare Pharmakologie (FMP), 13125 Berlin, Germany; NeuroCure, Charité-Universitätsmedizin Berlin, 10117 Berlin, Germany

This protocol describes how to record the single-channel activity of recombinant homomeric glycine receptors expressed in human embryonic kidney (HEK) cells. Cell-attached recordings readily reveal the large conductance (90 pS) and distinctive clusters of activations at high glycine concentration. This method for obtaining equilibrium recordings can be adapted to any ion channel receptor. The necessary extensions to outside-out patch for nonequilibrium recordings are also described, as are basic analyses of channel properties and activity.

MATERIALS

It is essential that you consult the appropriate Material Safety Data Sheets and your institution's Environmental Health and Safety Office for proper handling of equipment and hazardous materials used in this protocol.

RECIPES: Please see the end of this protocol for recipes indicated by <R>. Additional recipes can be found online at http://cshprotocols.cshlp.org/site/recipes.

Reagents

Extracellular solution <R>

> *To use in outside-out patch recording, substitute 20 mM NaCl in the extracellular solution with sodium gluconate (i.e., final solution is 100 mM NaCl, 20 mM sodium gluconate) to obtain outside and inside solutions with approximately symmetrical chloride, and add glycine (e.g., 1 mM).*

> *Balance osmolarity of the intracellular and extracellular solutions to 320 mOsm with sucrose as required.*

Human embryonic kidney (HEK) cells transfected with cDNA for the rat glycine α_1 receptor

> *We recommend transfection of HEK cells with glycine receptor cDNA using the calcium phosphate method, as previously described (Groot-Kormelink et al. 2002). Cells can be more densely seeded than for whole-cell experiments. For making outside-out recordings, it is useful to pull patches from connected cells. Substantial "blank" carrier DNA is included to keep receptor expression low. To check for transfection, it is standard to observe co-transfected GFP or that expressed from the same mRNA following an internal ribosomal entry site (IRES). In either case, looking for low expression, it will likely be pale cells that are optimal; therefore, the optical setup must have some dynamic range and a good way to see cells that are "less green." For well-expressing channels—those for which outside-out patches can contain hundreds of channels—it is usually necessary to work within 1 day of the transfection. The optimal level of fluorescence will depend on the channel/mutant in question and needs to be assessed anew for every untested construct.*

Hydrochloric acid (HCl; for etching pipette electrode)

---

[1]Correspondence: Plested@fmp-berlin.de; baranovic@fmp-berlin.de

Cite this protocol as *Cold Spring Harb Protoc*; doi:10.1101/pdb.prot091652

Intracellular solution for outside-out patch recording <R>

Use also as bath solution for inside-out patch recording.

Balance osmolarity of the intracellular and extracellular solutions to 320 mOsm with sucrose as required.

Pipette solution for cell-attached recordings (extracellular solution with 1 mM glycine added)

Store up to 6 mo at −20°C.

Silver chloride (powder)

Sylgard 184 (polydimethylsiloxane) silicone elastomer

Prepare according to manufacturer's instructions; mix very well. Aliquots of prepared solution may be stored for up to 2 mo at −20°C.

Equipment

Air table and Faraday cage

Borosilicate pipette glass (thick wall preferred)

Digitizer and computer (Digidata [Axon], CED 1401, ITC-18 [HEKA], or National Instruments 6000 series; the EPC 9 [HEKA] amplifier has an integral digitizer and is also acceptable)

Electronic filter (eight-pole Bessel with adjustable cutoff is recommended)

Graphing and analysis software (e.g., IGOR Pro [WaveMetrics]; see Step 15)

Heater mounted on dissecting microscope or polisher for curing Sylgard 184 elastomer (optional; see Step 8)

Inverted microscope with fluorescence

Microscopy can be done on an upright scope, but the solution meniscus around any immersion lens is a large source of noise.

Microforge for polishing and controlling the tip size (Narishige)

Micromanipulator

Oscilloscope (optional; see Steps 13 and 14)

Patch-clamp amplifier (Axopatch 200B [Axon] preferred; anything more modern than HEKA EPC 7 acceptable)

Perfusion system (optional; for excised patches only)

Pipette holder

The holder from G23 Instruments has excellent stability.

Pipette puller (Sutter P97/P1000 for borosilicate glass, P2000 for quartz glass)

Teflon-coated silver wire

Video monitor (optional; see Step 11)

METHOD

Prepare Electrode Holder and Pipettes

1. Even if it appears quite clean, scrupulously clean the holder, particularly any salt that comes from stray drops of solution.

   Excellent cleaning can be achieved by 10 min of sonication in water or 70% ethanol (dependent on the manufacturer's instructions) and subsequent blotting dry on a tissue before storage in a desiccator or in an oven maintained at 50°C.

2. Use silver wire with a Teflon coat for the pipette electrode, which allows good regularization of the noise level. Use a fresh scalpel or razor blade to gently cut away a short (~1-mm) length of the Teflon coat.

   This process can be done under a dissecting scope for greater precision.

   Silver is soft. Take great care at this stage to avoid cutting into the metal because nicks introduced at this stage will make the finished electrode prone to breakage.

Cite this protocol as *Cold Spring Harb Protoc*; doi:10.1101/pdb.prot091652

3. At the opposing end, do the same but expose several millimeters. Bend this section to contact onto the gold pin of the electrode holder.

4. Melt the silver chloride powder (e.g., in a small crucible over a Bunsen flame). Briefly dip the short exposed silver wire into the molten AgCl for <1 sec.

   Although Teflon melts at 300°C and silver chloride at 455°C, heat is conducted away by the silver wire before the Teflon melts, allowing a teardrop-shaped AgCl pellet to form on the end of the wire, overlapping the outside of the Teflon coat. This "pellet" must be narrow enough to fit down the shaft of the pipette. In our experience, the first blob of silver chloride should be pulled off by hand and the second will anneal nicely to the "cleaned" wire.

5. Etch the pellet for some 15–30 min in neat (undiluted) HCl to remove oxidation before use, otherwise substantial drift will result.

6. To delay subsequent tearing of the Teflon coat, fire-polish the ends of the pipettes.

   This can be done in bulk by rolling five or so pipettes between one's fingers in a Bunsen flame.

Achieve Good Recording Conditions

Solutions for various recording configurations are provided in the Reagents section. Choose the appropriate solutions for the bath and pipette according to the type of recording configuration to be used.

7. At the beginning of each day, fill a pipette with the desired solution and move the tip to immediately above the surface of the bath.

   This position represents the ideal noise that can be achieved and is the best way to measure noise. Typically, levels (in the 5-kHz bandwidth) between 100 and 200 fA RMS can be achieved with normal borosilicate glass and the Axon Instruments Axopatch 200B amplifier. Higher noise levels than this indicate some considerable pickup of noise.

8. For optimal noise performance, thickly coat the tapered shafts (but not the tip itself!) of borosilicate glass pipettes in Sylgard.

   Sylgard is a two-part elastomer that must be kept on ice during the experiment. It can be applied with a spare piece of pipette glass. Some pipette polishers (e.g., that from ALA Scientific) include a hot-air stream for curing Sylgard. Otherwise, a small heater coil may be used.

   Pipettes made of quartz glass need not be coated with Sylgard.

Perform Recording

9. For the lowest noise, place the ground electrode directly in the bath.

   Despite some concerns about silver toxicity, we always use this method when not performing permeability experiments with ion exchanges.

   Commercially available pellets come with a silver wire attached. This wire must be carefully shrouded in heat-shrink tubing to avoid the high noise that will result if bath solution comes into contact with it.

10. After polishing the tip of a patch pipette coated with Sylgard (see Step 8), fill the tip with ~10 μL pipette solution.

    Avoid overtightening the holder when mounting the pipette; otherwise the thread will loosen up over time, which can contribute to noise.

11. Bring the patch pipette to the surface of the cell. As for normal patch-clamp recording, monitor the resistance at the patch tip with a square wave of 5–10 mV amplitude (seal test).

    As the seal forms, the resistance should increase and thus the amplitude of the current due to the seal test should decrease. See Troubleshooting.

12. For recording glycine single channels, use the simplest recording configuration (i.e., cell-attached), which brings the lowest noise. Make the pipette angle as steep as possible.

    For best recordings, the solution level in the bath should be shallow—in the range of a few hundred microns.

13. Use the Axopatch 200B in patch mode, and select a gain of at least 100×—this engages a different circuit that introduces the least relative noise.

    If the leak is bad enough that this setting crops the channel activations, it is likely that the recording will not be of much use and can be stopped.

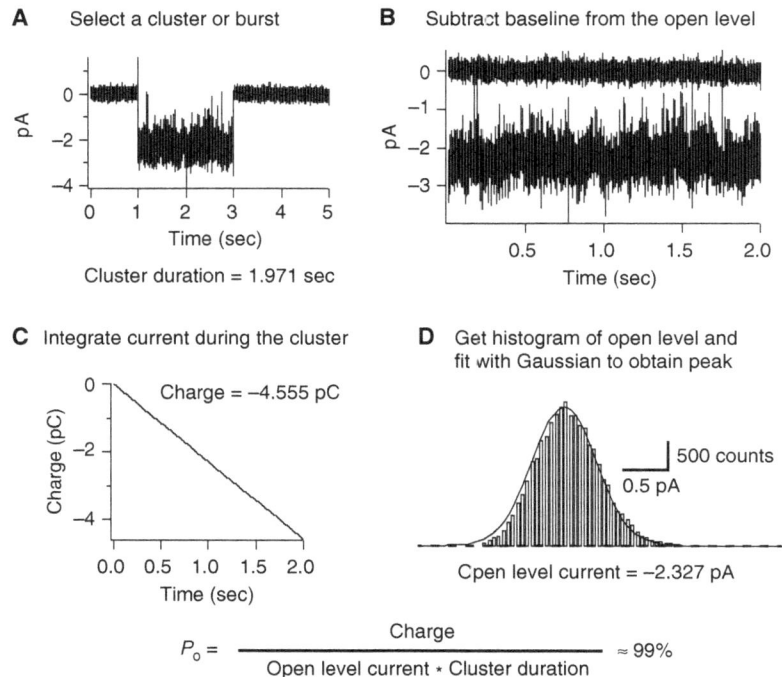

FIGURE 1. The charge-ratio approach for estimating open probability (P_{open}) during a burst or cluster.

For homomeric glycine α_1 channels, the level should be about 8–10 pA at +80 mV pipette potential (approximately −80 mV on the membrane). At 1 mm glycine, the predominant event should be long "clusters"—groups of openings that last for about a second on average (Beato et al. 2004).

Patches can be stable for tens of minutes at 100 mV, but some will not withstand more than a few seconds. See Troubleshooting.

14. Digitize data at 50 kHz or higher.

    The currents should not be filtered beyond 10 kHz at this stage. Although not essential, it is very useful indeed to split the amplifier output and use an eight-pole Bessel filter in front of an oscilloscope to see what the recording looks like at 1–3 kHz while it is being made. Any filtering that is necessary can be done in software, post hoc, using a Gaussian filter (Colquhoun and Sigworth 1995).

Data Analysis

15. To estimate channel conductance, make an all-points amplitude histogram after subtracting the baseline.

    A simple analysis can be conducted in IGOR Pro with the multipeak fitting add-in and with other packages as well. One should also determine and correct for the reversal potential by collecting unitary currents at a range of voltages.

16. Use a charge-ratio approach (Fig. 1) to assess activity (open probability) within a group of openings without idealization, because the Bessel filter preserves area.

    Here, the integral of the current during a group of openings can be compared to the charge expected if the channel had opened to its full conductance for 100% of the duration of the burst. This ratio gives the open probability but is sensitive to drifts and errors of conductance estimation. This method should not be applied to stretches of record that include gaps between clusters—because the clusters might originate from different channels, the shut duration between them does not relate to channel activity.

TROUBLESHOOTING

Problem (Step 11): Sealing onto the cell tightly is not possible—baseline is not obtained.
Solution: If expression is too high, it often appears that sealing is impossible, because the channels are activated by agonist in the pipette. Try cells with less expression. Well-expressing channels will be expressed in a matter of hours after transfection.

Cite this protocol as *Cold Spring Harb Protoc*; doi:10.1101/pdb.prot091652

Problem (Step 13): Channels endogenous to the overexpression host appear.

Solution: If this is a persistent problem, one might change the expression system or include specific inhibitors, if available. For glycine receptor recordings in HEK cells, we routinely include 20 mM tetraethylammonium chloride (TEA) to block background potassium channels.

Problem: Excised patches are not stable.

Solution: To obtain stable patches, it is necessary to use very small tips (15–25 MΩ). However, breaking through after sealing onto the cell gets progressively more difficult as the pipette tip gets smaller. The "Zap" control on the Axopatch amplifier can be useful here to break the membrane under the tip. Note that the seal should be fast for cell-attached recording. But for outside-out patch recordings, it is generally better to use patches for which the seals were obtained over the course of 30–60 sec.

Problem: Amplitudes are heterogeneous across experiments.

Solution: Note that in cell-attached mode, the membrane potential is not clamped. Individual HEK cells can show resting potentials from roughly −40 to −10 mV. The slope of the single-channel current–voltage relation can be used to determine the channel conductance. In the case of voltage-dependent gating, it is advisable that the holding voltage be adjusted to produce a similar unitary current across experiments. A brief spell with a very low filter cutoff (e.g., 1 kHz) at the beginning of the recording will help to pinpoint the open level.

RECIPES

Extracellular Solution

Reagent	Final concentration
NaCl	120 mM
KCl	4 mM
CaCl$_2$	2 mM
MgCl$_2$	2 mM
HEPES	10 mM
D-Glucose	10 mM

To block endogenous potassium channels, substitute 20 mM NaCl with 20 mM tetraethylammonium chloride (TEA·Cl) so that the solution contains 100 mM NaCl and 20 mM TEA·Cl.

Adjust pH to 7.4 with NaOH. Make fresh on the day of use. Store at room temperature during the day of use.

Intracellular Solution for Outside-Out Patch Recording

Reagent	Final concentration
KCl	107.1 mM
EGTA	11 mM
CaCl$_2$	1 mM
MgCl$_2$	1 mM
HEPES	10 mM
Tetraethylammonium chloride (TEA·Cl)	20 mM

Prepare this solution (adapted from Groot-Kormelink et al. 2002) by combining the reagents above and adjusting the pH to 7.2 with KOH. Store for up to 6 months at −20°C.

ACKNOWLEDGMENTS

This work was funded by the Deutsche Forschungsgemeinschaft (DFG) (PL619/1 and EXC-257) and European Research Council (ERC) ("Gluactive").

REFERENCES

Beato M, Groot-Kormelink PJ, Colquhoun D, Sivilotti LG. 2004. The activation mechanism of α_1 homomeric glycine receptors. *J Neurosci* **24:** 895–906.

Colquhoun D, Sigworth FJ. 1995. Fitting and statistical analysis of single-channel records. In *Single-channel recording* (ed. Sakmann B, Neher E), pp. 483–587. Plenum, New York.

Groot-Kormelink PJ, Beato M, Finotti C, Harvey RJ, Sivilotti LG. 2002. Achieving optimal expression for single channel recording: A plasmid ratio approach to the expression of α_1 glycine receptors in HEK293 cells. *J Neurosci Methods* **113:** 207–214.

CHAPTER 8

Measuring the Basic Physiological Properties of Synapses

Matthew A. Xu-Friedman[1]

Department of Biological Sciences, University at Buffalo, State University of New York, Buffalo, New York 14260

Studying synaptic physiology remains an important endeavor for understanding the mechanisms that underlie neurotransmitter release as well as synaptic development and refinement. It is also critical to understanding brain function. The basic conceptual framework for synaptic physiology was worked out by Katz and colleagues and has coalesced around three factors: the number of releasable vesicles (N), the probability of release (P), and quantal size (Q). Here, I discuss a few key experiments related to these quantities and how they are assessed.

BASIC PHYSIOLOGICAL PROPERTIES OF SYNAPSES

The basic mechanisms of synaptic transmission are well understood from many decades of physiological and molecular experiments. However, there are many synaptic proteins whose roles in modulating this process are not clear. In addition, much remains to be learned about synaptic function. The reasons any given synapse has the characteristics it does are largely unknown, particularly considering the sheer number of different types of synapses between all the different cell types in the brain. It is difficult to imagine understanding brain function without a reckoning of this diversity, daunting although it is. For those interested in function or mechanism, there are still many reasons to characterize synapses using electrophysiology. The accompanying protocol describes how to prepare brain slices for electrophysiology, with an emphasis on synaptic physiology in voltage clamp (see Protocol 1: Preparing Brain Slices to Study Basic Synaptic Properties [Xu-Friedman 2017]). This approach remains the gold standard for understanding the properties of individual neurons, as well as the connections between neurons.

Synapses are anatomically and molecularly specialized cellular compartments. The presynaptic neuron possesses clusters of vesicles that express an array of proteins that guide them through their cycle of activity—by mobilizing and docking them to the plasma membrane, filling them with the appropriate neurotransmitter, and triggering their fusion with the presynaptic membrane when presynaptic electrical activity drives a rise in intracellular calcium (Sudhof and Rothman 2009). On the other side of the synaptic cleft are postsynaptic specializations. In electron micrographs, these cluster into the postsynaptic density, which contains neurotransmitter receptors as well as signaling and structural elements (Sheng and Kim 2011).

THE EXCITATORY POSTSYNAPTIC CURRENT (EPSC)

A helpful framework for understanding the physiology of synapses was derived many decades ago, and reached a clear statement with the work of Bernard Katz. This framework considers the synaptic

---

[1]Correspondence: mx@buffalo.edu

Cite this introduction as *Cold Spring Harb Protoc*; doi:10.1101/pdb.top089680

potential as dependent on three factors: the number of releasable vesicles (N), the probability with which each vesicle releases its contents (P), and the quantal size of the current resulting from the release of neurotransmitter (Q). For simplicity, this introduction will focus on excitatory synapses. (Inhibitory are thought to be largely similar.) The excitatory postsynaptic current (EPSC) is then a product of these factors: $EPSC = NPQ$.

Quantal Size (Q)

The term "quanta" originated with Katz following the discoveries of Fatt and Katz (1952), in which they observed miniature versions of the normal, synaptic potential recorded at neuromuscular synapses (called endplates) evoked by motor neuron stimulation. These miniature endplate potentials (mEPPs) were only observed near the endplate and had similar reversal potential and sensitivity to receptor antagonists as normal-sized EPPs. It is noteworthy that the electron microscope had not yet provided images of the packages themselves, vesicles at presynaptic terminals.

Fatt and Katz (1952) showed that mEPPs, which occur spontaneously, were the individual building blocks of EPPs evoked by motor neuron stimulation. The individual blocks are presumably not distinguishable because so many of them are released simultaneously during a normal EPP. However, by lowering the concentration of calcium around the preparation, only a few quanta were released following each stimulation and the amplitudes of the evoked EPPs clustered at multiples of the mEPP size.

Measurement of Q can be done simply by patching a postsynaptic cell and recording long bouts of spontaneous electrical events. Many researchers distinguish between spontaneous synaptic currents, which may include action-potential-evoked synaptic currents, and true "minis" recorded in the presence of tetrodotoxin to prevent spontaneous activity.

Release Probability (P)

In the studies by Fatt and Katz (1952), the EPP amplitudes evoked in low calcium concentrations varied highly from trial to trial, such that release appeared random. Del Castillo and Katz (1954) analyzed the distribution of EPP amplitudes to compare them against random distributions. They began with the binomial model of release, which depends on two parameters, N and P; however, measuring N and P was experimentally intractable. Thus, they simplified their analysis to the Poisson distribution, which only requires a single parameter m, the average number of vesicles released per trial. This simplification is appropriate when N is large and P is small, as appears to be the case at the neuromuscular junction in low calcium. The parameter m can be estimated by dividing the average EPP by the average mEPP, allowing calculation of the expected frequency of single release events, double release events, etc. The match between theory and data were very good using this approach, called "quantal analysis," at the neuromuscular junction (Boyd and Martin 1956). These experiments indicate that neurotransmitter release is fundamentally a random process, likely obeying binomial statistics.

In contrast to the neuromuscular junction, quantal analysis in central synapses has usually failed to show such clear results. However, it is generally accepted that central synapses are similar in essence to the neuromuscular junction. It is often sufficient to get a relative measure of P by delivering pairs of stimuli at short intervals, and calculating the amplitude of the second EPSC relative to the first (the paired pulse ratio, or PPR). A low PPR reflects high P, and a high PPR reflects low P. The PPR does not yield a direct value of P, but is a sensitive measure of small changes in P.

There are many factors that could cause EPSC amplitudes to depart from binomial statistics in quantal analysis. One is that centrally, release sites are quite variable in size, leading to highly variable miniature EPSC amplitudes, so the distribution of EPSC amplitudes is too broad to recognize individual quanta. A second issue is that individual quanta likely do not summate linearly to produce the EPSC. The most likely reason for this is postsynaptic receptor saturation. Saturation occurs when the neurotransmitter from a single vesicle occupies a large fraction of receptors, such that additional neurotransmitter release has a diminished effect. Saturation can be shown by comparing the effects of low-affinity receptor antagonists on EPSCs under different conditions. Such antagonists block better

Cite this introduction as *Cold Spring Harb Protoc*; doi:10.1101/pdb.top089680

when the local glutamate concentration is low than when it is high. In an important experiment by Wadiche and Jahr (2001) at the climbing fiber synapse onto Purkinje cells in the cerebellum, the second in a pair of EPSCs was more strongly blocked relative to the first by a low-affinity antagonist. This is consistent with the glutamate concentration being lower for the second EPSC at the average release site. The most likely reason for this is that multiple vesicles could be released per release site. Furthermore, the amount of glutamate release is not normally proportional to EPSC amplitude in the absence of low-affinity antagonist, probably because of receptor saturation. Similar effects have been found at other synapses (Neher and Sakaba 2001; Foster et al. 2002; Chanda and Xu-Friedman 2010), suggesting receptor saturation is likely widespread.

Releasable Pool

Receptor saturation makes it difficult to assess presynaptic neurotransmitter release using postsynaptic receptors. One solution to this problem is the use of low-affinity antagonists. An alternative approach is to assay release using capacitance measurements of the presynaptic terminal as a direct measure of vesicle fusion. This technique is based on the fact that each vesicle, when it fuses with the plasma membrane, increases the total capacitance of the cell (Lindau and Neher 1988). Measuring capacitance requires patching the presynaptic terminal, so is only possible for large terminals. For example, capacitance was measured at the synaptic terminals of retinal bipolar cells in response to depolarization. For brief depolarization, the capacitance appeared to show an initial plateau at a capacitance equivalent to 1100 vesicles (Mennerick and Matthews 1996), and a second plateau equivalent to 5500 vesicles after much longer depolarization (von Gersdorff and Matthews 1994). These values appear to roughly match the numbers of vesicles counted in serial electron microscope reconstructions of retinal bipolar cell terminals, the smaller population closest to the presynaptic plasma membrane, and the larger population extending further into the synaptic specializations of these cells (von Gersdorff et al. 1996). This indicates that different physiological pools of vesicles have anatomical equivalents, and that they can be depleted.

There are additional approaches for characterizing N and P. One, mean-variance analysis, is based on the binomial model of release and the relationship between the EPSC mean and variance as measured at multiple release probabilities (i.e., calcium concentrations). This method is well described by Silver (2003). The second approach, the integration method, assesses the releasable pool by evoking enough EPSCs to deplete it and integrating their amplitudes, after correcting for synaptic recovery (Schneggenburger et al. 1999). This method has the virtue of simplicity, although it assumes a constant rate of recovery, which is almost certainly not the case (Dittman and Regehr 1998; Wang and Kaczmarek 1998; Sakaba and Neher 2001).

ACKNOWLEDGMENTS

I am grateful to T. Branco, S. Brenowitz, I. Duguid, and P. Kammermeier for their supervision of the Ion Channels and Synaptic Physiology course at Cold Spring Harbor Laboratory. I also thank T. Ngodup, H. Yang, Y. Yang, and X. Zhuang for their comments on this manuscript.

REFERENCES

Boyd IA, Martin AR. 1956. The end-plate potential in mammalian muscle. *J Physiol* **132:** 74–91.

Chanda S, Xu-Friedman MA. 2010. A low-affinity antagonist reveals saturation and desensitization in mature synapses in the auditory brainstem. *J Neurophysiol* **103:** 1915–1926.

Del Castillo J, Katz B. 1954. Quantal components of the end-plate potential. *J Physiol* **124:** 560–573.

Dittman JS, Regehr WG. 1998. Calcium dependence and recovery kinetics of presynaptic depression at the climbing fiber to Purkinje cell synapse. *J Neurosci* **18:** 6147–6162.

Fatt P, Katz B. 1952. Spontaneous subthreshold activity at motor nerve endings. *J Physiol* **117:** 109–128.

Foster KA, Kreitzer AC, Regehr WG. 2002. Interaction of postsynaptic receptor saturation with presynaptic mechanisms produces a reliable synapse. *Neuron* **35:** 1115–1126.

Lindau M, Neher E. 1988. Patch-clamp techniques for time-resolved capacitance measurements in single cells. *Pflugers Arch* **411:** 137–146.

Mennerick S, Matthews G. 1996. Ultrafast exocytosis elicited by calcium current in synaptic terminals of retinal bipolar neurons. *Neuron* **17:** 1241–1249.

Neher E, Sakaba T. 2001. Combining deconvolution and noise analysis for the estimation of transmitter release rates at the calyx of held. *J Neurosci* **21:** 444–461.

Sakaba T, Neher E. 2001. Calmodulin mediates rapid recruitment of fast-releasing synaptic vesicles at a calyx-type synapse. *Neuron* **32:** 1119–1131.

Schneggenburger R, Meyer AC, Neher E. 1999. Released fraction and total size of a pool of immediately available transmitter quanta at a calyx synapse. *Neuron* **23:** 399–409.

Sheng M, Kim E. 2011. The postsynaptic organization of synapses. *Cold Spring Harb Perspect Biol* **3:** a005678.

Silver RA. 2003. Estimation of nonuniform quantal parameters with multiple-probability fluctuation analysis: Theory, application and limitations. *J Neurosci Methods* **130:** 127–141.

Sudhof TC, Rothman JE. 2009. Membrane fusion: Grappling with SNARE and SM proteins. *Science* **323:** 474–477.

von Gersdorff H, Matthews G. 1994. Dynamics of synaptic vesicle fusion and membrane retrieval in synaptic terminals. *Nature* **367:** 735–739.

von Gersdorff H, Vardi E, Matthews G, Sterling P. 1996. Evidence that vesicles on the synaptic ribbon of retinal bipolar neurons can be rapidly released. *Neuron* **16:** 1221–1227.

Wadiche JI, Jahr CE. 2001. Multivesicular release at climbing fiber-Purkinje cell synapses. *Neuron* **32:** 301–313.

Wang LY, Kaczmarek LK. 1998. High-frequency firing helps replenish the readily releasable pool of synaptic vesicles. *Nature* **394:** 384–388.

Xu-Friedman MA. 2017. Preparing brain slices to study basic synaptic properties. *Cold Spring Harb Protoc* doi: 10.1101/pdb.prot093328.

Cite this introduction as *Cold Spring Harb Protoc*; doi:10.1101/pdb.top089680

Preparing Brain Slices to Study Basic Synaptic Properties

Matthew A. Xu-Friedman[1]

Department of Biological Sciences, University at Buffalo, State University of New York, Buffalo, New York 14260

This protocol describes how to prepare brain slices for electrophysiology, with an emphasis on synaptic physiology in voltage clamp. This approach remains the gold standard for understanding the properties of individual neurons, as well as the connections between neurons.

MATERIALS

It is essential that you consult the appropriate Material Safety Data Sheets and your institution's Environmental Health and Safety Office for proper handling of equipment and hazardous materials used in this protocol.

RECIPES: Please see the end of this protocol for recipes indicated by <R>. Additional recipes can be found online at http://cshprotocols.cshlp.org/site/recipes.

Reagents

Agar (2%) (as needed; see Step 12)

Artificial cerebrospinal fluid (ACSF) (1×) <R>

> *During experiments, 1× ACSF must be constantly bubbled with carbogen (95% O_2, 5% CO_2), or the pH will turn basic.*

Bleach (as needed, for rechloriding silver wire)

CsF/CsCl recording solution <R>

Cyanoacrylate glue

Isoflurane

Rodents for brain dissection

> *Optimal slices are obtained from rodents in the age range of 2 wk old or younger. Cell survival declines markedly at older ages, and increased myelination reduces visibility. Even so, recordings are possible at all ages.*

Sucrose cutting solution <R>

> *Use an ice-cold slurry for Step 6 and Step 11. Prewarm solution to 30°C for Step 13.*

> *During experiments, sucrose cutting solution must be constantly bubbled with carbogen (95% O_2, 5% CO_2), or the pH will turn basic.*

Equipment

Air table with Faraday cage, for supporting microscope

---

[1]Correspondence: mx@buffalo.edu

Cite this protocol as *Cold Spring Harb Protoc*; doi:10.1101/pdb.prot093328

Amplifier, for voltage clamp recording

Beaker (150-mL)

Coverslips, precoated with poly-L-lysine (0.01% solution)

Alternatively, a harp (U-shaped platinum wire with fine nylon strands across) can be used to secure slices in the recording chamber (Step 17).

Cutting chamber

The cutting chamber is usually supplied with the slicing machine (see below), and has a platform (chuck) to which the brain is glued. To prepare the cutting chamber before use, pour sucrose cutting solution into the bottom of the chamber, leaving the chuck uncovered, and then freeze the chamber at −20°C. Be cautious with plastic cutting chambers, not all of which are freezable.

Data acquisition system, for capturing responses to computer

Dissection chamber (small Petri dish or similar, with Sylgard-lined bottom), chilled in freezer to −20°C

Electrode holders for recording and stimulating micropipettes

These include chlorided silver wire, side ports for applying pressure, and gaskets for maintaining pressure.

Forceps

In-line solution heater, for performing recordings at physiological temperature

Use physiological temperature in the recording chamber when possible. Sealing and breaking in are generally easier, and cell properties are more physiologically relevant. However, recordings may be shorter.

Microelectrode puller and capillary glass tubing, for preparing micropipettes for recording and stimulating

We use a five-stage pull on a Sutter puller (see Step 18), with a 3-mm × 3-mm box-type filament and thick-wall capillary glass (1.5-mm OD, 0.86-mm ID).

Micromanipulator, for holding recording and stimulating electrodes

Oscilloscope

Peristaltic pump, for delivering 1× ACSF to preparation

Razor blade

Recording chamber

The recording chamber will hold the brain slice during recording and allows 1× ACSF to flow over and bathe the slice. The chamber sits on the movable microscope stage between the objective and the condenser. The chamber consists of three connected areas, one which accepts a tube from the peristaltic pump carrying 1× ACSF, a second (roughly 1 × 1 cm) where the slice will sit to be viewed through the microscope objective, and a third where a vacuum drains the 1× ACSF from the recording area. The bottom of the recording chamber is optical glass, such as a coverslip. Recording chambers can be bought commercially or made out of plastic. Chambers should be inspected regularly for leaks, to avoid 1× ACSF getting into the microscope condenser.

Recording setup

Recovery chamber (grid cut from "egg crate" diffusers for fluorescent lights, with nylon mesh glued to bottom, placed in a close-fitting container)

Scissors, fine

Scissors, large (or guillotine)

Spatula

Stimulus isolator, for isolating stimulating electrode from setup and delivering stimuli

Tubing, with a 1-mL syringe, for applying positive or negative pressure

Devices for monitoring pressure are available, but these are not strictly necessary.

Upright microscope with water-immersion 40× or 60× objectives

Vibrating slicer, for cutting brain slices

Use a fresh blade for slicing. Stainless steel blades are usually adequate, but ceramic and sapphire blades may improve quality.

Water bath at 30°C

Cite this protocol as *Cold Spring Harb Protoc*; doi:10.1101/pdb.prot093328

METHOD

Dissecting the Brain

1. Anesthetize the rodent with isoflurane.
2. Decapitate with large scissors (or guillotine if necessary).
3. Cut open the skull with fine scissors, starting from the spinal cord, taking care not to injure the brain area of interest.
4. Peel back the skull with forceps.
5. Urge the brain out of the braincase with scissors, cutting the cranial nerves underneath to free it.
6. Drop the brain in a 150-mL beaker containing an ice-cold slurry of sucrose cutting solution bubbled with carbogen.
7. Transfer the brain to a chilled dissection chamber. Cut the tissue with a razor blade to prepare the specific brain area of interest. Make sure that one face of the tissue block is flat in preparation for gluing to the chuck in the desired orientation. Remove as much of the meninges as is practical.

Slicing the Brain

8. Daub 1–2 drops of cyanoacrylate glue on the chuck of the cutting chamber.
9. Transfer the brain from the dissection chamber to the spot of glue using a small spatula.
10. Attach the cutting chamber to the slicer.
11. Fill the cutting chamber with ice-cold sucrose cutting solution until the brain is submerged. Bubble the solution continuously with carbogen.
12. Raise the preparation to the blade, and cut at slow speed (0.05 mm/sec) with rapid vibration (80 Hz). Cut 100- to 300-μm-thick slices, depending on what is convenient to handle.

    Advancement can be faster for unimportant brain areas (e.g., up to 0.20 mm/sec).

    The chunk of brain to be cut for slices should be relatively firm to resist the push of the blade. If cutting particularly floppy or tall chunks, it is best to brace the tissue from behind with a small block of 2% agar.

13. Transfer slices of the area of interest to a recovery chamber containing prewarmed sucrose cutting solution. Hold the recovery chamber for 20 min in a water bath at 30°C, bubbling the cutting solution continuously with carbogen.

    Do not incubate more than 20 min in sucrose after slicing.

    Some users skip this sucrose incubation and proceed directly to incubation in 1× ACSF (Step 14).

14. Transfer the slices to 1× ACSF in the recovery chamber and hold for an additional 40 min in a water bath at 30°, bubbling continuously with carbogen.
15. Remove the recovery chamber from the water bath, and keep the slices at room temperature until recording.

Aligning the Electrode

16. Perfuse the recording chamber with 1× ACSF at 1–4 mL/min using the peristaltic pump. Bubble continuously with carbogen.
17. Adhere a brain slice to a coverslip coated in poly-L-lysine, and place in the recording chamber under the microscope. Alternatively, place a slice in the recording chamber and weigh the slice down with a harp.
18. Pull a recording patch pipette using a five-stage pull. After pulling, inspect the pipette tips under a high-power microscope to make sure they are the correct size.

Upon installing a new filament, we conduct a ramp test, make a 1-line program with heat = ramp, pull = 0, time = 250, and adjust the velocity for a reliable five-stage pull. Next, we create a five-step program with pull = 0, time = 250, velocity = the empirically determined velocity. The heat values in the first three steps are ramp, and the last two decrement by 10–20. The final heat value is adjusted to yield patch pipettes of the preferred size, with size assayed by resistance, for example, 1 MΩ for large cells and 2 or 3 MΩ for small cells (see Step 23). Smaller electrodes should be avoided, except for dendritic or axonal patching. Large electrodes provide the best recording quality.

It is unnecessary to pull electrodes until slice quality has been assessed. Also, electrodes should be pulled in small batches, to avoid overheating the puller, which can change the efficacy of the puller program.

19. Fill the tip of the pulled pipette with CsF/CsCl recording solution. Insert the pipette in the electrode holder, and attach to the micromanipulator.

20. At low power (e.g., 4×) on the upright microscope, minimize the field diaphragm (below the condenser) and confirm that the condenser is properly in focus.

21. Adjust the pipette position so the tip is centered in the light spot, hovering just above the brain slice.

22. Lower the recording pipette into the bath solution of the recording chamber, applying constant, positive pressure through a tube connected to the pressure port on the electrode holder.

    These pressures are most easily applied by mouth, for example using a 1-mL syringe as the mouthpiece.

23. Set the amplifier to the voltage-clamp configuration, and check the electrode resistance on an oscilloscope using a −5 mV pulse (Fig. 1).

    The patch pipette resistance should be as low as possible, ideally 1 MΩ for large cells and 2 or 3 MΩ for small cells.

24. Adjust the pipette offset on the amplifier to 0 pA.

    If the pipette offset requires constant adjustment in the bath, the silver wire in the electrode should be re-chlorided. This is most easily accomplished by scraping off the old coating and immersing the wire in bleach for ~20 min.

Forming a Seal

25. Switch the microscope objective to a high power (40× or 60×) and immerse in the bath.

26. Put the slice in the focal plane of the objective by adjusting the focus toward increasing brightness.

27. Locate a good cell to patch.

    Somata should be round with subtle borders. Sharply defined or irregular borders reflect ill health.

    See Troubleshooting.

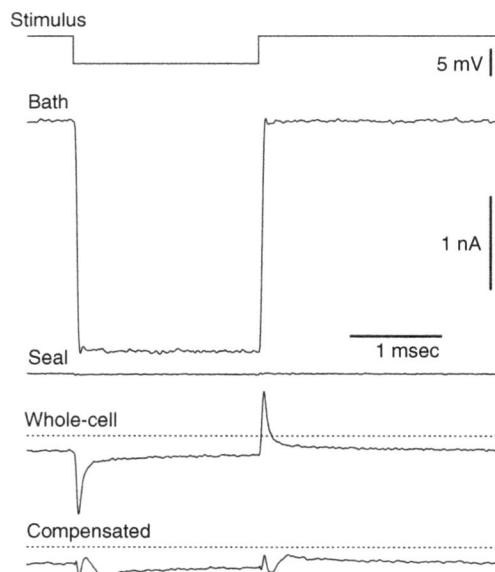

FIGURE 1. Voltage-clamp traces during patching. The stimulus (−5 mV pulse) is shown at the *top*. The following four traces (*top* to *bottom*) are the recorded currents when the electrode is in the bath, after forming a giga-seal, after breaking in to whole-cell voltage clamp, and after turning on series resistance compensation. Dashed lines in the whole-cell and compensated traces indicate zero current. The zero line is omitted from the bath and seal traces for clarity.

Cite this protocol as *Cold Spring Harb Protoc*; doi:10.1101/pdb.prot093328

28. Lower the recording electrode into the slice just above the selected cell. Use positive pressure to clear a path to the cell.

29. When the pipette tip is close to the cell, reduce the positive pressure so that the path just remains open.

30. Lower the electrode tip and press it into the center of the soma, until a bright ring or indentation is visible.

31. Very quickly, release the positive pressure and apply small negative pressure, while monitoring electrode resistance on the oscilloscope.

32. When the pipette resistance passes 100 MΩ, release the negative pressure and allow a giga-seal to form (Fig. 1).

    Ideally, this only takes a few seconds.

33. Lower the holding potential to the desired level (e.g., −70 mV) using the amplifier or computer interface.

34. Adjust the pipette capacitance compensation to remove rapid capacitance transients.

35. When holding current drops below the acceptable value (−20 to −4 pA, depending on the cell), rupture the patch by applying a pulse of suction.

    This is typically done by making a gentle kissing gesture into the syringe connected to the pressure port. The pressure of the kiss is increased if there is no break-in.

36. Once break-in occurs, assess the acquired data for patch quality using the leak current and peak of the capacitance transient (Fig. 1).

    Leak may range from 10 to 200 pA, depending on the cell. Poor leak cannot be fixed, and the cell should be abandoned if it is too high. Access resistance is limited by the size of the electrode, and should produce a transient peak of 200 pA (for a 2–3 MΩ pipette) to 1000 pA (for a 1 MΩ pipette). Poor access can be fixed by applying transient or sustained positive or negative pressure.

    See Troubleshooting.

37. Use whole-cell compensation and series-resistance compensation (Fig. 1), consulting the amplifier user manual for the proper methods.

    Spontaneous synaptic currents can be recorded at this stage.

Synaptic Stimulation

38. Prepare a pipette for stimulating. Insert the pipette in the electrode holder, and attach to the micromanipulator.

    The simplest stimulating electrodes are glass micropipettes that are pulled to 2–4 μm in diameter, filled with 1× ACSF, and referenced to bath ground. Larger pipettes tend to stimulate more broadly, and smaller pipettes stimulate more focally. Other options are theta glass or concentric ring electrodes, where current is delivered from one chamber into the other. These require special holders.

39. Lower the stimulating electrode into the slice some distance from the recorded cell, being careful not to move the surface of the slice too much and disrupt the recording.

40. Using the computer interface, trigger short pulses of current through the stimulus isolator, for example 0.2 msec duration and 4–20 μA.

    Synaptic currents should follow the stimulus artifact reliably. If there are no synaptic currents, the stimulating current should be increased, or the stimulating electrode position moved.

TROUBLESHOOTING

Problem (Step 27): Cell survival is poor.
Solution: Cell survival may be improved by performing transcardial perfusion before brain dissection, by cutting at physiological temperature, or by using different cutting or recovery solutions (e.g.,

NMDG-based solutions). It is often unclear why slices may be bad on a given day, and sometimes the only solution is to keep trying.

Problem (Step 36): Access resistance is unstable.
Solution: To enhance the stability of access resistance throughout a long experiment, try the following: use large pipettes, keep the electrode centered over the cell, lift the electrode above the cell after break-in, and apply constant positive pressure during recording.

RECIPES

Artificial Cerebrospinal Fluid (ACSF) (1×)

Reagent	Amount to add	Final concentration
Sodium L-lactate	0.448 g	4 mM
Sodium pyruvate	0.220 g	2 mM
Sodium L-ascorbate	0.079 g	0.4 mM
Glucose	4.5 g	25 mM

Add a small amount of H_2O to a 1-L volumetric flask. Add the above reagents and swirl to dissolve. Add 100 mL of Artificial Cerebrospinal Fluid (ACSF) (10×) <R>. Fill the flask to ~950 mL with H_2O. Add 3.95 mL of Calcium Chloride Stock Solution <R> for a final concentration of 1.5 mM $CaCl_2$. Top off the flask with H_2O to 1 L, and bubble the solution with carbogen for 10 to 20 min (which should resolve any cloudiness from the addition of calcium). Transfer the solution to a 1-L bottle and cap. Store 1× ACSF at 4°C between experiments, and use for up to 1 or 2 d. Do not use the solution if it takes on a pinkish or yellowish color. (1× ACSF can be prepared with different calcium concentrations by changing the added amount of calcium chloride stock solution, and compensating by changing the $MgCl_2$ so that the total divalent concentration is 2.5 mM. We find it difficult to raise the calcium concentration above 3 mM.)

Artificial Cerebrospinal Fluid (ACSF) (10×)

Reagent	Amount to add	Final concentration in 10× ACSF	Final concentration in 1× ACSF
NaCl	73.05 g	1.25 M	125 mM
$NaHCO_3$	21.843 g	260 mM	26 mM
NaH_2PO_3	1.725 g	12.5 mM	1.25 mM
KCl	1.864	25 mM	2.5 mM

In a 1-L volumetric flask, combine the above reagents. Shake to prevent poorly soluble monobasic crystals from forming. Add $MgCl_2$ from a concentrated stock solution to a final concentration of 10 mM (in 10× ACSF). (For example, add 5 mL of a 2 M $MgCl_2$ stock.) Add H_2O to the 1-L line and stir until all reagents are dissolved. Store 10× ACSF in a 1-L flask at 4°C for up to 2 wk. Use this 10× stock solution daily to prepare Artificial Cerebrospinal Fluid (ACSF) (1×) <R>. Do not use if crystals form. (10× ACSF can also be prepared without $MgCl_2$ for experiments requiring different calcium concentrations.)

Calcium Chloride Stock Solution

To prepare calcium chloride stock solution, add 58 g $CaCl_2$ to 1 L of H_2O. Measure the osmolarity using an osmometer: if below 1000 mmol/kg, add more $CaCl_2$; if above 1000, add H_2O. (The osmolarity of 1000 mmol/kg is ~0.383 M. Preparation of this calibrated solution is necessary because calcium chloride is hygroscopic, so is not reliably measured by weight. Slight errors in calcium concentration have an outsized effect on synaptic physiology.) Store the solution at 4°C. Solution stored under these conditions appears to be stable for several months.

Cite this protocol as *Cold Spring Harb Protoc*; doi:10.1101/pdb.prot093328

CsF/CsCl Recording Solution

Reagent	Amount	Final concentration
CsF	1.329 g	35 mM
CsCl	4.209 g	100 mM
EGTA	0.951 g	10 mM
HEPES	0.596 g	10 mM

In a 250-mL volumetric flask, combine the above reagents. Add ~200 mL of H_2O, stirring constantly. Adjust the pH to 7.3 using CsOH solution. Measure the osmolarity of the CsF/CsCl recording solution using an osmometer. Adjust to ~300 mOsm by adding H_2O. Verify the pH after adjustment of osmolarity, adding CsOH or HCl if necessary. Store the solution at 4°C for up to several months. (This recording solution is optimized for simple voltage-clamp recordings. Different recording solutions are appropriate for different experimental goals. QX-314 [2 mM] may be added to block sodium channels and improve clamp.)

Sucrose Cutting Solution

Reagent	Amount	Final concentration
NaCl	4.441 g	76 mM
$NaHCO_3$	2.1 g	26 mM
Sucrose	25.673 g	75 mM
Glucose	4.504 g	25 mM
KCl	0.186 g	2.5 mM
NaH_2PO_4	0.172 g	1.25 mM

In 1-L volumetric flask, combine the above reagents. Shake to prevent monobasic crystals from clumping. Fill the flask to ~950 mL with H_2O. Add $MgCl_2$ from a concentrated stock solution to a final concentration of 7 mM. (For example, add 3.5 mL of a 2 M $MgCl_2$ stock.) Add 1.32 mL of Calcium Chloride Stock Solution <R> for a final concentration of 0.5 mM $CaCl_2$. Top off the flask with H_2O to 1 L, and bubble the solution with carbogen for 10–20 min (which should resolve any cloudiness from the addition of calcium). Transfer the solution to a 1-L bottle and cap. Store the solution for up to 1 wk at 4°C.

CHAPTER 9

Cellular and Synaptic Properties of Local Inhibitory Circuits

Court Hull[1,2]

[1]*Department of Neurobiology, Duke University, Durham, North Carolina 27710*

Inhibitory interneurons play a key role in sculpting the information processed by neural circuits. Despite the wide range of physiologically and morphologically distinct types of interneurons that have been identified, common principles have emerged that have shed light on how synaptic inhibition operates, both mechanistically and functionally, across cell types and circuits. This introduction summarizes how electrophysiological approaches have been used to illuminate these key principles, including basic interneuron circuit motifs, the functional properties of inhibitory synapses, and the main roles for synaptic inhibition in regulating neural circuit function. It also highlights how some key electrophysiological methods and experiments have advanced our understanding of inhibitory synapse function.

INTRODUCTION

The concept of inhibition was not widely recognized by many early pioneers of neuroscience who relied on anatomical and behavioral approaches. Ramon and Cajal, for example, despite his tremendous contributions to neuronal theory and circuit organization, never acknowledged a role for inhibition in his writings. In large part, the modern concept of inhibition arose from the work of Charles Scott Sherrington and his seminal experiments on spinal reflexes. Sherrington was the first to formally describe the inhibition of decerebrate muscle rigidity that results from a variety of tactile stimuli (Sherrington 1898), as well as the concept reciprocal inhibitory innervation that produces relaxation of antagonist muscle groups (Sherrington 1906). However, it was only through the advent of new electrophysiological approaches that the key principles of synaptic inhibition were ultimately elucidated.

Inspired by the findings of Sherrington and the emerging concept of inhibition in the central nervous system, several investigators used in vivo extracellular electrophysiology to provide key experimental evidence describing both the mechanism and role of inhibitory synaptic transmission in spinal circuits during the mid-20th century. In particular, the work of Lloyd (1941), Renshaw (1941), and others (for review, see Callister and Graham 2010) demonstrated that evoked excitatory potentials in ventral horn motor neurons could be inhibited by anterograde electrical stimulation of sensory neurons in the dorsal root, or retrograde stimulation of motor neuron axons themselves. In each case, the authors noted a brief 2–3 msec delay between the onset of evoked excitation and inhibition in the recorded potentials. Such delayed latency measurements between excitation and inhibition suggested the presence of intermediate interneurons in the circuit. Even today in the era of optogenetics, latency measurements are a valuable tool to dissect circuit connectivity and can

[2]Correspondence: Hull@neuro.duke.edu

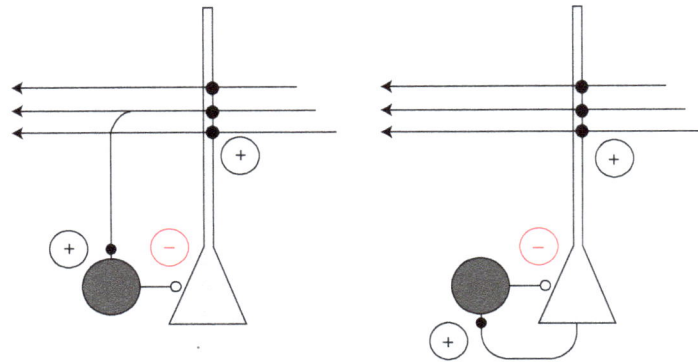

FIGURE 1. Schematic of feedforward and feedback inhibitory circuits. (*Left*) A feedforward interneuron (dark gray filled circle) is recruited by the same axons that excite a principal cell (light gray pyramid). (*Right*) A feedback interneuron is recruited by the activity of the principal cell itself. Positive signs (black) denote synaptic excitation; negative signs (red) denote synaptic inhibition.

provide essential information about whether optically or electrically evoked inhibition results from direct interneuron activation or a multisynaptic pathway.

The results of Renshaw and others also established the basis for our understanding of the two fundamental inhibitory circuit motifs that have since been documented across the nervous system: feedforward and feedback inhibitory circuits. Feedforward inhibition occurs when the same excitatory pathway that impinges on a principal neuron also activates inhibitory neurons that synapse onto the same principal cell population (Fig. 1, left). In the spinal cord, feedforward inhibition results from dorsal root stimulation because these sensory neurons both directly excite motor neurons, and excite inhibitory interneurons that synapse onto the same motor neuron population. Conversely, feedback inhibition results when the activity of a given principal neuron recruits inhibitory interneurons that feed back onto the same target population (Fig. 1, right). In the spinal cord example, feedback inhibition results from axon collaterals from motor neurons that recruit inhibitory neurons synapsing onto the very same motor neurons.

Despite the growing recognition of inhibitory circuit function in the mid-20th century, the mechanism of neural inhibition remained highly controversial. While Otto Loewi's famous experiments had already established the idea that neuronal communication was a chemical process, many early investigators including Eccles believed that inhibition was a distinct, non-chemical process mediated by electrical transmission (Hultborn 2006). The key experiment demonstrating that synaptic inhibition was indeed a chemical process required the advent of a crucial new technique: intracellular recording. By impaling motor neurons with a glass electrode, Eccles and colleagues were, for the first time, able to measure the neuronal membrane potentials evoked by stimulating different synaptic pathways. In so doing, they demonstrated the first intracellularly recorded inhibitory postsynaptic potential or IPSP (Brock et al. 1952). The central observation in these recordings was that evoked inhibitory potentials had the opposite polarity of excitatory potentials. That is, rather than depolarizing the neuronal membrane potential, IPSPs hyperpolarized the motor neurons, taking them further from spike threshold. Because there was no possible mechanism for electrical transmission to invert the sign of a signal and convert an excitatory impulse to a hyperpolarizing potential, Eccles correctly concluded that synaptic inhibition was a chemical process requiring a distinct neurotransmitter.

BASIC PROPERTIES OF INHIBITORY SYNAPSES

We now recognize that the primary inhibitory neurotransmitter in the spinal cord is glycine, whereas γ-aminobutyric acid (GABA) is the most common inhibitory transmitter throughout the rest of the

Cite this introduction as *Cold Spring Harb Protoc*; doi:10.1101/pdb.top095281

central nervous system. Both GABA and glycine can bind to ionotropic ligand-gated receptors (i.e., GABA$_A$ and glycine receptors) that mediate chloride conductance; the influx of chloride ions across the neuronal membrane produces the hyperpolarizing potentials that underlie IPSPs at most physiological membrane potentials. GABA can also bind metabotropic GABA$_B$ receptors that couple to G-proteins; these receptors typically act to either open a potassium conductance or inhibit calcium channels involved in vesicular release. Hence, GABA$_B$ receptors are also inhibitory, but can act on slower time scales than GABA$_A$ receptors. Effects mediated by each of these receptors can be dissociated pharmacologically, as GABA$_A$ receptors are blocked by the selective antagonist gabazine (SR-95531), whereas GABA$_B$ receptors can be blocked by CGP-54626.

Importantly, GABA receptors can be tailored to suit specialized circuit functions. For example, by expressing different subunit combinations, GABA$_A$ receptors can have different kinetics, open probability, desensitization properties, and sensitivity to modulators such as benzodiazepines (Mody and Pearce 2004; Farrant and Nusser 2005; Jacob et al. 2008). One such example of specialized GABA$_A$ receptors is found in the cerebellum, where granule cells contain GABA$_A$ receptors that express the α_6 subunit (Brickley et al. 1996; Wall and Usowicz 1997; Rossi and Hamann 1998). These GABA$_A$ receptors do not desensitize in the continuous presence of GABA, allowing them to mediate a tonic inhibitory current that is distinct from the typical transient currents mediated by desensitizing GABA$_A$ receptors. These, and certain other GABA$_A$ receptor subtypes, can be distinguished pharmacologically. For example, α_6-subunit-containing GABA$_A$ receptors are sensitive to furosemide, but not to benzodiazepine modulation by diazepam (Hamann et al. 2002). One useful approach that has often been used to test the properties of different GABA receptors is to pull outside-out patches of membrane from a neuron of interest using standard glass capillary recording electrodes (Jones and Westbrook 1995). With such patches, it is possible to rapidly apply agonists and antagonists by using a gravity-fed flow pipette attached to a piezoelectric translator. This setup allows for microsecond switching of solutions, and thus provides the ability to record receptor-mediated currents that are shaped predominately by the kinetics of the receptors themselves, without the influence of synaptic release dynamics, diffusion, transmitter uptake, or other factors that can influence the time course of synaptic currents.

INTERNEURON IDENTIFICATION AND FUNCTION IN NEURAL CIRCUITS

In the central nervous system, inhibitory interneurons display a wide diversity of morphologies, biophysical properties, molecular markers, and functional roles (Markram et al. 2004; Monyer and Markram 2004). Hence, the most common way of classifying inhibitory cells involves a combination of these parameters, and whole-cell recording techniques along with transgenic approaches provide a valuable way to identify and manipulate individual interneuron subtypes. For example, interneurons typically classified as "basket cells" are so-called because they target the somatic compartment of their postsynaptic partners (Hu et al. 2014). By adding biocytin to a whole-cell recording electrode, it is possible to recover the morphology of a recorded neuron by performing *post hoc* histology to visualize its characteristic axonal arborization (Sik et al. 1995). Basket cells also typically express the calcium binding protein parvalbumin and hence are often referred to as "PV cells." This marker has proved a valuable tool for establishing transgenic lines that, for example, express either green fluorescent protein or channelrhodopsin-2, allowing targeted recordings or optical manipulation of these interneurons (Cardin 2012). In addition, many basket cells exhibit high firing frequencies in response to depolarization, and little adaptation of their spike rates over time. Hence, they are also often classified as "fast-spiking" cells, and current-clamp recordings can provide such physiological characterization for initial identification during recordings. In contrast, interneurons called "Martinotti cells" target the dendrites of their postsynaptic targets, express the peptide hormone somatostatin, and show slower, adapting spike rates in response to depolarization (Wang et al. 2004). However, it is important to recognize that these parameters often do not clearly divide interneuron types into unique categories, and interneuron classification remains a hotly debated topic (DeFelipe et al. 2013).

The role of inhibitory interneurons in synaptic processing can depend on both the circuit motif in which they are engaged and the unique functional and anatomical properties of the interneurons in question. For example, feedforward inhibitory circuits often regulate spike timing by establishing a narrow period of excitability referred to as an "integration time window" (Pouille and Scanziani 2001). This window is established by the latency between the arrival of direct synaptic excitation and disynaptic inhibition engaged by the same feedforward pathway. In contrast, feedback inhibitory circuits are typically involved in regulating population-level spike synchrony (Cobb et al. 1995), and regulating global levels of principal cell activity by normalizing inhibition according to the spiking of the principal cells themselves (Isaacson and Scanziani 2011). Such actions of inhibitory interneurons can serve to regulate the strength of excitatory synaptic transmission in both an additive and divisive manner by, respectively, moving the membrane potential away from spike threshold, and increasing the conductance of the neuronal membrane according to Ohm's law.

Whole-cell electrophysiological approaches, particularly in combination with an in vitro brain slice preparation, have provided a key experimental platform for testing the role of distinct interneuron subtypes in neuronal circuit function. Protocol 1: Measuring Feedforward Inhibition and Its Impact on Local Circuit Function (Hull 2016) describes methods for testing the effects of synaptic inhibition using cerebellar circuits as a model system.

REFERENCES

Brickley SG, Cull-Candy SG, Farrant M. 1996. Development of a tonic form of synaptic inhibition in rat cerebellar granule cells resulting from persistent activation of GABAA receptors. *J Physiol* 497: 753–759.

Brock LG, Coombs JS, Eccles JC. 1952. The recording of potentials from motoneurones with an intracellular electrode. *J Physiol* 117: 431–460.

Callister RJ, Graham BA. 2010. Early history of glycine receptor biology in mammalian spinal cord circuits. *Front Mol Neurosci* 3: 13.

Cardin JA. 2012. Dissecting local circuits in vivo: Integrated optogenetic and electrophysiology approaches for exploring inhibitory regulation of cortical activity. *J Physiol Paris* 106: 104–111.

Cobb SR, Buhl EH, Halasy K, Paulson O, Somogyi P. 1995. Synchronization of neuronal activity in hippocampus by individual GABAergic interneurons. *Nature* 378: 75–78.

DeFelipe J, López-Cruz PL, Benavides-Piccione R, Bielza C, Larrañaga P, Anderson S, Burkhalter A, Cauli B, Fairén A, Feldmeyer D, et al. 2013. New insights into the classification and nomenclature of cortical GABAergic interneurons. *Nat Rev Neurosci* 14: 202–216.

Farrant M, Nusser Z. 2005. Variations on an inhibitory theme: Phasic and tonic activation of GABA$_A$ receptors. *Nat Rev Neurosci* 6: 215–229.

Hamann M, Rossi DJ, Attwell D. 2002. Tonic and spillover inhibition of granule cells control information flow through cerebellar cortex. *Neuron* 33: 625–633.

Hu H, Gan J, Jonas P. 2014. Interneurons. Fast-spiking, parvalbumin$^+$ GABAergic interneurons: From cellular design to microcircuit function. *Science* 345: 1255263.

Hull C. 2016. Measuring feedforward inhibition and its impact on local circuit function. *Cold Spring Harb Protoc* doi: 10.1101/pdb.prot095828.

Hultborn H. 2006. Spinal reflexes, mechanisms and concepts: From Eccles to Lundberg and beyond. *Prog Neurobiol* 78: 215–232.

Isaacson JS, Scanziani M. 2011. How inhibition shapes cortical activity. *Neuron* 72: 231–243.

Jacob TC, Moss SJ, Jurd R. 2008. GABA$_A$ receptor trafficking and its role in the dynamic modulation of neuronal inhibition. *Nat Rev Neurosci* 9: 331–343.

Jones MV, Westbrook GL. 1995. Desensitized states prolong GABA$_A$ channel responses to brief agonist pulses. *Neuron* 15: 181–191.

Lloyd DPC. 1941. A direct central inhibitory action of dromically conducted impulses. *J Neurophsyiol* 4: 184–190.

Markram H, Toledo-Rodriguez M, Wang Y, Gupta A, Silberberg G, Wu C. 2004. Interneurons of the neocortical inhibitory system. *Nat Rev Neurosci* 5: 793–807.

Mody I, Pearce RA. 2004. Diversity of inhibitory neurotransmission through GABA$_A$ receptors. *Trends Neurosci* 27: 569–575.

Monyer H, Markram H. 2004. Interneuron diversity series: Molecular and genetic tools to study GABAergic interneuron diversity and function. *Trends Neurosci* 27: 90–97.

Pouille F, Scanziani M. 2001. Enforcement of temporal fidelity in pyramidal cells by somatic feed-forward inhibition. *Science* 293: 1159–1163.

Renshaw B. 1941. Influence of discharge of motoneurons upon excitation of neighboring motoneurons. *J Neurophsyiol* 4: 167–183.

Rossi DJ, Hamann M. 1998. Spillover-mediated transmission at inhibitory synapses promoted by high affinity α_6 subunit GABA$_A$ receptors and glomerular geometry. *Neuron* 20: 783–795.

Sherrington CS. 1898. Decerebrate rigidity, and reflex coordination of movements. *J Physiol* 22: 319–332.

Sherrington CS. 1906. *The integrative action of the nervous system.* Yale University Press, New Haven and London.

Sik A, Penttonen M, Ylinen A, Buzsáki G. 1995. Hippocampal CA1 interneurons: An in vivo intracellular labeling study. *J Neurosci* 15: 6651–6665.

Wall MJ, Usowicz MM. 1997. Development of action potential-dependent and independent spontaneous GABA$_A$ receptor-mediated currents in granule cells of postnatal rat cerebellum. *Eur J Neurosci* 9: 533–548.

Wang Y, Toledo-Rodriguez M, Gupta A, Wu C, Silberberg G, Luo J, Markram H. 2004. Anatomical, physiological and molecular properties of Martinotti cells in the somatosensory cortex of the juvenile rat. *J Physiol* 561: 65–90.

Cite this introduction as *Cold Spring Harb Protoc*; doi:10.1101/pdb.top095281

Measuring Feedforward Inhibition and Its Impact on Local Circuit Function

Court Hull[1,2]

[1]Department of Neurobiology, Duke University, Durham, North Carolina 27710

This protocol describes a series of approaches to measure feedforward inhibition in acute brain slices from the cerebellar cortex. Using whole-cell voltage and current clamp recordings from Purkinje cells in conjunction with electrical stimulation of the parallel fibers, these methods demonstrate how to measure the relationship between excitation and inhibition in a feedforward circuit. This protocol also describes how to measure the impact of feedforward inhibition on Purkinje cell excitability, with an emphasis on spike timing.

MATERIALS

It is essential that you consult the appropriate Material Safety Data Sheets and your institution's Environmental Health and Safety Office for proper handling of equipment and hazardous material used in this protocol.

RECIPES: Please see the end of this protocol for recipes indicated by <R>. Additional recipes can be found online at http://cshprotocols.cshlp.org/site/recipes.

Reagents

Artificial cerebrospinal fluid (ACSF)
Carbogen (95% O_2, 5% CO_2)
Cesium methanesulfonate recording solution <R> (optional; for voltage clamp experiments)
Gabazine, 25 mM stock solution (Tocris Bioscience, 1262)
NBQX, 50 mM stock solution (Tocris Bioscience, 1044)
Potassium gluconate recording solution <R> (optional; for current clamp experiments)

Equipment

Air table, equipped with Faraday cage
Amplifier (used for voltage and current clamp recording)
Capillaries, glass
Data acquisition system
Dissection tools (i.e., scissors, forceps, scalpel, spatula)
In-line solution heater (for maintaining solutions at near-physiological temperatures)

[2]Correspondence: Hull@neuro.duke.edu

Microelectrode puller (for preparing glass capillary recording and stimulating electrodes)

Micromanipulators (for holding recording and stimulating electrodes)

Microscope, upright, equipped with water-immersion 40× objective

Osmometer (for calibrating solutions)

Peristaltic pump (for delivering ACSF and pharmacological reagents to the slice chamber)

Petri dish, with Sylgard bottom

Slice recovery chamber

Stimulus isolator (for isolating stimulating electrode from electrical noise sources)

Vibrating slicer (for cutting brain slices)

METHOD

Preparing the Tissue Sections

1. Prepare ACSF and acute rodent cerebellar slices as described in Chapter 8 Protocol 1: Preparing Brain Slices to Study Basic Synaptic Properties (Xu-Friedman 2016) with the following considerations.

   i. When dissecting the cerebellum, perform the initial cut below the cerebellum at the spinal cord level. Continue around the skull until the bone is loose enough to remove with forceps.

   ii. Gently lift the brain with a flat spatula. Block off the cerebellum and proceed as described in the slicing protocol.

   iii. For transverse sections of cerebellar cortex, arrange the tissue such that the blade cuts in the direction of the parallel fibers (i.e., mediolaterally).

   Glue a small piece of agar behind the tissue to act as a backstop for support during cutting.

Evoking Feedforward Inhibition under Voltage Clamp

These experiments require a cesium-based internal solution with QX 314 chloride to allow voltage clamping at depolarized potentials.

2. Obtain a whole-cell recording from Purkinje cells as described in Chapter 8 Protocol 1: Preparing Brain Slices to Study Basic Synaptic Properties (Xu-Friedman 2016).

3. Place a stimulating electrode in the parallel fibers 200–500 µm away from the recorded Purkinje cell.

   i. Use a glass micropipette the same size as that used for Purkinje cell recording for the stimulating electrode.

   This stimulating electrode should be filled with ACSF and contain a bare wire attached to one pole of the stimulus isolation unit.

   ii. To minimize stimulus artifacts, attach a second wire from the other pole of the isolation unit to the air table (or place it in the bath).

   In the latter case, the immersed wire should be silver so as not to introduce metal ions that could interfere with recordings.

   iii. Ensure that the stimulus isolation unit resides within the Faraday cage and is set to "current" mode.

   Take care to avoid any power cables in the unit's vicinity or near the wires traveling to the stimulus electrode, as these will carry electrical noise into the recording chamber.

4. Begin by voltage clamping the Purkinje cell at −70 mV.

Cite this protocol as *Cold Spring Harb Protoc*; doi:10.1101/pdb.prot095828

5. Stimulate the parallel fiber pathway with single pulses at 0.1 Hz. Gradually increase stimulation intensity until a small (e.g., 100–300 pA) excitatory postsynaptic current (EPSC) appears.

   Stimulation in the range of a few to 10 sec of µA at 0.2-msec pulse width is usually sufficient to drive parallel fiber activity.

6. To measure feedforward excitation, adjust the membrane potential to find the reversal potential for inhibition (typically −70 to −80 mV with a standard low chloride [e.g., 3–5 mM] internal solution).

   When adjusted properly, the EPSC should have a mono-exponential decay.

7. Collect 15–20 sweeps at the reversal potential for inhibition.

8. For instructive purposes, adjust the membrane potential to a somewhat depolarized value (e.g., approximately −50 mV).

   An outward current should develop with a 1–2-msec delay following the onset of excitation.

   i. To isolate the feedforward inhibition, slowly adjust the membrane potential toward 0 mV.

   Purkinje cells have many active conductances and are notoriously difficult to voltage clamp well, and an outward holding current will develop as the neuron is depolarized.

   ii. To identify the EPSC reversal potential, continue to depolarize the neuron while stimulating the parallel fibers until only an outward evoked current remains.

   Because of the liquid junction potential, this will typically occur at ∼+10 mV. Importantly, this outward inhibitory postsynaptic current (IPSC) should occur with a 1–2 msec delay relative to the previously evoked EPSC (Fig. 1).

   Purkinje cells receive input from molecular layer interneurons, which are spontaneously active. At depolarized potentials, a high rate of spontaneous IPSCs will also be present. These will not contaminate your measurement of evoked IPSCs as long as multiple trials are averaged.

9. After 15–20 sweeps have been collected at the reversal potential for excitation, assess the system further using one of the following methods.

To block synaptic inhibition

This method pharmacologically verifies (and corrects for errors in) the accuracy of the chosen reversal potentials.

   i. Apply 5 µM gabazine (a selective GABA$_A$ receptor antagonist) via bath perfusion while the Purkinje cell is still voltage clamped at 0 mV.

   All evoked and spontaneous inhibition should be blocked. Any small remaining evoked current is the result of misestimation of the EPSC reversal potential.

   ii. Collect several sweeps under these conditions to correct the initially recorded IPSC by subtracting the remaining current.

   iii. While still in the presence of gabazine, gradually change the membrane potential back to the reversal potential of inhibition measured previously.

   iv. Collect several sweeps for averaging.

   This measurement represents the pure, uncontaminated EPSC, which can be compared with that obtained before blocking the GABA$_A$ receptors.

   v. Apply 5–10 µM NBQX to the bath in addition to gabazine.

   This will block all remaining current, and provide an average trace with only the stimulation artifact. This can also be subtracted from previous recordings to cosmetically eliminate the electrical artifact.

To block excitation

   vi. Apply 5–10 µM NBQX while recording at the EPSC reversal potential.

   This manipulation should block all evoked inhibition, but leave spontaneous inhibition intact. This measurement is important to confirm that all evoked inhibition resulted from recruitment of molecular layer interneurons by the parallel fibers and not by direct electrical depolarization of the interneurons via the stimulus electrode.

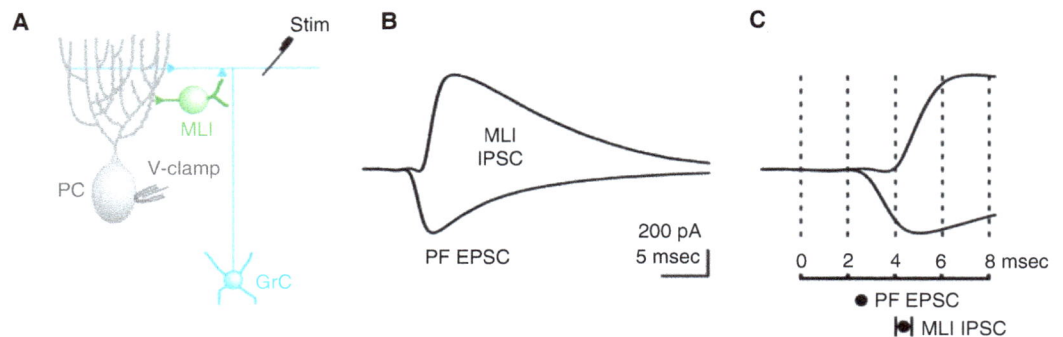

FIGURE 1. Schematic of recording setup and example of evoked feedforward excitation and inhibition. (*A*) Recording configuration showing a voltage-clamped Purkinje cell (PC) and electrical stimulation of parallel fibers (stim) emanating from granule cells (GrCs), as well as from the molecular layer interneurons (MLIs), providing feedforward inhibition. (*B*) A feedforward IPSC recorded at 0 mV, an EPSC, and a direct parallel fiber (PF) EPSC recorded at −75 mV. (*C*) A higher temporal resolution of the recording in *B* shows an average 1.8-msec latency between the direct EPSC and the feedforward IPSC (*n* = 12). (Adapted from Hull and Regehr 2012.)

Testing the Impact of Feedforward Inhibition on Spike Timing

These experiments require a potassium-based internal solution to allow spiking and perform current clamp measurements.

10. Achieve a whole-cell recording from a Purkinje cell as described in Step 2.

11. Place a stimulating electrode in the parallel fibers as described in Step 3.

12. Switch to current clamp mode, but do not inject any current.

    The Purkinje cell will spike spontaneously; this should occur at a rate of 20–50 Hz.
    See Troubleshooting.

13. Apply just enough negative holding current to stop the cell from spiking.

    This should result in a membrane potential near −60 mV.

14. Stimulate the parallel fibers at 0.1 Hz, adjusting the stimulus intensity until action potentials are generated in ∼30%–60% of trials.

15. Apply 5 μM gabazine via bath perfusion while continuing to stimulate at the same frequency and intensity.

    By blocking feedforward inhibition, the probability of action potential generation should increase. Furthermore, the jitter in action potential timing should increase as the integration time window broadens in the absence of inhibition.

16. Optionally, insert a second stimulation electrode to activate an independent set of parallel fibers.

    In this configuration, it is possible to stimulate a second pathway with variable delay with respect to the first pathway, and thus measure the duration of the integration time window in the presence and absence of feedforward inhibition.

TROUBLESHOOTING

Problem (Step 12): Spiking is not stable; the rate continues to increase.
Solution: The recording should be rejected for poor quality. A poor quality recording can be confirmed by switching back to voltage clamp, and checking whether significant holding current (>100 pA) is present at −60 mV.

Cite this protocol as *Cold Spring Harb Protoc*; doi:10.1101/pdb.prot095828

RECIPES

Cesium Methanesulfonate Recording Solution

140 mM cesium methanesulfonate (Sigma-Aldrich, C1426)
0.5 mM EGTA (Sigma-Aldrich, 03777)
15 mM HEPES (Sigma-Aldrich, H4034)
2 mM MgATP (Sigma-Aldrich, A9187)
0.3 mM NaGTP (Sigma-Aldrich, G8877)
10 mM phosphocreatine-Tris$_2$ (Sigma-Aldrich, P1937)
2 mM *QX 314* chloride (Tocris Bioscience, 2313)
2 mM tetramethylammonium chloride (Sigma-Aldrich, 86616)

Dissolve the above listed chemicals with ~70 mL of distilled water in a 100-mL volumetric flask. Remove the stir bar; add water to fill to exactly 100 mL. Stir again, and adjust the pH to 7.25 using CsOH. If necessary, adjust the osmolarity to ~305 mOsm with water. Aliquot solutions into 1–1.5 mL Eppendorf tubes. Store at −80°C until use (within 1 yr).

Potassium Gluconate Recording Solution

0.5 mM EGTA (Sigma-Aldrich, 03777)
0.5 mM GTP (Sigma-Aldrich, G8877)
10 mM HEPES (Sigma-Aldrich, H4034)
3 mM MgATP (Sigma-Aldrich, A9187)
3 mM KCl
5 mM phosphocreatine-Na$_2$ (Sigma-Aldrich, P7936)
5 mM phosphocreatine-Tris$_2$ (Sigma-Aldrich, P1937)
150 mM potassium gluconate (Sigma-Aldrich, P1847)

Dissolve the above listed chemicals with ~70 mL of distilled water in a 100-mL volumetric flask. Remove the stir bar; add water to fill to exactly 100 mL. Stir again, and adjust the pH to 7.25 using NaOH. If necessary, adjust the osmolarity to ∼305 mOsm with water. Aliquot solutions into 1–1.5 mL Eppendorf tubes. Store at −80°C until use (within 1 yr).

REFERENCES

Hull C, Regehr WG. 2012. Identification of an inhibitory circuit that regulates cerebellar Golgi cell activity. *Neuron* **73:** 149–158.

Xu-Friedman MA. 2016. Preparing brain slices to study basic synaptic properties. *Cold Spring Harb Protoc* doi: 10.1101/pdb.prot093328.

CHAPTER 10

Fast Cholinergic Synaptic Transmission in the Mammalian Central Nervous System

Michael Beierlein[1,2]

[1]*Department of Neurobiology and Anatomy, McGovern Medical School at UTHealth, Houston, Texas 77030*

Release of acetylcholine (ACh) in the brain controls several cognitive processes, and a number of disorders including Alzheimer's and Parkinson's diseases are associated with a loss of cholinergic function. Despite the importance of ACh signaling in modulating information processing in thalamocortical circuits, understanding the dynamics of cholinergic function has long been limited by a lack of in vitro model systems. Recent studies employing both electrical as well as optogenetic stimulation techniques have overcome this challenge, resulting in the identification of multiple forms of fast cholinergic signaling throughout the mammalian brain. Here we highlight a simple strategy making use of extracellular electrical stimulation techniques that allows for the study of cholinergic synaptic inputs onto neurons in the thalamic reticular nucleus (TRN).

CHOLINERGIC SIGNALING IN THE THALAMOCORTICAL SYSTEM

Cholinergic neurons in the basal forebrain (BF) and the brain stem control a variety of computations in the thalamocortical system, including arousal, sensory processing, and learning (Hasselmo and Sarter 2011). On a cellular level, the actions of ACh are mediated by both nicotinic and muscarinic receptors (nAChRs and mAChRs) expressed in pre- and postsynaptic elements (Arroyo et al. 2014; Munoz and Rudy 2014). Activation of these receptors can lead to changes in neuronal excitability and neuronal firing rate, trigger different forms of short- or long-term plasticity, or regulate activity in local neuronal circuits. Despite the importance of ACh release in controlling information processing, our understanding of the synaptic mechanisms that govern cholinergic signaling has long remained incomplete. Many brain structures including neocortex, thalamus, and hippocampus are the targets of extensive cholinergic afferent projections (Zaborszky 2002; Woolf and Butcher 2011; Wu et al. 2014). Furthermore, neurons in these areas express distinct types of nAChRs and mAChRs in considerable densities, often in a cell-type and layer-specific manner (Poorthuis et al. 2013; Munoz and Rudy 2014). Until recently, cholinergic control has been primarily investigated using exogenous agonists, which does not necessarily mimic the dynamics of receptor activation by synaptically released ACh (Unal et al. 2015). A near-absence of in vitro models for cholinergic transmission in the thalamocortical system seemed to agree with anatomical work that failed to find good evidence for the existence of ultrastructurally defined synapses. Instead, the considerable distances reported between sites of ACh release and postsynaptic receptors appeared to be consistent with a slower and more

[2]Correspondence: michael.beierlein@uth.tmc.edu

Cite this introduction as *Cold Spring Harb Protoc*; doi:10.1101/pdb.top095083

diffuse form of cholinergic signaling, a process often referred to as volume transmission (Descarries et al. 1997; Picciotto et al. 2012).

Recent studies making use of optogenetics or more conventional stimulation techniques have made it clear that this view might be overly simplistic by demonstrating that short-latency cholinergic synaptic transmission is prominent throughout the mammalian thalamocortical system (Munoz and Rudy 2014) and can be readily examined using acute brain slice preparations (Arroyo et al. 2012; Sun et al. 2013; Pita-Almenar et al. 2014; Hay et al. 2016).

EXAMINING CHOLINERGIC SYNAPTIC SIGNALING IN VITRO

Much of our recent insights into mechanisms and function of cholinergic synaptic signaling can be attributed to the ability to selectively activate cholinergic neurons or synapses by specifically targeting expression of light-gated cation channel channelrhodopsin-2 (ChR2) to neurons that express choline acetyltransferase (ChAT), the enzyme responsible for acetylcholine synthesis. ChR2 delivery and expression can be accomplished employing multiple strategies, including viral approaches with or without the use of Cre-expressing driver lines (Fenno et al. 2011; Madisen et al. 2012), or the generation of transgenic mouse lines that enable constitutive expression of ChR2 (Arenkiel et al. 2007; Zhao et al. 2011).

A selective activation of cholinergic afferents using optogenetics can have significant advantages over more conventional stimulation techniques, depending on the neuronal system under study and the specific research question being addressed. The cell bodies and axonal afferents of cholinergic neurons are extensively intermingled with other transmitter systems. Therefore, even with careful placement of the stimulating electrode, a selective activation of cholinergic afferents using extracellular electrical stimulation is usually not feasible. In addition, cholinergic synaptic release occurs at a considerable distance from the presynaptic cell body, making dual recordings from synaptically connected neurons impractical. Optogenetic techniques are therefore critical to study the functional impact of cholinergic synaptic activity in experiments that do not allow for the pharmacological block of other transmitter systems. Similarly, selective optical activation of cholinergic afferents enables the study of co-release of ACh and either glutamate or GABA, which appears to be a prominent form of signaling for several types of cholinergic neurons (Ren et al. 2011; Saunders et al. 2015; Tritsch et al. 2016). Lastly, use of wide-field optical stimuli makes it straightforward to identify and activate relatively sparse cholinergic afferents onto a target structure, which is difficult to accomplish using more focal electrical stimuli.

Despite its limitations, extracellular electrical stimulation remains a viable and easy-to-implement alternative to examine cholinergic synaptic signaling in a number of model systems, including the spinal cord (d'Incamps and Ascher 2014), hippocampus (Gu and Yakel 2011), and thalamus (Sun et al. 2013). Brief currents applied with extracellular electrodes lead to the generation of action potentials with a presynaptic waveform and do not perturb the intracellular milieu. Therefore, this technique remains important for studies aimed to quantify distinct forms of synaptic plasticity that are highly sensitive to subtle changes in presynaptic Ca^{2+} influx. Furthermore, electrical stimulation allows for the generation of very high presynaptic firing rates. Maximum firing rates achieved with optogenetic stimulation typically reach ~50 Hz (Jackman et al. 2014), though the development of ChR2 variants with faster kinetics will likely help to increase these values. In addition, the localized nature of electrical stimulation and the ability to precisely control current intensity and duration enables activation of individual presynaptic axons (minimal stimulation). Furthermore, in our hands, electrical stimuli allow for the stable activation of cholinergic synaptic afferents over many trials (>200), which remains a challenge in experiments employing ChR2.

We have recently adapted extracellular electrical stimulation to characterize fast cholinergic synaptic transmission in the TRN, the target of extensive cholinergic innervation from neurons in the basal forebrain and the brainstem (Sun et al. 2013). (See also the accompanying Protocol 1: Examining Cholinergic Synaptic Signaling in the Thalamic Reticular Nucleus (TRN) [Dasgupta et al. 2016]).

Cite this introduction as *Cold Spring Harb Protoc*; doi:10.1101/pdb.top095083

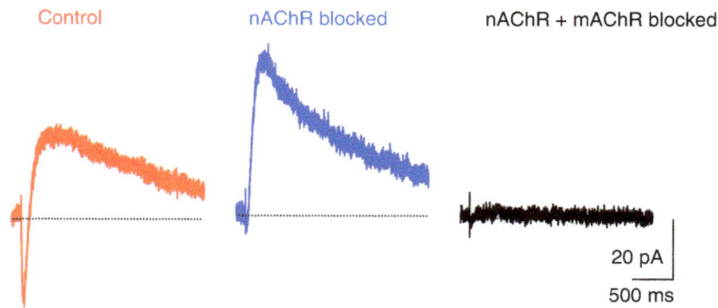

FIGURE 1. Synaptic release of ACh elicits biphasic postsynaptic response in TRN neurons. Biphasic postsynaptic response in a TRN neuron (red trace), with inward current blocked by a nAChR antagonist (blue trace). The remaining outward current is blocked by a mAChR antagonist (black trace). (Adapted from Sun et al. 2013.)

Stimulation of these afferents elicits fast-latency, biphasic postsynaptic responses in the TRN, mediated by the activation of $\alpha4\beta2$ nAChRs and M2 mAChRs (Fig. 1). ACh release at these synapses is tightly regulated by autoinhibition mediated by presynaptic M2 mAChRs (Fig. 2A). Cholinergic afferents contact virtually every single TRN neuron, and lead to large-amplitude postsynaptic responses that remain stable over many trials. Cholinergic synaptic inputs to TRN therefore serve as an excellent model system to investigate the dynamic properties and functional consequences of ACh release in the mammalian brain (Sun et al. 2013, 2016; Pita-Almenar et al. 2014). However, it is important to point out that the demonstration of short-latency cholinergic synaptic transmission does not exclude the existence of cholinergic signaling modes that are considerably slower and more diffuse, in the TRN or elsewhere in the brain (Picciotto et al. 2012). Both nAChRs and mAChRs are thought to be expressed extrasynaptically, on presynaptic terminals of glutamatergic and GABAergic terminals, and on axons, often at considerable distances from potential

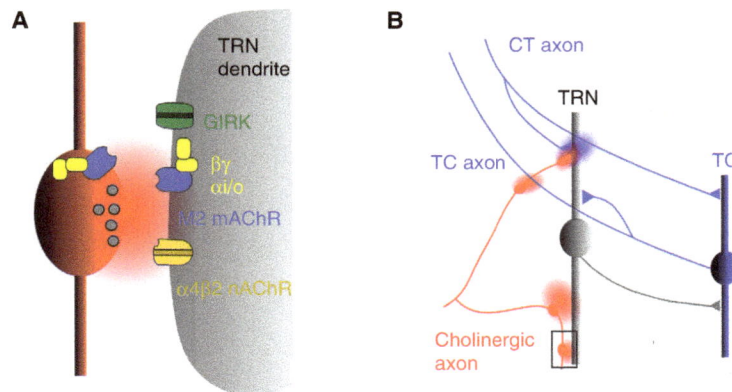

FIGURE 2. Cholinergic signaling in the TRN. (A) Schematic illustrating the cellular mechanisms underlying cholinergic synaptic signaling in the TRN. ACh release activates postsynaptic ionotropic nAChRs (likely $\alpha4\beta2$) and metabotropic $G_{i/o}$-coupled M2 mAChRs, which are expressed both pre- and postsynaptically. M2 mAChR activation in the postsynaptic membrane triggers the liberation of the $\beta\gamma$ subunit complex, in turn leading to the opening of G protein-coupled inwardly rectifying K^+ (GIRK) channels. Presynaptic M2 mAChRs are responsible for autoinhibition of ACh release, likely generated by $\beta\gamma$ subunit complex-mediated inhibition of presynaptic Ca^{2+} channels. (B) Diagram illustrates thalamic circuitry and potential forms of cholinergic signaling in the TRN. Cholinergic afferents (red) contact distal dendrites of TRN neurons and generate biphasic cholinergic responses, likely mediated by ultrastructurally defined synapses (area outlined by square and shown in more detail in A). In turn, glutamate release from corticothalamic (CT) axons can modulate cholinergic synaptic strength on distinct time scales (Sun et al. 2016). In addition, ACh released from *en passant* varicosities could act over larger distances and activate extrasynaptic receptors expressed on TRN dendrites, nAChRs expressed by thalamocortical (TC) axons or receptors expressed at presynaptic CT terminals. Figure modified from (Beierlein 2014).

sites of ACh release (Fig. 2B). The exact conditions leading to the recruitment of these receptors remain to be determined.

ACKNOWLEDGMENTS

Our research is supported by funds from NS077989 National Institute of Neurological Disorders and Stroke (NINDS) to M.B.

REFERENCES

Arenkiel BR, Peca J, Davison IG, Feliciano C, Deisseroth K, Augustine GJ, Ehlers MD, Feng G. 2007. In vivo light-induced activation of neural circuitry in transgenic mice expressing channelrhodopsin-2. *Neuron* **54**: 205–218.

Arroyo S, Bennett C, Aziz D, Brown SP, Hestrin S. 2012. Prolonged disynaptic inhibition in the cortex mediated by slow, non-α7 nicotinic excitation of a specific subset of cortical interneurons. *J Neurosci* **32**: 3859–3864.

Arroyo S, Bennett C, Hestrin S. 2014. Nicotinic modulation of cortical circuits. *Front Neural Circuits* **8**: 30.

Beierlein M. 2014. Synaptic mechanisms underlying cholinergic control of thalamic reticular nucleus neurons. *J Physiol* **592**: 4137–4145.

Dasgupta R, Seibt F, Sun Y-G, Beierlein M. 2016. Examining cholinergic synaptic signaling in the thalamic reticular nucleus (TRN). *Cold Spring Harb Protoc* doi: 10.1101/pdb.prot095836.

Descarries L, Gisiger V, Steriade M. 1997. Diffuse transmission by acetylcholine in the CNS. *Prog Neurobiol* **53**: 603–625.

d'Incamps BL, Ascher P. 2014. High affinity and low affinity heteromeric nicotinic acetylcholine receptors at central synapses. *J Physiol* **592**: 4131–4136.

Fenno L, Yizhar O, Deisseroth K. 2011. The development and application of optogenetics. *Annu Rev Neurosci* **34**: 389–412.

Gu Z, Yakel JL. 2011. Timing-dependent septal cholinergic induction of dynamic hippocampal synaptic plasticity. *Neuron* **71**: 155–165.

Hasselmo ME, Sarter M. 2011. Modes and models of forebrain cholinergic neuromodulation of cognition. *Neuropsychopharmacol* **36**: 52–73.

Hay YA, Lambolez B, Tricoire L. 2016. Nicotinic transmission onto layer 6 cortical neurons relies on synaptic activation of non-α7 receptors. *Cereb Cortex* **26**: 2549–2562.

Jackman SL, Beneduce BM, Drew IR, Regehr WG. 2014. Achieving high-frequency optical control of synaptic transmission. *J Neurosci* **34**: 7704–7714.

Madisen L, Mao T, Koch H, Zhuo JM, Berenyi A, Fujisawa S, Hsu YW, Garcia AJ III, Gu X, Zanella S, et al. 2012. A toolbox of Cre-dependent optogenetic transgenic mice for light-induced activation and silencing. *Nat Neurosci* **15**: 793–802.

Munoz W, Rudy B. 2014. Spatiotemporal specificity in cholinergic control of neocortical function. *Curr Opin Neurobiol* **26**: 149–160.

Picciotto MR, Higley MJ, Mineur YS. 2012. Acetylcholine as a neuromodulator: Cholinergic signaling shapes nervous system function and behavior. *Neuron* **76**: 116–129.

Pita-Almenar JD, Yu D, Lu HC, Beierlein M. 2014. Mechanisms underlying desynchronization of cholinergic-evoked thalamic network activity. *J Neurosci* **34**: 14463–14474.

Poorthuis RB, Bloem B, Schak B, Wester J, de Kock CP, Mansvelder HD. 2013. Layer-specific modulation of the prefrontal cortex by nicotinic acetylcholine receptors. *Cereb Cortex* **23**: 148–161.

Ren J, Qin C, Hu F, Tan J, Qiu L, Zhao S, Feng G, Luo M. 2011. Habenula "cholinergic" neurons co-release glutamate and acetylcholine and activate postsynaptic neurons via distinct transmission modes. *Neuron* **69**: 445–452.

Saunders A, Granger AJ, Sabatini BL. 2015. Corelease of acetylcholine and GABA from cholinergic forebrain neurons. *Elife* **4**. doi: 10.7554/eLife.06412

Sun YG, Pita-Almenar JD, Wu CS, Renger JJ, Uebele VN, Lu HC, Beierlein M. 2013. Biphasic cholinergic synaptic transmission controls action potential activity in thalamic reticular nucleus neurons. *J Neurosci* **33**: 2048–2059.

Sun YG, Rupprecht V, Zhou L, Dasgupta R, Seibt F, Beierlein M. 2016. mGluR1 and mGluR5 Synergistically control cholinergic synaptic transmission in the thalamic reticular nucleus. *J Neurosci* **36**: 7886–7896.

Tritsch NX, Granger AJ, Sabatini BL. 2016. Mechanisms and functions of GABA co-release. *Nat Rev Neurosci* **17**: 139–145.

Unal CT, Pare D, Zaborszky L. 2015. Impact of basal forebrain cholinergic inputs on basolateral amygdala neurons. *J Neurosci* **35**: 853–863.

Woolf NJ, Butcher LL. 2011. Cholinergic systems mediate action from movement to higher consciousness. *Behav Brain Res* **221**: 488–498.

Wu H, Williams J, Nathans J. 2014. Complete morphologies of basal forebrain cholinergic neurons in the mouse. *Elife* **3**: e02444.

Zaborszky L. 2002. The modular organization of brain systems. Basal forebrain: The last frontier. *Prog Brain Res* **136**: 359–372.

Zhao S, Ting JT, Atallah HE, Qiu L, Tan J, Gloss B, Augustine GJ, Deisseroth K, Luo M, Graybiel AM, et al. 2011. Cell type-specific channelrhodopsin-2 transgenic mice for optogenetic dissection of neural circuitry function. *Nat Methods* **8**: 745–752.

Examining Cholinergic Synaptic Signaling in the Thalamic Reticular Nucleus

Rajan Dasgupta,[1,2,3] Frederik Seibt,[1,3] Yan-Gang Sun,[1] and Michael Beierlein[1,2,4]

[1]Department of Neurobiology and Anatomy, McGovern Medical School at UTHealth, Houston, Texas 77030;
[2]University of Texas Graduate School of Biomedical Sciences at Houston, Houston, Texas 77030

This protocol describes how to obtain monosynaptic cholinergic responses in neurons of the thalamic reticular nucleus (TRN) by making use of extracellular stimulation techniques. These methods are easy to implement and allow for the study of various forms of cholinergic synaptic plasticity and modulation. For many synapses throughout the mammalian brain, short-term plasticity is mediated by endocannabinoids released from postsynaptic neurons that activate presynaptic type I cannabinoid receptors (CB1Rs), resulting in the inhibition of presynaptic Ca^{2+} channels and a reduction of release probability. Neurons in the TRN are known to liberate endocannabinoids that can control transmitter release at GABAergic terminals. However, expression of CB1Rs on cholinergic terminals contacting the TRN has not been demonstrated. Here we outline strategies aimed to record stable postsynaptic responses and to quantify changes in cholinergic synaptic strength, using presynaptic modulation of acetylcholine (ACh) release by a CB1R agonist as an illustrative example.

MATERIALS

It is essential that you consult the appropriate Material Safety Data Sheets and your institution's Environmental Health and Safety Office for proper handling of equipment and hazardous material used in this protocol.

RECIPES: Please see the end of this protocol for recipes indicated by <R>. Additional recipes can be found online at http://cshprotocols.cshlp.org/site/recipes.

Reagents

Acute rodent thalamic brain slices (see Step 1)

AM-251 (Tocris Bioscience)

Artificial cerebrospinal fluid (ACSF) with antagonists for glutamatergic and GABAergic receptors <R>

To isolate cholinergic synaptic inputs, fast glutamatergic and GABAergic synaptic transmission need to be blocked pharmacologically using specific receptor antagonists.

TRN neurons in slices derived from younger animals (postnatal days 13–21) are easier to visualize, because myelination in the TRN resulting from thalamocortical and corticothalamic fibers is less pronounced relative to slices obtained from more mature animals. For animals <P13, cholinergic synaptic release is not fully developed, making stable recordings of large-amplitude postsynaptic responses more challenging.

Internal solution for whole-cell recording, cesium-based

---

[3]These authors contributed equally to this work.
[4]Correspondence: michael.beierlein@uth.tmc.edu

Copyright © Cold Spring Harbor Laboratory Press; all rights reserved
Cite this protocol as *Cold Spring Harb Protoc*; doi:10.1101/pdb.prot095836

Synaptic release of ACh will recruit both nicotinic ACh receptors (nAChRs) and muscarinic ACh receptors (mAChRs), resulting in a biphasic postsynaptic response (Sun et al. 2013). Cesium-based internal solutions block potassium conductances, including the G-protein-coupled inwardly rectifying potassium (GIRK) channels recruited by M2 mAChR activation. Therefore, these solutions should be employed in experiments aimed to quantify changes in presynaptic release of ACh, using the isolated nAChR–mediated response as a readout.

Poly-L-lysine (Sigma-Aldrich)

WIN 55,212-2 (Tocris Bioscience)

Equipment

CCD camera, IR-sensitive (e.g., Hamamatsu)

Coverslips

Heater, inline, equipped with a temperature controller (e.g., Warner Instruments)

Microscope, upright, fixed-stage, equipped for infrared differential interference contrast visualization (e.g., BX51WI, Olympus)

Peristaltic pump (e.g., Gilson [Minipuls 3])

Pipette puller

Slice submersion chamber, low profile (e.g., Warner Instruments [RC-26GLP])

Theta glass (e.g., World Precision Instruments [TST150-6])

METHOD

Preparing Brain Slices

Because cholinergic synaptic responses are most readily evoked using stimulus electrodes placed near the recorded neuron, all slice preparations that allow for the identification of TRN boundaries are suitable.

1. Prepare acute brain slices (300–400 µm) containing the TRN from mice or any other suitable species according to standard techniques and animal welfare guidelines. Slice orientation should be chosen based on the specific experimental requirements:

    i. Thalamocortical slices preserve connectivity between the thalamus and the neocortex (Agmon and Connors 1991).

    ii. Horizontal thalamic slices allow for better connectivity between the TRN and the adjacent thalamic relay nucleus (Sohal et al. 2003; Pita-Almenar et al. 2014).

    iii. Parasagittal slices preserve cholinergic axonal afferents from the basal forebrain (Pita-Almenar et al. 2014).

Recording Station Setup

Any standard setup used for slice electrophysiology that allows for visualization of the recording process is suitable for these experiments. All tubing associated with the recording setup should be replaced frequently to avoid buildup of contaminants, particularly following wash-in of exogenous agonists and antagonists.

2. Fill submersion chamber with ACSF. Using a pump, perfuse the chamber with ACSF at a flow rate of 3–4 mL/min.

3. Adjust the inline heater to raise the bath temperature in the slice chamber to near-physiological levels (32°C–34°C).

4. Prepare the recording electrodes. Fill with a cesium-based internal solution to isolate nAChR-mediated excitatory postsynaptic currents (nEPSCs).

5. Prepare the stimulation electrodes. Fill with ACSF.

    Either monopolar electrodes using patch pipettes or bipolar electrodes pulled from theta glass can be used for extracellular stimulation. Bipolar electrodes using theta glass tend to generate brief stimulus artifacts that do not contaminate the synaptic response but require more careful positioning to obtain a sufficiently large postsynaptic response. For both types of electrodes, the tip diameter should be 5–15 µm.

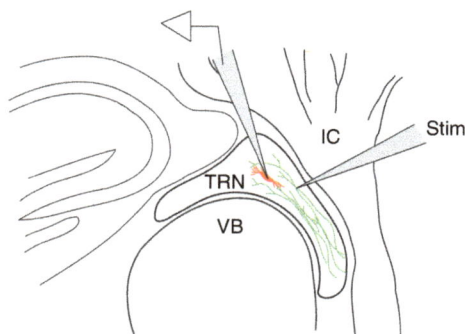

FIGURE 1. Investigating cholinergic synaptic transmission in the TRN in vitro. The schematic shows a horizontal slice containing the TRN, with the approximate location of the recording and stimulus (Stim) electrodes. VB, ventrobasal nucleus of the thalamus; IC, internal capsule.

Recording

6. Place brain slices on glass coverslips coated with poly-L-lysine before transfer to the recording bath.

7. Transfer the slice containing the TRN to the recording chamber.

    For thalamocortical slices (400 μm) obtained from P13–21 mice, one or two slices per hemisphere will contain somatosensory TRN, which surrounds the thalamus like a shell.

8. Position the stimulating electrode directly in the TRN (Fig. 1A) between 100 and 200 μm from the recorded cell.

    Avoid placing the electrode too deep into the tissue, as it might be necessary to move the electrode after a whole-cell recording configuration has been achieved.

9. Position the recording eletrode near the target neuron and obtain whole cell recording.

    Almost all TRN neurons receive cholinergic afferent projections. However, in our experience, recordings from neurons located within the wider part of the TRN will yield postsynaptic responses with very high likelihood (Fig. 1).

10. Adjust stimulus intensity and position:

    i. Following establishment of whole-cell configuration, begin stimulating at 0.1 Hz, with an intensity of ~10 μA. Stimulus duration should be kept at 200 μsec.

    ii. Increase stimulus intensities in small steps (~5 μA) until a postsynaptic response can be detected.

    Response amplitudes for nEPSCs should be between 30 and 50 pA, with larger responses up to 100 pA.
    See Troubleshooting.

11. Obtain a stable baseline:

    i. For experiments using single stimuli, interstimulus intervals (ISI) should be at least 12 sec to minimize fatigue of the synaptic response.

    See Troubleshooting.

    ii. If brief trains (>10 stimuli) are used, increase ISIs to ~40 sec.

12. Characterize the modulation of cholinergic synaptic strength.

    Figure 2 shows recordings of nEPSCs before and after wash-in of the CB1R agonist WIN 55,212-2, followed by wash-in of the CB1R antagonist AM-251.

TROUBLESHOOTING

Problem (Step 10.ii): Stimulation does not evoke a postsynaptic response.

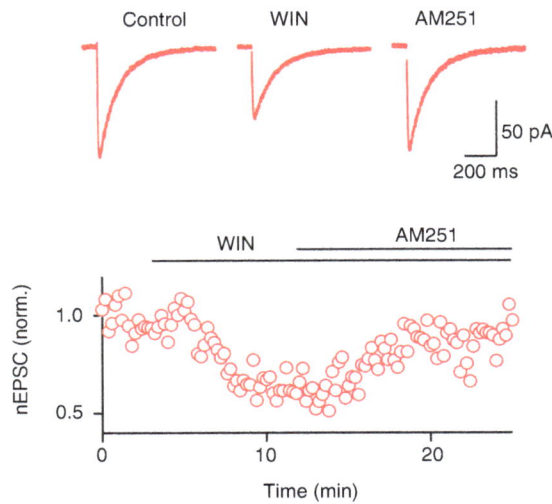

FIGURE 2. Presynaptic expression of CB1Rs at cholinergic synapses contacting TRN neurons. Neurons in the TRN are recorded in voltage clamp using a cesium-based internal solution to isolate nEPSCs, and responses are evoked using single pulses applied every 15 sec. (*Top*) Average postsynaptic nAChR-mediated response recorded in control conditions (*left*), in the presence of 5 µM of the CB1R agonist WIN 55,212-2 (WIN; *middle*) and following application of 10 µM of the CB1R antagonist AM-251 (*right*). (*Bottom*) Time course of response amplitude (normalized to control) for the same neuron shown above (YG Sun and M Beierlein, unpubl.).

Solution: Effective stimulation might not occur using glass pipettes. Confirm that axonal afferents in the vicinity of the electrode are reliably recruited. This can most easily be accomplished in the absence of any antagonists for glutamatergic and GABAergic synaptic transmission. Under these conditions placing the stimulating electrode into the internal capsule near the TRN to activate both thalamoreticular as well as corticoreticular inputs should lead to a large postsynaptic glutamatergic EPSC in TRN neurons.

Alternatively, the recorded neuron might not be within TRN. The border between TRN and the adjacent ventrobasal nucleus of the thalamus (VB) can be difficult to detect in some slice orientations, particularly in younger animals, and in our hands, no fast cholinergic responses can be detected in VB neurons. Confirm that neurons targeted for recordings are clearly within TRN.

It is also possible that cholinergic afferents to the recorded neuron might not be recruited. Adjust the position of the stimulus electrode while maintaining distances of 100–200 µm to the recorded cell until a response is detected. This is most easily accomplished by using a dry objective to guide movement and placement of the electrode tip just below the slice surface. Access resistance of the recorded neuron should be monitored continuously to verify minimal movement of the slice during this procedure.

Problem (Step 10.ii): Postsynaptic response shows large-amplitude fluctuations from trial-to-trial.
Solution: Stimulation could be at the threshold of axon recruitment. Trial-to-trial variability of nicotinic postsynaptic responses is very low (Sun et al. 2013). Large fluctuations might indicate recruitment of an additional axon in some trials. In that case, a slight increase or decrease of stimulus intensity should reduce response variability.

Problem (Step 10.ii and Step 11.i): Postsynaptic response amplitude decreases by >10% over a small number of trials (e.g., <30).
Solution: Synaptic fatigue might be occurring, caused by a high stimulation frequency. In that case, increasing ISIs should minimize response rundown. Furthermore, it is possible that stimulus intensities are too high and lead to axonal fatigue. Consider adjusting the position of the stimulus electrode or reducing stimulus intensities. Alternatively, the slice could be unhealthy; in our experience, cholinergic afferent systems are particularly sensitive to deteriorating slice health.

Problem (Step 10.ii): Stimulus artifact overlaps with synaptic response.
Solution: The stimulation electrode might be too close to the recorded cell. Bipolar electrodes generate brief artifacts and are therefore particularly suitable for extracellular stimulation in the vicinity of the recorded neuron. Alternatively, it is possible that the recorded neuron is being stimulated

Cite this protocol as *Cold Spring Harb Protoc*; doi:10.1101/pdb.prot095836

directly. The dendritic tree of many TRN neurons is roughly parallel to the border between the internal capsule and the TRN. Consider moving the stimulus electrode in a perpendicular direction to minimize the direct activation of dendrites.

Problem (Step 10.ii): Recorded currents show nonsynaptic components.

Solution: Synaptic inputs recruit intrinsic conductances. Cholinergic afferents contact the distal dendrites of TRN neurons, which express high densities of low-threshold T-type Ca^{2+} channels in their dendrites. Voltage clamp recordings of large-amplitude cholinergic responses can therefore be associated with significant space clamp errors. In the extreme, this can result in the generation of escape currents, indicative of the activation of voltage-gated Ca^{2+} channels. Consider recording responses of smaller amplitude (~50 pA). Recording electrodes should have resistances 2–4 MΩ and recordings should have stable access resistances of <20 MΩ to minimize space clamp errors.

Another possibility is that depolarization of neighboring neurons contributes to recorded currents. TRN neurons are interconnected via Cx36-dependent gap junctions (Landisman et al. 2002). Therefore, action potential activity in nearby neurons frequently leads to small-amplitude spikelets in the recorded neuron. However, most of the current transmitted by electrical synapses under these conditions will reflect slow synaptic depolarizations and low-threshold Ca^{2+} spikes in local neurons, resulting from the low-pass filtering characteristics of electrical synapses. Because cholinergic afferents to TRN show considerable divergence (Pita-Almenar et al. 2014), activation of nearby neurons is almost impossible to prevent. Bath-application of the specific T-type Ca^{2+} channel blocker TTA-P2 (Dreyfus et al. 2010) eliminates both low-threshold Ca^{2+} spikes and fast action potentials in electrically coupled neurons, while having little effect on presynaptic ACh release (Sun et al. 2013). A better strategy involves eliminating gap junctional coupling using mefloquine (Cruikshank et al. 2004), although any nonspecific effects on cholinergic signaling have not been characterized to date for this compound.

RECIPE

Artificial Cerebrospinal Fluid (ACSF) with Antagonists for Glutamatergic and GABAergic Receptors

2 mM $CaCl_2$
10 mM Glucose
2.5 mM KCl
2 mM $MgCl_2$
126 mM NaCl
26 mM $NaHCO_3$
1.25 mM NaH_2PO_4
5 μM CGP 55845 ($GABA_B$ receptor blocker)
10 μM NBQX (AMPA receptor blocker)
50 μM Picrotoxin ($GABA_A$ receptor blocker)
5 μM R-CPP (NMDA receptor blocker)

Saturate with 95% O_2/5% CO_2.

ACKNOWLEDGMENTS

Our research is supported by funds from NS077989 National Institute of Neurological Disorders and Stroke (NINDS) to M.B. R.D. is supported by a Zilkha Family Discovery Fellowship in Neuroengineering.

REFERENCES

Agmon A, Connors BW. 1991. Thalamocortical responses of mouse somatosensory (barrel) cortex in vitro. *Neuroscience* **41:** 365–379.

Cruikshank SJ, Hopperstad M, Younger M, Connors BW, Spray DC, Srinivas M. 2004. Potent block of Cx36 and Cx50 gap junction channels by mefloquine. *Proc Natl Acad Sci* **101:** 12364–12369.

Dreyfus FM, Tscherter A, Errington AC, Renger JJ, Shin HS, Uebele VN, Crunelli V, Lambert RC, Leresche N. 2010. Selective T-type calcium channel block in thalamic neurons reveals channel redundancy and physiological impact of $I_{Twindow}$. *J Neurosci* **30:** 99–109.

Landisman CE, Long MA, Beierlein M, Deans MR, Paul DL, Connors BW. 2002. Electrical synapses in the thalamic reticular nucleus. *J Neurosci* **22:** 1002–1009.

Pita-Almenar JD, Yu D, Lu HC, Beierlein M. 2014. Mechanisms underlying desynchronization of cholinergic-evoked thalamic network activity. *J Neurosci* **34:** 14463–14474.

Sohal VS, Keist R, Rudolph U, Huguenard JR. 2003. Dynamic GABA$_A$ receptor subtype-specific modulation of the synchrony and duration of thalamic oscillations. *J Neurosci* **23:** 3649–3657.

Sun YG, Pita-Almenar JD, Wu CS, Renger JJ, Uebele VN, Lu HC, Beierlein M. 2013. Biphasic cholinergic synaptic transmission controls action potential activity in thalamic reticular nucleus neurons. *J Neurosci* **33:** 2048–2059.

Cite this protocol as *Cold Spring Harb Protoc*; doi:10.1101/pdb.prot095836

In Vitro Investigation of Synaptic Plasticity

Therese Abrahamsson,[1,4] Txomin Lalanne,[1,2,4] Alanna J. Watt,[3] and P. Jesper Sjöström[1,5]

[1]*Centre for Research in Neuroscience, Department of Neurology and Neurosurgery, The Research Institute of the McGill University Health Centre, Montréal General Hospital, Montréal, Québec H3G 1A4, Canada;* [2]*Integrated Program in Neuroscience, McGill University, Montréal, Quebec H3A 2B4, Canada;* [3]*Department of Biology, Bellini Life Sciences Building, McGill University, Montréal, Québec H3G 0B1, Canada*

A classical in vitro model for investigation of information storage in the brain is based on the acute hippocampal slice. Here, repeated high-frequency stimulation of excitatory Schaeffer collaterals making synapses onto pyramidal cells in the hippocampal CA1 region leads to strengthening of evoked field-recording responses—long-term potentiation (LTP)—in keeping with Hebb's postulate. This model remains tremendously influential for its reliability, specificity, and relative ease of use. More recent plasticity studies have explored various other brain regions including the neocortex, which often requires more laborious whole-cell recordings of synaptically connected pairs of neurons, to ensure that the identities of recorded cells are known. In addition, with this experimental approach, the spiking activity can be controlled with millisecond precision, which is necessary for the study of spike-timing-dependent plasticity (STDP). Here, we provide protocols for in vitro study of hippocampal CA1 LTP using field recordings, and of STDP in synaptically connected pairs of layer-5 pyramidal cells in acute slices of rodent neocortex.

IN VITRO SYNAPTIC PLASTICITY: A MODEL OF LEARNING IN THE BRAIN

It is widely believed that the storing of information in the brain is accomplished by alterations in synaptic strength among connected neurons. This view is typically attributed to the Canadian neuroscientist Donald Hebb (1949), although many before him had advanced similar propositions for how learning in the brain could be accomplished (Markram et al. 2011). In Hebb's postulate, he states that a way of storing information would be to increase synaptic strength of already connected cells if they are repeatedly and persistently activated simultaneously (Hebb 1949), a concept that has been summarized as "cells that fire together, wire together" (Shatz 1992).

The first evidence for Hebbian plasticity was reported by Bliss and Lømo (1973). They stimulated inputs to the dentate gyrus of rabbit hippocampus in vivo and found that response amplitudes would not only potentiate but also remain potentiated several hours after brief trains of high-frequency stimulation. As high-frequency stimulation drives postsynaptic cells, the persistent synaptic strengthening was in agreement with Hebb's postulate. However, a causal relationship between synaptic plasticity and learning still remains to be formally established in mammals. Tremendous progress has been made to link synaptic plasticity in the amygdala to fear conditioning (Stevens 1998; Pape and Pare 2010; Johansen et al. 2011; Nabavi et al. 2014). But memories are often distributed across multiple synapses, which makes it hard to establish clear-cut causal links between plasticity and behavior.

---

[4]These authors contributed equally to this work.

[5]Correspondence: jesper.sjostrom@mcgill.ca

Cite this introduction as *Cold Spring Harb Protoc*; doi:10.1101/pdb.top087262

Here, we focus on two in vitro synaptic plasticity models that rely on acute slices from the rodent brain: field-recording long-term potentiation (LTP) in the hippocampal CA1 region (see Protocol 1: Long-Term Potentiation by Theta-Burst Stimulation Using Extracellular Field Potential Recordings in Acute Hippocampal Slices [Abrahamsson et al. 2016]) and spike-timing-dependent plasticity (STDP) in paired recordings of layer-5 pyramidal cells of visual cortex (Sjöström et al. 2001; see Protocol 2: Using Multiple Whole-Cell Recordings to Study Spike-Timing-Dependent Plasticity in Acute Neocortical Slices [Lalanne et al. 2016]). We provide tips and tricks as well as troubleshooting information. These protocols are easy to adapt to other brain regions or activity paradigms. Additional background on each of these methods is provided below.

LONG-TERM POTENTIATION BY THETA-BURST STIMULATION USING EXTRACELLULAR FIELD RECORDINGS IN ACUTE HIPPOCAMPAL SLICES

There are several advantages to using extracellular field recordings for the study of LTP. First, it is a relatively simple method that is suitable even for beginner electrophysiologists with little background or expertise. Second, it is a relatively noninvasive technique that does not disrupt the internal milieu of the neurons. This is in contrast to whole-cell recordings, where the experimenter runs the risk of washing out substances that are essential for LTP as the cell is dialyzed (Malinow and Tsien 1990; Isaac et al. 1996). Third, it is possible to generate stable recordings for a long period of time, up to several hours, which is technically more challenging with whole-cell recordings (Malinow and Tsien 1990; Watt et al. 2004; Sjöström and Häusser 2006). Finally, response variability is lowered with field potential measurements, which sample the activity of large numbers of neurons simultaneously. This averaging helps produce robust data sets rapidly. Figure 1 in Protocol 1: Long-Term Potentiation by Theta-Burst Stimulation Using Extracellular Field Potential Recordings in Acute Hippocampal Slices (Abrahamsson et al. 2016) provides a comparison between field excitatory postsynaptic potentials (fEPSPs) and EPSPs from whole-cell recordings.

Theta-burst stimulation is a highly influential LTP induction paradigm that is commonly used because it resembles physiological theta activity and because it is quite robust in brain regions as different as neocortex and hippocampus (Kirkwood et al. 1993). Another standard LTP induction paradigm uses uninterrupted high-frequency stimulation—tetanization—rather than theta-burst stimulation (e.g., the original LTP study by Bliss and Lømo [1973]).

USING QUADRUPLE WHOLE-CELL RECORDINGS TO STUDY SPIKE-TIMING-DEPENDENT PLASTICITY IN ACUTE NEOCORTICAL SLICES

Using extracellular field recordings, electrophysiologists have made great strides in the study of synaptic plasticity in hippocampus (see Protocol 1: Long-Term Potentiation by Theta-Burst Stimulation Using Extracellular Field Potential Recordings in Acute Hippocampal Slices) [Abrahamsson et al. 2016]). With the neocortex, however, it has not been quite as straightforward. Although a considerable amount has been learned about neocortical plasticity by stimulating in the white matter or in layer 4 (Kirkwood et al. 1993, 1995; Kirkwood and Bear 1994), neocortical extracellular stimulation experiments often suffer from the shortcoming that it is difficult to know which synapse types were recorded from.

With paired recordings, however, the experimenter knows precisely what neuronal types are being stimulated and recorded (Miles and Poncer 1996; Debanne et al. 2008). Paired recordings are thus particularly suited for the study of neocortical circuits, where multiple cell types exist side by side and where plasticity is known to be synapse- and cell type–specific (Buchanan et al. 2012; Blackman et al. 2013; Larsen and Sjöström 2015). To benefit maximally from paired recordings, they should ideally be combined with morphological reconstruction and classification, either from biocytin histology or 3D imaging stacks obtained with two-photon laser-scanning microscopy (2PLSM) (Blackman et al. 2014; Ferreira et al. 2014). In addition, paired recordings provide pharmacological access to both the pre- and postsynaptic cell, thus enabling wash-in of drugs or dyes into the transmitting or recipient neuron

Cite this introduction as *Cold Spring Harb Protoc*; doi:10.1101/pdb.top087262

(Kaiser et al. 2004; Koester and Johnston 2005; Rodriguez-Moreno and Paulsen 2008; Buchanan et al. 2012). Finally, paired recordings also enable precise timing of spikes in connected neurons, which is absolutely essential for the STDP experimental paradigm (Markram et al. 1997; Sjöström et al. 2001).

Unfortunately, neocortical connectivity is sparse—typically only 10%–50% of neighboring excitatory cells are monosynaptically connected (Song et al. 2005; Lefort et al. 2009; Ko et al. 2011)—which makes paired recordings slow and painstaking. Fortunately, the number of connections tested scales favorably with the number of cells recorded: With n neighboring cells simultaneously patched, the number of connections tested is $n(n-1)$. As n increases, more manipulators are required, resulting in considerable spatial and financial constraints. Still, several studies have been reported with seven to 12 simultaneous whole-cell recordings (Lefort et al. 2009; Perin et al. 2011). Because quadruple recordings sample 12 possible connections simultaneously with reasonable spatial constraints and at a relatively realistic cost, we suggest that $n = 4$ recordings represent an ideal choice for many electrophysiology laboratories. However, care should be taken when identifying the cells that are connected. With multiple simultaneous patching, the nonconnected cells are typically also labeled with the same dye and visualized, which can make it difficult to tell individual neurons apart. One possible way to avoid this problem is to use different dyes in the different recording pipettes (Kaiser et al. 2004; Koester and Johnston 2005).

REFERENCES

Abrahamsson T, Lalanne T, Watt AJ, Sjöström PJ. 2016. Long-term potentiation by theta-burst stimulation using extracellular field potential recordings in acute hippocampal slices. *Cold Spring Harb Protoc* doi: 10.1101/pdb.prot091298.

Blackman AV, Abrahamsson T, Costa RP, Lalanne T, Sjöström PJ. 2013. Target cell-specific short-term plasticity in local circuits. *Front Synaptic Neurosci* 5: 11.

Blackman AV, Grabuschnig S, Legenstein R, Sjöström PJ. 2014. A comparison of manual neuronal reconstruction from biocytin histology or 2-photon imaging: Morphometry and computer modeling. *Front Neuroanat* 8: 65.

Bliss TV, Lømo T. 1973. Long-lasting potentiation of synaptic transmission in the dentate area of the anaesthetized rabbit following stimulation of the perforant path. *J Physiol* 232: 331–356.

Buchanan KA, Blackman AV, Moreau AW, Elgar D, Costa RP, Lalanne T, Tudor Jones AA, Oyrer J, Sjöström PJ. 2012. Target-specific expression of presynaptic NMDA receptors in neocortical microcircuits. *Neuron* 75: 451–466.

Debanne D, Boudkkazi S, Campanac E, Cudmore RH, Giraud P, Fronzaroli-Molinieres L, Carlier E, Caillard O. 2008. Paired-recordings from synaptically coupled cortical and hippocampal neurons in acute and cultured brain slices. *Nat Protoc* 3: 1559–1568.

Ferreira TA, Blackman AV, Oyrer J, Jayabal S, Chung AJ, Watt AJ, Sjöström PJ, van Meyel DJ. 2014. Neuronal morphometry directly from bitmap images. *Nat Methods* 11: 982–984.

Hebb DO. 1949. *The organization of behavior.* Wiley, New York.

Isaac JT, Hjelmstad GO, Nicoll RA, Malenka RC. 1996. Long-term potentiation at single fiber inputs to hippocampal CA1 pyramidal cells. *Proc Natl Acad Sci* 93: 8710–8715.

Johansen JP, Cain CK, Ostroff LE, LeDoux JE. 2011. Molecular mechanisms of fear learning and memory. *Cell* 147: 509–524.

Kaiser KM, Lübke J, Zilberter Y, Sakmann B. 2004. Postsynaptic calcium influx at single synaptic contacts between pyramidal neurons and bitufted interneurons in layer 2/3 of rat neocortex is enhanced by back-propagating action potentials. *J Neurosci* 24: 1319–1329.

Kirkwood A, Bear MF. 1994. Hebbian synapses in visual cortex. *J Neurosci* 14: 1634–1645.

Kirkwood A, Dudek SM, Gold JT, Aizenman CD, Bear MF. 1993. Common forms of synaptic plasticity in the hippocampus and neocortex in vitro. *Science* 260: 1518–1521.

Kirkwood A, Lee HK, Bear MF. 1995. Co-regulation of long-term potentiation and experience-dependent synaptic plasticity in visual cortex by age and experience. *Nature* 375: 328–331.

Ko H, Hofer SB, Pichler B, Buchanan KA, Sjöström PJ, Mrsic-Flogel TD. 2011. Functional specificity of local synaptic connections in neocortical networks. *Nature* 473: 87–91.

Koester HJ, Johnston D. 2005. Target cell-dependent normalization of transmitter release at neocortical synapses. *Science* 308: 863–866.

Lalanne T, Abrahamsson T, Sjöström PJ. 2016. Using multiple whole-cell recordings to study spike-timing-dependent plasticity in acute neocortical slices. *Cold Spring Harb Protoc* doi: 10.1101/pdb.prot091306.

Larsen RS, Sjöström PJ. 2015. Synapse-type-specific plasticity in local circuits. *Curr Opin Neurobiol* 35: 127–135.

Lefort S, Tomm C, Floyd Sarria JC, Petersen CC. 2009. The excitatory neuronal network of the C2 barrel column in mouse primary somatosensory cortex. *Neuron* 61: 301–316.

Malinow R, Tsien RW. 1990. Presynaptic enhancement shown by whole-cell recordings of long-term potentiation in hippocampal slices. *Nature* 346: 177–180.

Markram H, Lübke J, Frotscher M, Sakmann B. 1997. Regulation of synaptic efficacy by coincidence of postsynaptic APs and EPSPs. *Science* 275: 213–215.

Markram H, Gerstner W, Sjöström PJ. 2011. A history of spike-timing-dependent plasticity. *Front Synaptic Neurosci* 3: 1–24.

Miles R, Poncer JC. 1996. Paired recordings from neurones. *Curr Opin Neurobiol* 6: 387–394.

Nabavi S, Fox R, Proulx CD, Lin JY, Tsien RY, Malinow R. 2014. Engineering a memory with LTD and LTP. *Nature* 511: 348–352.

Pape HC, Pare D. 2010. Plastic synaptic networks of the amygdala for the acquisition, expression, and extinction of conditioned fear. *Physiol Rev* 90: 419–463.

Perin R, Berger TK, Markram H. 2011. A synaptic organizing principle for cortical neuronal groups. *Proc Natl Acad Sci* 108: 5419–5424.

Rodriguez-Moreno A, Paulsen O. 2008. Spike timing-dependent long-term depression requires presynaptic NMDA receptors. *Nat Neurosci* 11: 744–745.

Shatz CJ. 1992. The developing brain. *Sci Am* 267: 60–67.

Sjöström PJ, Häusser M. 2006. A cooperative switch determines the sign of synaptic plasticity in distal dendrites of neocortical pyramidal neurons. *Neuron* 51: 227–238.

Sjöström PJ, Turrigiano GG, Nelson SB. 2001. Rate, timing, and cooperativity jointly determine cortical synaptic plasticity. *Neuron* 32: 1149–1164.

Song S, Sjöström PJ, Reigl M, Nelson S, Chklovskii DB. 2005. Highly nonrandom features of synaptic connectivity in local cortical circuits. *PLoS Biol* 3: e68.

Stevens CF. 1998. A million dollar question: Does LTP = memory? *Neuron* 20: 1–2.

Watt AJ, Sjöström PJ, Häusser M, Nelson SB, Turrigiano GG. 2004. A proportional but slower NMDA potentiation follows AMPA potentiation in LTP. *Nat Neurosci* 7: 518–524.

Long-Term Potentiation by Theta-Burst Stimulation Using Extracellular Field Potential Recordings in Acute Hippocampal Slices

Therese Abrahamsson,[1] Txomin Lalanne,[1,2] Alanna J. Watt,[3] and P. Jesper Sjöström[1,4]

[1]Centre for Research in Neuroscience, Department of Neurology and Neurosurgery, The Research Institute of the McGill University Health Centre, Montréal General Hospital, Montréal, Québec H3G 1A4, Canada; [2]Integrated Program in Neuroscience, McGill University, Montréal, Quebec H3A 2B4, Canada; [3]Department of Biology, Bellini Life Sciences Building, McGill University, Montréal, Québec H3G 0B1, Canada

This protocol describes how to carry out theta-burst long-term potentiation (LTP) with extracellular field recordings in acute rodent hippocampal slices. This method is relatively simple and noninvasive and provides a way to sample many neurons simultaneously, making it suitable for applications requiring higher throughput than whole-cell recording.

MATERIALS

It is essential that you consult the appropriate Material Safety Data Sheets and your institution's Environmental Health and Safety Office for proper handling of equipment and hazardous material used in this protocol.

RECIPE: Please see the end of this protocol for recipes indicated by <R>. Additional recipes can be found online at http://cshprotocols.cshlp.org/site/recipes.

Reagents

Acute hippocampal rodent brain slices

Many different species can be used (e.g., rats, mice, guinea pigs). The slices should be 200–500-μm thick. In thicker slices, gas and metabolite exchange become compromised, as diffusion time increases exponentially with distance. Gas exchange is better with an interface slice recording chamber (Reid et al. 1988), but high flow rate may alleviate this problem in a submerged slice recording chamber (Hajos et al. 2009). Brain tissue can be sliced in different orientations (e.g., coronally or horizontally), which may alter experiments because specific structures are preserved in some orientations but not in others (Skrede and Westgaard 1971; Marcaggi and Attwell 2007; Bartlett et al. 2011; Sherwood et al. 2012). It is typically easier to record from juvenile slices because young slices stay healthier for longer (see Troubleshooting). As a consequence, thicker slices or entire brain structures of neonatal rodent brain can be used for in vitro studies (Khalilov et al. 1997). Anecdotally, we find that irrespective of age, rat slices are easier to work with than mouse slices.

Artificial cerebrospinal fluid (ACSF) for synaptic plasticity studies <R>
Carbogen gas (95% O_2/5% CO_2)

This gas mixture keeps the slices oxygenated and the pH at ~7.4 during storage and recording.

---

[4]Correspondence: jesper.sjostrom@mcgill.ca

Equipment

Amplifier

An amplifier (e.g., Molecular Devices MultiClamp 700B, HEKA EPC 10, Dagan BVC-700A, or A-M Systems Model 1700 or 3000) is required to boost electrode signals. Patch-clamp amplifiers can act as field-recording amplifiers, thus providing a saving for those laboratories that already have them. However, field potential amplifiers are considerably cheaper if the sole intended purpose is field potential recordings.

Computer with acquisition and analysis software

There is a plethora of very good recording software packages available. Many of these can be downloaded for free (e.g., ePhus, Strathclyde Electrophysiology Software, or NeuroMatic), although they may require a software framework that is not free, such as IGOR Pro (WaveMetrics) or MATLAB (MathWorks). Off-the-shelf electrophysiology software packages include AxoGraph, WinLTP, and pCLAMP. The choice of software is tremendously important, as it may determine what kinds of experiments can be performed and which acquisition boards must be used. A flexible alternative is to use custom-made programs written in, for example, IGOR Pro, MATLAB, or LabVIEW. This simplifies data acquisition and analysis, but requires computer programming skills.

Data acquisition boards

The choice of data acquisition board depends critically on the choice of software (see above). Relatively inexpensive boards of a wide variety can be purchased from National Instruments (e.g., PCI-6229, PCI-6221-37, or USB-6003). Boards dedicated to electrophysiology can be purchased from Molecular Devices (e.g., Digidata 1550) or HEKA (e.g., InstruTech ITC-18). Dedicated boards come with advantages such as integration with specific amplifiers, but this may be of less importance for field potential recordings. Many types of boards can be used with general data analysis software frameworks such as IGOR Pro (WaveMetrics) or MATLAB (MathWorks), although custom scripts may have to be written or downloaded.

Electrodes and electrode holders

Stimulation and recording electrodes can be made from borosilicate glass capillaries using a pipette puller. Capillaries of various thicknesses and diameters are available; we use those with an outer diameter of 1.5 mm and an inner diameter of 0.86 mm. We have used Sutter Instruments' P-97 and P-1000 pullers, the Narishige Group's PC-10 puller, and AutoMate Scientific's Zeitz DMZ puller, and they all work very well—the chief differences are degree of automation, versatility, and price. Both recording and stimulation pipettes may be filled with ACSF, in which case the tip size should typically be 2–50 µm. It may be difficult to obtain a large enough tip diameter, in which case gently breaking the pipette tip against tissue paper may help. Capillaries with an internal filament are easier to fill. The recording pipette can also be filled with 1–4 M NaCl to increase the electrode conductance. In this case, however, the tip size has to be much smaller (maximally 1 µm) to avoid NaCl leakage, which may damage tissue and which may result in nonphysiologically high Cl− concentration locally. A chlorided silver wire attached to an electrode holder is placed inside the recording pipette. We know of two methods for chloriding electrode wires: Either the wire is immersed in bleach for ~10 min (rinse carefully in distilled water afterward) or a positive current is passed through the electrode dipped in ~1 M NaCl using, for example, a 4.5-V battery until a white layer appears. For stimulation, a variety of different types of electrodes can be used (e.g., insulated tungsten electrodes [World Precision Instruments or FHC Inc.]). These have lower resistance and capacitance but larger tip diameter.

Faraday cage

A grounded metal cage used to reduce electric noise can be bought (e.g., from Harvard Apparatus or World Precision Instruments) or can be made by attaching a metal mesh to a frame. A Faraday cage is often not necessary.

Ground electrodes

Two ground electrodes should be placed in the recording chamber: one connected to the amplifier and the other to the stimulus isolator. For the recording ground, use a chlorided silver wire. For stimulation, use a bare silver wire to avoid buildup of electrode polarization. If bipolar metal electrodes are used for stimulation, there is no need for a stimulation ground.

Micromanipulators

For field recordings, inexpensive mechanical micromanipulators (e.g., Narishige Group NMN-21, Scientifica LBM-7, or Sutter Instruments MP85) are normally adequate. Motorized micromanipulators cost more but simplify experiments—a nonexhaustive list includes PatchStar and MicroStar by Scientifica, Mini and Micro by Luigs & Neumann, and MP-285 and MP-225 by Sutter Instruments.

Microscope

There are several microscopes suitable for electrophysiology on the market, such as the Olympus BX-50, Scientifica SliceScope, Leica DM series, and Zeiss Axio Examiner. Important properties to consider include size, price, choice of contrast enhancement (differential interference, oblique illumination, Dodt contrast), and level of automation. For field-recording experiments, an inexpensive stereoscope may provide an excellent training environment.

Oscilloscope

We recommend having a dedicated oscilloscope to simplify debugging. Digital oscilloscopes of low sampling rate are not expensive (e.g., Tektronix TDS 200 series or PicoScope).

Perfusion system

To perfuse the slice with oxygenated ACSF, use a peristaltic pump (e.g., Gilson Minipulse 3 or World Precision Instruments Peri-Star) or a siphon combined with suction for removal. Peristaltic pumps can be used to recycle expensive drug-containing perfusion solutions to keep the overall ACSF volume low. Add a drip chamber to the inlet to visualize flow rate, to reduce electrical noise levels when using a recirculating pump, and to prevent perfusion line bubble formation.

Recording chamber

These are commercially available (e.g., from Harvard Apparatus or Scientifica) but can also be constructed in-house by making a hole in the bottom of a Petri dish and attaching a coverslip using silicon adhesive.

Slice holder ("harp")

Use a flattened U-shaped platinum wire with individual nylon strings superglued in place, forming the shape of a harp. Pantyhose is a good source of nylon strings. Human hair provides an alternative. Platinum is soft and easy to shape and is also inert, which reduces the risk of toxicity. Slice anchors are also commercially available (e.g., from Harvard Apparatus or Warner Instruments).

Stimulus isolator

To activate input fibers, a stimulus isolator ("stim box") is needed. This unit takes a command signal from a computer or other external source, isolates it to minimize electrical noise, and injects current via the stimulation electrode into the preparation. There are many equivalent units on the market—we have had good experience with Dagan BSI-950, Digitimer DS2A, and AMPI ISO-Flex. Some isolators accept a command signal to control pulse amplitude, whereas others only accept a TTL (transistor–transistor logic)-style gating pulse, with pulse amplitude controlled by a dial. Some have constant-current or constant-voltage modes, and some can be switched between mono- and biphasic pulses. We recommend biphasic pulses (two contiguous monophasic pulses of opposite sign), as they minimize electrochemical reactions at the electrode tip, thus improving long-term stability.

Temperature controller

To obtain physiological ACSF temperature, use an in-line heater (e.g., Harvard Apparatus CL-100 or Scientifica SM-4500). Measure temperature in the recording chamber, because it drops in the tubing system.

Transfer pipette

Carefully break off a few centimeters of the narrow part of a glass Pasteur pipette and attach a rubber bulb. The wide end is then used to transfer the slice. Alternatively, cut off the tip of a plastic transfer pipette.

METHOD

Setup

1. Start the perfusion to circulate oxygenated ACSF into and out of the recording chamber.

   Perfusion flow rate should be >1 drop per second (>2 mL/min). For recordings nominally at physiological temperatures, recording chamber temperature should be 31°C–34°C, because at higher temperatures, slice quality suffers and evoked responses decay. Plasticity experiments are typically performed at nominally physiological temperature, although LTP may be observed at room temperature as well (Feldman 2000).

2. Attach the recording electrode to the electrode holder. Attach the stimulation electrode to one of the micromanipulators and connect it to the stimulus isolator.

   Before starting recordings, check noise levels (see Troubleshooting).

Recording

3. Transfer a slice to the recording chamber using a transfer pipette. Secure the slice in place with a slice holder.

   Reposition the slice so that the electrodes can be placed without touching the edges of the recording chamber. Slice holder strings should not run between stimulation and recording electrodes.

4. Position the stimulation and recording electrodes in the *stratum radiatum* (Fig. 1A).

   Electrodes should penetrate the slice ~50 μm without damaging tissue appreciably. A white halo can typically be seen surrounding the pipette tip as it enters the tissue.

Cite this protocol as *Cold Spring Harb Protoc*; doi:10.1101/pdb.prot091298

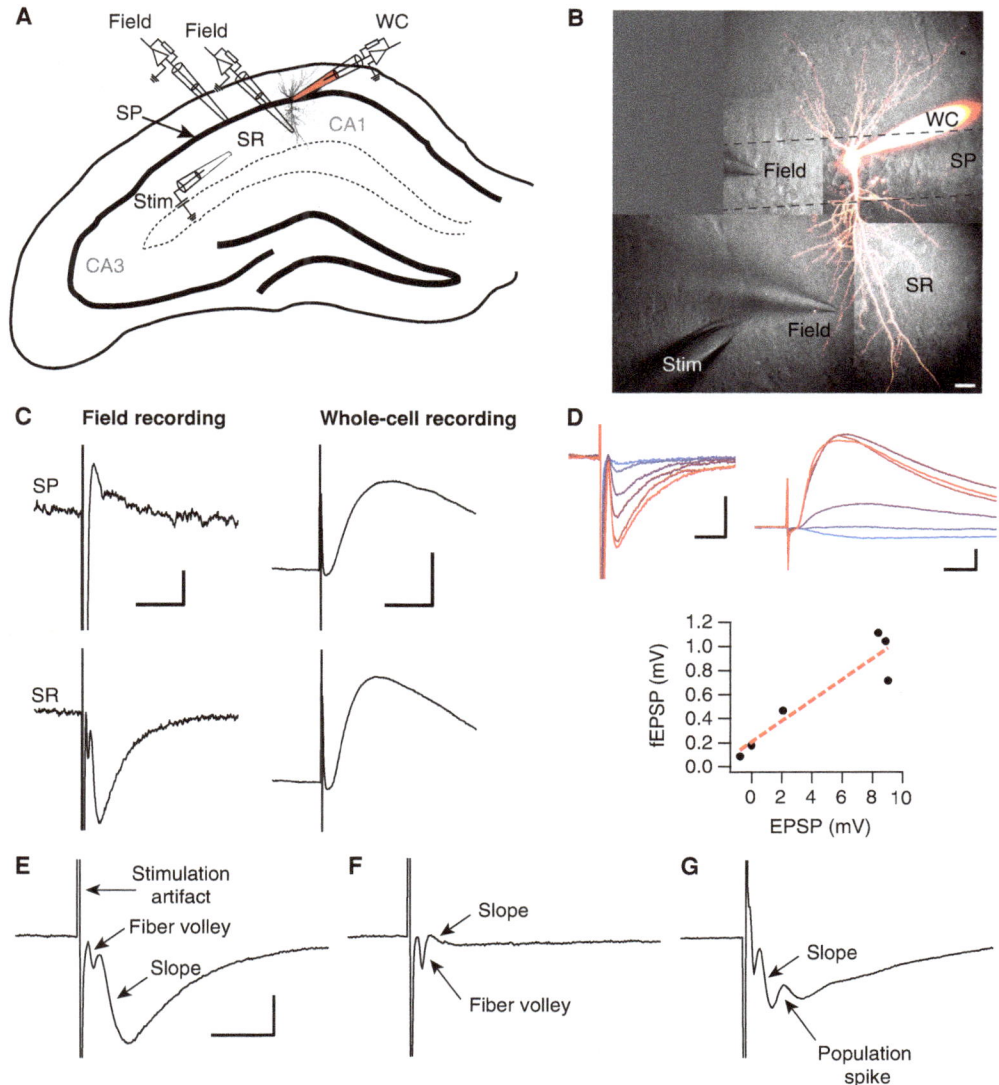

FIGURE 1. Factors determining field excitatory postsynaptic potential (fEPSP) recordings. (*A*) Schematic drawing of a hippocampal slice showing positioning of stimulation and recording electrodes. A stimulation electrode ("stim") is placed in the *stratum radiatum* (SR) of the CA1 area, to activate axons originating from pyramidal cells in CA3. For comparison, a whole-cell recording electrode (WC) is placed in the cell body layer, *stratum pyramidale* (SP). The field-recording electrode ("field") is moved between SP and SR positions. (*B*) Two-photon microscopy maximal intensity projection recorded pyramidal cell loaded with Alexa 594 is superimposed on infrared Dodt contrast images. Positions of stimulation and field-recording electrodes are shown, similar to the schematic in *A* (scale bar: 20 µm). (*C*) fEPSPs recorded in SP (*top left*) and SR (*bottom left*) have reversed polarities (scale bars: 200 µV, 10 msec). EPSP recorded in whole-cell configuration remains the same (*right*, scale bars: 2 mV, 10 msec). Note that the SR fEPSP (*bottom left*) has faster dynamics and the opposite sign compared with the intracellular EPSP (*bottom right*). (*D*) fEPSPs (*top left*) and whole-cell EPSPs (*top right*) recorded simultaneously at six different stimulation intensities (30, 40, 50, 70, 80, and 100 V, indicated by lines ranging from blue to red) are significantly correlated (*bottom*; Pearson's $r = 0.933$, $P < 0.05$), indicating equivalency in terms of measuring synaptic strength (scale bars: 500 µV, 5 msec [*top left*]; 2 mV, 5 msec [*top right*]). (*E*) Example of a good fEPSP recording. The stimulus artifact, fiber volley, and slope are indicated. The fEPSP is large compared with the fiber volley, which indicates a healthy slice (scale bar: 500 µV, 5 msec). (*F*) A fEPSP that is much smaller than the fiber volley suggests that the slice is of poor quality; despite stimulating many axons, the response is very small. (*G*) A fEPSP that has a population spike (positive deflection) riding on the PSP interfering with the measurement of the peak. This occurs when the recorded neurons fire.

5. Set the stimulation strength for the baseline recording.

   Adjust stimulation amplitude and duration to obtain the desired amplitude of the field excitatory postsynaptic potential (fEPSP). Typical stimulation values are 10–300 μA or 20–100 V, and 20–200 μsec, but these vary enormously depending on, for example, the stimulation electrode and type of brain tissue. A monopolar glass electrode with a 1- to 2-μm tip should reliably evoke responses in hippocampal P12 (postnatal day 12) slices with biphasic ±30–100-V pulses of 200-μsec duration. Maximize fEPSP amplitude by adjusting electrode positions and stimulation strength (Fig. 1A–D). A fiber volley should ideally precede the fEPSP (Fig. 1E), although it may not be evident if capacitive coupling between the stimulation and recording electrode generates a large stimulation artifact (Stangl and Fromherz 2008). The fiber volley represents presynaptic action potentials; variations in fiber volley amplitude are commonly used as a measure of stability, whereas a fEPSP that is smaller than the fiber volley indicates an unhealthy slice (Fig. 1F; see Troubleshooting). A larger distance between stimulation and recording electrodes separates the fiber volley from the fEPSP better, but at the expense of fEPSP amplitude. Increase stimulation strength until a population spike appears and (Fig. 1G), and then reduce it until fEPSP is halved.

6. Record responses every 10–60 sec to obtain a ~30-min baseline period (Fig. 2).

   Experiments without a stable baseline should be rejected, which may result in a 10%–50% rejection rate. To avoid bias, it is important to consistently apply the same stability criteria across all experiments including controls—for example, baseline fEPSP amplitude should not significantly correlate with time (Pearson's r at $P < 0.05$ level). Baseline length should also reflect experiment duration: recordings lasting several hours may require an hour-long baseline. To further ensure stability, an additional control stimulation electrode is often used (see Troubleshooting).

   Most amplifiers come with a set of built-in filters. To reduce noise, recorded signals should be filtered with a 2-kHz low-pass filter. A 0.03-Hz high-pass filter removes any offsets, and a notch filter eliminates line frequency noise. External adaptive filters (e.g., The HumBug, Quest Scientific) work better than built-in notch filters but are expensive. Proper systematic grounding of key equipment is the most efficient method to reduce noise but can be quite time-consuming.

7. Induce LTP (Fig. 2).

   For theta-burst LTP, deliver trains of 4–5 pulses at 100 Hz every 200 msec, repeated 10–75 times (Larson and Lynch 1986; Larson et al. 1986; Staubli and Lynch 1987; Kirkwood et al. 1993; Kirkwood and Bear 1994). For tetanization, ~100 pulses are typically delivered at 100 Hz, repeated two to four times (Bliss and Lømo 1973; Lynch et al. 1977; Dunwiddie and Lynch 1978). Theta-burst stimulation generally elicits LTP more reliably than tetanization (see Introduction: In Vitro Investigation of Synaptic Plasticity [Abrahamsson et al. 2016] and Larson and Lynch 1986).

8. Resume baseline recordings (Fig. 2).

   Overall duration of the recording depends on which phases of LTP are studied. Early LTP may last up to 3 h after induction, whereas late LTP should be measured 3–4 h after induction (Frey and Morris 1998; Kandel 2001). Regardless, plasticity is typically measured 30–60 min after the induction.

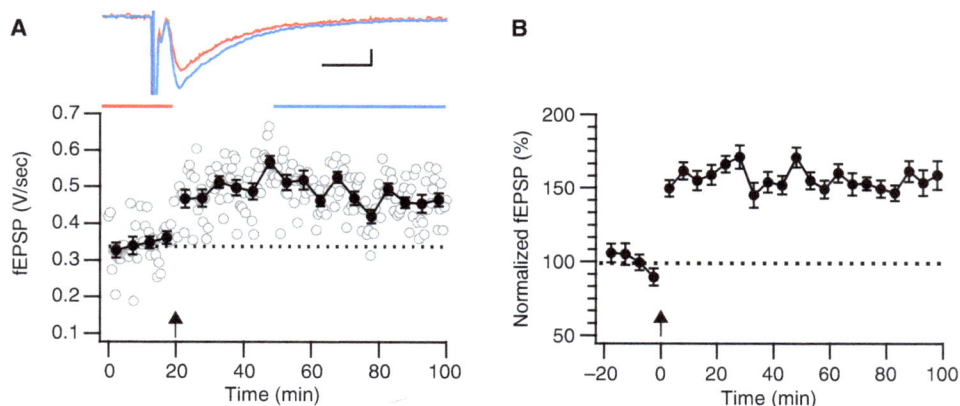

FIGURE 2. Theta-burst LTP. (A) Sample theta-burst induction experiment showing robust LTP. During the induction (arrow), 12 trains of 4 pulses at 100 Hz were delivered at 5 Hz, and this was repeated three times at 0.1 Hz. Error bars indicate standard error of the mean (s.e.m.). *Inset top*: fEPSPs averaged over indicated time periods before (red) and after (blue) induction (scale bars: 100 μV, 5 msec). (B) This ensemble average of five such theta-burst experiments indicates how robust this LTP paradigm is. Before averaging, experiments were normalized to their individual baseline periods, so error bars (s.e.m.) reflect the variability around the mean, not absolute amplitude variability across experiments.

Cite this protocol as *Cold Spring Harb Protoc*; doi:10.1101/pdb.prot091298

9. Analyze the data.

The size of the responses is measured by determining the slope of the fEPSP (i.e., by calculating the linear regression of the rising phase of the fEPSP [see Fig. 1E]). It is more appropriate to measure slope than amplitude, because field recordings represent a time derivative of extracellular current sources and sinks (Johnston and Wu 1994). In addition, spiking may contaminate peak amplitude measurements, leading to underestimation of the amplitude.

TROUBLESHOOTING

Problem: The fEPSP is small despite using high stimulation strength.
Solution: Consider the following.

- Electrodes may not be positioned correctly. Move the stimulation electrode vertically in the *stratum radiatum* to maximize fEPSP amplitude.

- Slice may be unhealthy. fEPSP responses that are much smaller than the fiber volley indicate an unhealthy slice (Fig. 1E, F). Slice health is crucial for LTP, and there are several ways to achieve and maintain healthier slices. Make sure the osmolality and pH of the ACSF are correct: The pH should be ~7.4, whereas the osmolality depends on the animal species being used (Bourque 2008). We recommend 320 mM for rat slices, adjusted with D-glucose. An essential step in the preparation of slices is the dissection of the brain, which should be relatively quick. Aim for <15 min from decapitation to the end of the slicing procedure. The decapitation and removal of the skin and skull should take <1 min. It may be easier to generate healthy slices from juvenile animals, but LTP mechanisms vary with developmental stage (Isaac et al. 1995; Liao et al. 1995; Yasuda et al. 2003; Massey and Bashir 2007; Abrahamsson et al. 2008; Larsen et al. 2010). With adult animals, slice quality can be improved by cardiac perfusion with ice-cold ACSF to cool the brain rapidly (Moyer and Brown 1998).

Problem: Response amplitudes are not stable.
Solutions: Consider the following.

- Electrodes may be drifting. If the slice or the electrodes are gradually moving, responses tend to run up or down. The harp should hold the slice firmly in place to minimize slice movement. Minimizing electrode drift can be more challenging. Drift can be monitored using a microscope but can be very difficult to notice along the microscope *z*-axis. Causes of drift include taut manipulator wires pulling on manipulator housing, ambient room temperature changes (e.g., air conditioning), and electrodes settling in their holder.

- Gas exchange in the slice may be compromised. Dropping oxygen levels reduce fEPSPs amplitude but not the fiber volley. In addition, CO_2 accumulation suppresses the fEPSP (Lee et al. 1996). Increase perfusion rate to enhance slice oxygenation. Use Teflon or Tygon tubing to prevent gas leakage (Hajos et al. 2009).

- Perfusion temperature may be changing. Check heater settings. Bubbles in the perfusion line cause temperature transients; reduce bubble formation by filling the drip chamber a little at the beginning of the experiment. Pharmacological manipulations via perfusate changes also tend to cause bubble formation and temperature changes.

- Slice may be unhealthy. Take a new slice from the incubation chamber or dissect new slices.

- An additional control input may be needed to monitor fEPSP stability. One input will undergo LTP but the other will not. Slice deterioration, oxygen deprivation, and perfusion temperature fluctuations affect both inputs. The two stimulation electrodes are typically placed on opposite sides of the recording electrode, on different levels in the *stratum radiatum*. To ensure that the afferent pathways are independent, the two inputs are activated separately and simultaneously and then the algebraic sum of the separate fEPSPs is compared with the simultaneously evoked fEPSP.

Problem: Recording sweeps are disrupted by epileptiform activity.

Solution: Hyperexcitable acute slices display epileptiform activity. Epileptiform activity can be observed as multiple regenerative population spikes in the decay phase of the fEPSP. To prevent this, make a cut between CA3 and CA1. Increase the total concentration of Ca^{2+} and Mg^{2+} in the ACSF to 4 mM. Hyperexcitability can be a consequence of perfusing inhibitory blockers such as picrotoxin or bicuculline. If inhibitory blockade is essential, try applying drugs locally in the slice (Feldman 2000).

Problem: Recordings suffer from electrical noise.

Solution: Electrical noise is a major problem in electrophysiology, but noise debugging is an art form that we regrettably cannot cover exhaustively here. The Faraday cage should help reduce noise, but the key is proper grounding. The input to the amplifier should be floating, which means that the perfusion ground should not be shared with the equipment chassis ground—the latter is there to protect the user from electrical shock, whereas the former is a signal reference. There should be no ground loops, which means that equipment should only be grounded once. This is achieved by methodically using a star grounding pattern: The grounding leads should all connect to the same grounding point, thus forming a star shape. To debug for noise, measure root-mean-square noise level with the acquisition system and systematically turn equipment on and off until the identity of the poorly grounded culprit(s) is revealed. Noise frequency can give hints to the source: Line frequency noise may be caused by transformers, cathode ray tubes, or lights, whereas double line frequency noise can be caused by fluorescent light tubes. High-frequency noise can result from equipment such as computers, monitors, or nearby lasers and can be filtered out with a low-pass filter. Nearby light sources often cause noise. Shielding using grounded aluminum foil may provide a temporary solution but is a last resort that we do not generally recommend.

DISCUSSION

Investigating LTP using extracellular field recordings is a technique that has been used for decades since LTP was first discovered (Bliss and Lømo 1973). Besides theta-burst stimulation, there are several other types of LTP induction protocols—for example, tetanization (Bliss and Lømo 1973; Lynch et al. 1977; Dunwiddie and Lynch 1978), spike-timing-dependent plasticity (STDP; see Protocol 2: Using Multiple Whole-Cell Recordings to Study Spike-Timing-Dependent Plasticity in Acute Neocortical Slices [Lalanne et al. 2016] and Sjöström et al. 2001), and pairing of presynaptic spiking with postsynaptic depolarization (Artola et al. 1990). Note that different induction protocols may recruit different plasticity mechanisms that vary with synapse type, brain region, and age of the animal (Sjöström et al. 2008). These parameters have to be taken into account when comparing results from different studies. For example, whether LTP is expressed pre- or postsynaptically in hippocampus has been hotly debated (Malenka and Bear 2004; Kullmann 2012), but this disagreement can partly be caused by comparisons across different plasticity paradigms.

Extracellular field recordings are ideally used in laminated structures such as hippocampus, where pre- and postsynaptic cells are well separated. Here, this method enables rapid and relatively easy sampling of a large number of synaptic connections. Extracellular field recordings do have some drawbacks compared with patch-clamp recordings, however. Patch-clamp recordings offer cell specificity, the possibility of filling cells with drugs or dyes, etc. (see Protocol 2: Using Multiple Whole-Cell Recordings to Study Spike-Timing-Dependent Plasticity in Acute Neocortical Slices [Lalanne et al. 2016]; for additional background on both methods, also see Introduction: In Vitro Investigation of Synaptic Plasticity [Abrahamsson et al. 2016]). Ultimately, the choice of method depends on the experiment at hand. For example, it may make little sense to carry out high-throughput drug screening with technically demanding whole-cell recordings; field recordings would be more appropriate. The field recording technique will therefore remain an indispensable tool in the neuroscientist's toolbox for the foreseeable future.

Cite this protocol as *Cold Spring Harb Protoc*; doi:10.1101/pdb.prot091298

RECIPES

Artificial Cerebrospinal Fluid (ACSF) for Synaptic Plasticity Studies

Prepare the following 10× stock solution in double-distilled water (ddH$_2$O). Store it at 4°C for a maximum of 1 wk.

NaCl	1250 mM
KCl	25 mM
NaH$_2$PO$_4$	12.5 mM
NaHCO$_3$	260 mM

On the day of the experiment, dilute the 10× solution 10-fold, bubble for 10 min with carbogen (95% O$_2$/5% CO$_2$), and supplement with the following.

MgCl$_2$	1 mM
CaCl$_2$	2 mM
Glucose	~26 mM

ACSF—in particular Ca^{2+} and Mg^{2+} concentrations—may be varied depending on what is to be studied. For example, higher Ca^{2+} concentration favors long-term potentiation (LTP), whereas lower Ca^{2+} concentration promotes long-term depression (LTD). Adjust osmolality to ~320 mOsm with D-glucose for rat and ~338 mOsm for mouse.

ACKNOWLEDGMENTS

We thank Ian Duguid, Karri Lamsa, Eric Hanse, David Stellwagen, Elvis Cela, and Jérôme Maheux for help and useful discussions, as well as Pauline Vitte and all the students in BIOL389 at McGill University in 2015, for helpful input and for testing this protocol. This work was funded by Canada Foundation for Innovation (CFI) LOF 28331 (P.J.S.), Canadian Institutes of Health Research (CIHR) OG 126137 (P.J.S.), CIHR OG 130570 (A.J.W), CIHR NIA 288936 (P.J.S.), Natural Sciences and Engineering Research Council of Canada (NSERC) DG 418546-2 (P.J.S.), by a Research Institute of the McGill University Health Centre (RI MUHC) studentship award (T.L.), and by the Biology Department of McGill University.

REFERENCES

Abrahamsson T, Gustafsson B, Hanse E. 2008. AMPA silencing is a prerequisite for developmental long-term potentiation in the hippocampal CA1 region. *J Neurophysiol* 100: 2605–2614.

Abrahamsson T, Lalanne T, Watt AJ, Sjöström PJ. 2016. In vitro investigation of synaptic plasticity. *Cold Spring Harb Protoc* doi: 10.1101/pdb.top087262.

Artola A, Bröcher S, Singer W. 1990. Different voltage-dependent thresholds for inducing long-term depression and long-term potentiation in slices of rat visual cortex. *Nature* 347: 69–72.

Bartlett TE, Lu J, Wang YT. 2011. Slice orientation and muscarinic acetylcholine receptor activation determine the involvement of N-methyl D-aspartate receptor subunit GluN2B in hippocampal area CA1 long-term depression. *Mol Brain* 4: 41.

Bliss TV, Lømo T. 1973. Long-lasting potentiation of synaptic transmission in the dentate area of the anaesthetized rabbit following stimulation of the perforant path. *J Physiol* 232: 331–356.

Bourque CW. 2008. Central mechanisms of osmosensation and systemic osmoregulation. *Nat Rev Neurosci* 9: 519–531.

Dunwiddie T, Lynch G. 1978. Long-term potentiation and depression of synaptic responses in the rat hippocampus: Localization and frequency dependency. *J Physiol* 276: 353–367.

Feldman DE. 2000. Timing-based LTP and LTD at vertical inputs to layer II/III pyramidal cells in rat barrel cortex. *Neuron* 27: 45–56.

Frey U, Morris RG. 1998. Synaptic tagging: implications for late maintenance of hippocampal long-term potentiation. *Trends Neurosci* 21: 181–188.

Hajos N, Ellender TJ, Zemankovics R, Mann EO, Exley R, Cragg SJ, Freund TF, Paulsen O. 2009. Maintaining network activity in submerged hippocampal slices: Importance of oxygen supply. *Eur J Neurosci* 29: 319–327.

Isaac JT, Nicoll RA, Malenka RC. 1995. Evidence for silent synapses: Implications for the expression of LTP. *Neuron* 15: 427–434.

Johnston D, Wu SMS. 1994. *Foundations of cellular neurophysiology*. The MIT Press, Cambridge, MA.

Kandel ER. 2001. The molecular biology of memory storage: A dialogue between genes and synapses. *Science* 294: 1030–1038.

Khalilov I, Esclapez M, Medina I, Aggoun D, Lamsa K, Leinekugel X, Khazipov R, Ben-Ari Y. 1997. A novel in vitro preparation: The intact hippocampal formation. *Neuron* 19: 743–749.

Kirkwood A, Bear MF. 1994. Hebbian synapses in visual cortex. *J Neurosci* 14: 1634–1645.

Kirkwood A, Dudek SM, Gold JT, Aizenman CD, Bear MF. 1993. Common forms of synaptic plasticity in the hippocampus and neocortex in vitro. *Science* 260: 1518–1521.

Kullmann DM. 2012. The Mother of All Battles 20 years on: Is LTP expressed pre- or postsynaptically? *J Physiol* 590: 2213–2216.

LaLanne T, Abrahamsson T, Sjöström PJ. 2016. Using multiple whole-cell recordings to study spike-timing-dependent plasticity in acute neocortical slices. *Cold Spring Harb Protoc* doi: 10.1101/pdb.prot091306.

Larsen RS, Rao D, Manis PB, Philpot BD. 2010. STDP in the developing sensory neocortex. *Front Synaptic Neurosci* 2: 9.

Larson J, Lynch G. 1986. Induction of synaptic potentiation in hippocampus by patterned stimulation involves two events. *Science* 232: 985–988.

Larson J, Wong D, Lynch G. 1986. Patterned stimulation at the θ frequency is optimal for the induction of hippocampal long-term potentiation. *Brain Res* 368: 347–350.

Lee J, Taira T, Pihlaja P, Ransom BR, Kaila K. 1996. Effects of CO_2 on excitatory transmission apparently caused by changes in intracellular pH in the rat hippocampal slice. *Brain Res* 706: 210–216.

Liao D, Hessler NA, Malinow R. 1995. Activation of postsynaptically silent synapses during pairing-induced LTP in CA1 region of hippocampal slice. *Nature* 375: 400–404.

Lynch GS, Dunwiddie T, Gribkoff V. 1977. Heterosynaptic depression: A postsynaptic correlate of long-term potentiation. *Nature* 266: 737–739.

Malenka RC, Bear MF. 2004. LTP and LTD: An embarrassment of riches. *Neuron* 44: 5–21.

Marcaggi P, Attwell D. 2007. Short- and long-term depression of rat cerebellar parallel fibre synaptic transmission mediated by synaptic crosstalk. *J Physiol* 578: 545–550.

Massey PV, Bashir ZI. 2007. Long-term depression: Multiple forms and implications for brain function. *Trends Neurosci* 30: 176–184.

Moyer JR Jr, Brown TH. 1998. Methods for whole-cell recording from visually preselected neurons of perirhinal cortex in brain slices from young and aging rats. *J Neurosci Methods* 86: 35–54.

Reid KH, Edmonds HL Jr, Schurr A, Tseng MT, West CA. 1988. Pitfalls in the use of brain slices. *Prog Neurobiol* 31: 1–18.

Sherwood JL, Amici M, Dargan SL, Culley GR, Fitzjohn SM, Jane DE, Collingridge GL, Lodge D, Bortolotto ZA. 2012. Differences in kainate receptor involvement in hippocampal mossy fibre long-term potentiation depending on slice orientation. *Neurochem Int* 61: 482–489.

Sjöström PJ, Turrigiano GG, Nelson SB. 2001. Rate, timing, and cooperativity jointly determine cortical synaptic plasticity. *Neuron* 32: 1149–1164.

Sjöström PJ, Rancz EA, Roth A, Häusser M. 2008. Dendritic excitability and synaptic plasticity. *Physiol Rev* 88: 769–840.

Skrede KK, Westgaard RH. 1971. The transverse hippocampal slice: A well-defined cortical structure maintained in vitro. *Brain Res* 35: 589–593.

Stangl C, Fromherz P. 2008. Neuronal field potential in acute hippocampus slice recorded with transistor and micropipette electrode. *Eur J Neurosci* 27: 958–964.

Staubli U, Lynch G. 1987. Stable hippocampal long-term potentiation elicited by "theta" pattern stimulation. *Brain Res* 435: 227–234.

Yasuda H, Barth AL, Stellwagen D, Malenka RC. 2003. A developmental switch in the signaling cascades for LTP induction. *Nat Neurosci* 6: 15–16.

Using Multiple Whole-Cell Recordings to Study Spike-Timing-Dependent Plasticity in Acute Neocortical Slices

Txomin Lalanne,[1,2] Therese Abrahamsson,[1] and P. Jesper Sjöström[1,3]

[1]Centre for Research in Neuroscience, Department of Neurology and Neurosurgery, The Research Institute of the McGill University Health Centre, Montréal General Hospital, Montréal, Québec H3G 1A4, Canada; [2]Integrated Program in Neuroscience, McGill University, 3801 University Street, Montréal, Quebec H3A 2B4, Canada

This protocol provides a method for quadruple whole-cell recording to study synaptic plasticity of neocortical connections, with a special focus on spike-timing-dependent plasticity (STDP). It also describes how to morphologically identify recorded cells from two-photon laser-scanning microscopy (2PLSM) stacks.

MATERIALS

It is essential that you consult the appropriate Material Safety Data Sheets and your institution's Environmental Health and Safety Office for proper handling of equipment and hazardous materials used in this protocol.

RECIPES: Please see the end of this protocol for recipes indicated by <R>. Additional recipes can be found online at http://cshprotocols.cshlp.org/site/recipes.

Reagents

Acute neocortical rodent brain slices

> *For information on our experience with different slice types and preparation methods, see Protocol 1: Long-Term Potentiation by Theta-Burst Stimulation Using Extracellular Field Potential Recordings in Acute Hippocampal Slices (Abrahamsson et al. 2016b) and Davie et al. (2006).*

Artificial cerebrospinal fluid (ACSF) for synaptic plasticity studies <R>
Carbogen gas (95% O_2/5% CO_2)

> *This gas mixture keeps the slices oxygenated and the pH at ~7.4 during storage and recording.*

Internal solution for whole-cell recording <R>

Equipment

Antivibration air table (e.g., Newport, TMC, Thorlabs)
Borosilicate capillary glass tubing (Harvard Apparatus [G150F-4])
Contrast enhancement (e.g., Luigs & Neumann DGC tube, Scientifica Dodt contrast, or Olympus DIC)

> *Dodt contrast can also be custom-built from Thorlabs parts.*

Data acquisition board (National Instruments [PCI-6229]; see Fig. 1C)
Electrode holders and silver wire (e.g., Harvard Apparatus [2037760664]; see Fig. 1A,B)

[3]Correspondence: jesper.sjostrom@mcgill.ca

A

Headstage

Signal cable

Rod

Patch pipette
Electrode holder

Micromanipulator
Pressure tubing

B

Objective

Bath ground

Slice

Electrode holder

Pressure tubing

Patch pipette

Temperature probe

Slice holder

Heater

C

Data acquisition boards

Amplifiers

Oscilloscope

Video screen

Micromanipulator controllers

Stage controller

Control units

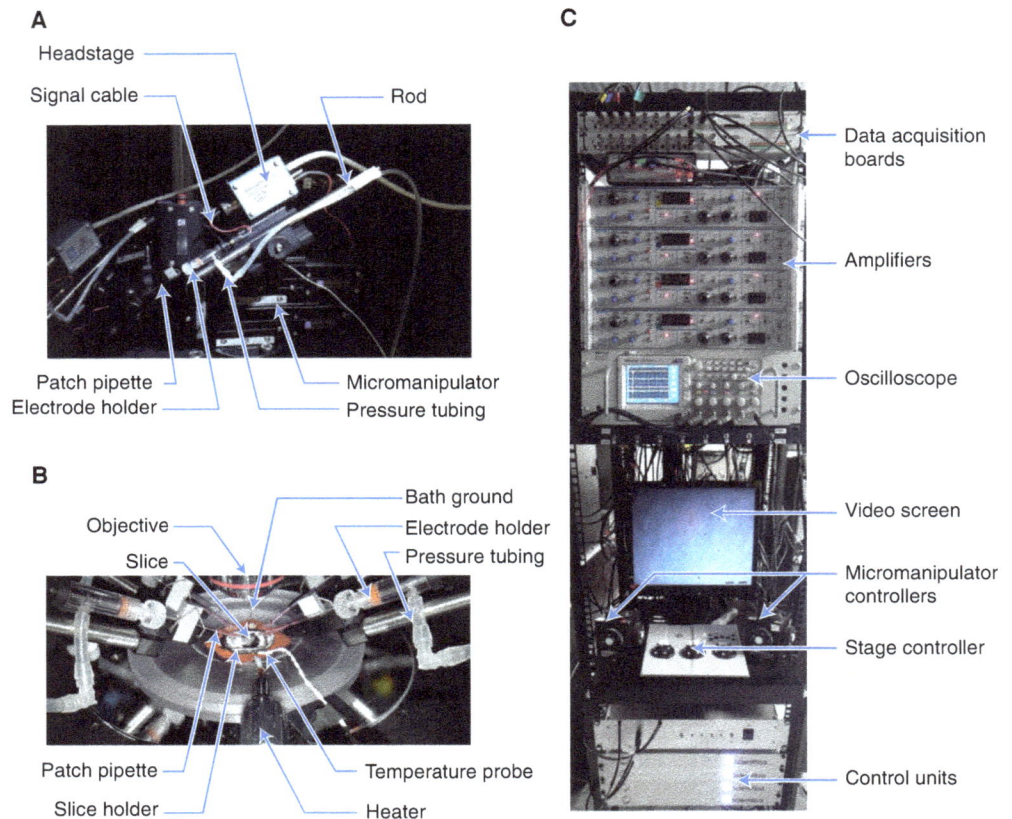

FIGURE 1. Recording setup. (*A*) Motorized micromanipulator with patch pipette. With this design, the rod is slid away from the chamber for pipette exchange; other designs may rotate or swing backward. The mode of pipette exchange determines the total number of manipulators that can be fit. (*B*) Perfusion chamber during quadruple whole-cell recording. Pipettes are filled with enough internal solution to touch the electrode wire. Glass pipettes are clamped to the rod to minimize movements. The temperature probe and the bath ground should be fully immersed in ACSF. (*C*) Acquisition boards provide communication between computer and amplifiers. Recording channels and manipulators are color-coded (see oscilloscope screen) for simplicity. Video screen shows a Dodt-contrast-enhanced image of the slice. Microscope stage, objective, and manipulators are remotely controlled to minimize the risk of disrupting ongoing recordings.

Electrode-tracking software (e.g., Scientifica [LinLab], WaveMetrics [IGOR Pro], or MathWorks [MATLAB] for custom-written programs)

Electronic manometer (optional; see Step 3)

Faraday cage (e.g., custom-made, Luigs & Neumann, Scientifica)

Infrared (IR)-sensitive charge-coupled device (CCD) camera (e.g., Watec [WAT902H], TILL Photonics [VX55])

Inline heater, sensor, and temperature controller (e.g., Scientifica [HPT-2A], Warner Instruments [SH-27B])

Micromanipulator (e.g., Scientifica [MicroStar; see Fig. 1A], Luigs & Neumann [MLE/MRE 3axes Mini25])

Motorized microscope with XY stage (e.g., Scientifica [SliceScope Pro], Luigs & Neumann [Infrapatch 380])

Oscilloscope (e.g., Tektronix [TDS2024C]; see Fig. 1C], Pico Technology [3406A/B])

Patch-clamp amplifier (Dagan Corporation [BVC-700A; see Fig. 1C], Molecular Devices [MultiClamp 700B or Axopatch 200B])

Patch-pipette filler (e.g., Advanced Instruments [MF28G67-5], Eppendorf ["microloader," 930001007])

Pipettes (prepared with pipette puller, e.g., Narishige [PC-10], Sutter Instruments [P-97, P-1000], Harvard Apparatus [PMP-102], AutoMate Zeitz DMZ)

Pressure tubing (see Fig. 1A,B; Tygon)

Recording acquisition software (e.g., pCLAMP [Molecular Devices] or AxoGraph X [AxoGraph], or IGOR Pro [WaveMetrics] or MATLAB [MathWorks] for custom-written programs such as Neuromatic or ePhus)

Recording chamber (see Protocol 1: Long-Term Potentiation by Theta-Burst Stimulation Using Extracellular Field Potential Recordings In Acute Hippocampal Slices [Abrahamsson et al. 2016b])

Slice holder (Harvard Apparatus, Warner Instruments, or custom-made from platinum wire with nylon strands held in place by cyanoacrylate glue [see Fig. 1B])

Software for morphological reconstructions (e.g., Neuromantic [http://www.reading.ac.uk/ neuromantic/], Neurolucida [MBF Bioscience], Imaris [Bitplane])

Software for morphometry (e.g., L-Measure [http://cng.gmu.edu:8080/Lm/], Fiji [http://fiji.sc/Fiji])

Stabilizing pipette rod holder (Scientifica; see Fig. 1A)

Three-way stopcocks (e.g., Cole-Parmer [EW-30600-23], VWR [89134-220])

Vacuum system or pump (e.g., Charles Austen [Dymax 5], Masterflex [HV7791620], Gilson [MINIPULS 3])

Water immersion objective (Olympus [40×: LUMPLFLN40XW; see Fig. 1B; 60×: LUMPLFLN60XW])

METHOD

Set Up the Experiment

1. Prepare ACSF and internal solution and cut neocortical brain slices (see Protocol 1: Long-Term Potentiation by Theta-Burst Stimulation Using Extracellular Field Potential Recordings in Acute Hippocampal Slices [Abrahamsson et al. 2016b] and Davie et al. 2006). Ensure that ACSF is circulating in the recording chamber and that the temperature is 31°C–34°C.

    ACSF should circulate at a rate of at least ~1 drop/sec (~2 mL/min).

2. Hold down the slice in the recording chamber (Fig. 1B) with the slice holder. To ensure that layer 5 (L5) pyramidal cells (PCs) were not damaged during dissection, visualize their apical dendrites as far as L1 if possible. Select the cells to patch, keeping in mind that connectivity is highest for cells located closer than 100-µm apart (Holmgren et al. 2003; Perin et al. 2011) and for neurons deep in the slice (Ko et al. 2011).

    Cells that appear smooth are usually healthier than those that are of high contrast.

3. Fill pipettes with internal solution and insert them into the electrode holders. Ensure pressure tubing and wires are attached to electrode holders. Clamp pipettes with a rod (Fig. 1A) to stabilize recordings. Apply positive pressure with a 20-mL syringe or by mouth. Close a three-way stop-cock to maintain positive pressure.

    An electronic manometer attached to the tubing can be helpful in monitoring pipette pressure.

Perform Patch-Clamping

4. Place all four electrodes just above the region of interest in the slice. Null the amplifier offsets and measure pipette resistances. Use a software solution to semiautomatic electrode movements (e.g., the *Follow* function in LinLab software from Scientifica or custom scripts).

    Computer assistance saves time and reduces the risk of damaging pipettes and/or the tissue.

5. Patch the first cell.

    i. Approach the cell to be patched. Voltage-clamp the pipette to 0 mV and apply a −5-mV test pulse running at 30–40 Hz with 50% duty cycle to monitor pipette resistance with an oscilloscope. Verify the positive pressure—as you advance the pipette through the slice, the positive pressure should push tissue aside. Approach the cell slowly, ideally along the diagonal axis of the pipette while circumventing other cells.

> *Do not go straight through other cells, as this makes the tip dirty, which makes the formation of a GΩ seal difficult or impossible.*
>
> *If the test pulse readout suddenly drops, the tip has become blocked, either by dirt inside the pipette or by brain tissue.*

ii. When the pipette tip is located ~10 μm from the cell, reduce positive pressure: Open the three-way stopcock and then reapply and hold positive pressure quickly by mouth. Advance the pipette tip a few microns into the cell until a dimple forms. Quickly release the pressure and gently apply light negative pressure to gradually form a GΩ seal without rupturing it (see Step 7).

iii. As seal resistance increases beyond ~100 MΩ, switch the holding voltage from 0 to −70 mV, as this helps establish the GΩ seal. Once the GΩ seal is formed, open the stopcock and release to atmospheric pressure.

6. Repeat Step 5 for the other three electrodes.

> *As the next pipette is brought down into the tissue with positive pressure, the tissue moves, thus requiring continual readjustments of the previous electrodes. Ideally, the tip of the pipettes should follow any movement of the cell so that the pipette tip remains at the same position relative to the cell as when it was first patched.*

7. Rupture the four patches.

i. Once four GΩ seals have been established, go into whole-cell configuration. Doing this in quick succession on all four cells ensures that intracellular components necessary for plasticity induction are not dialyzed unequally from the different cells (Malinow and Tsien 1990; Sjöström et al. 2001).

ii. Gradually apply gentle suction until the patch is ruptured while monitoring the oscilloscope test pulse.

> *Patch rupture is evidenced by a sudden increase in test pulse current step. The negative pressure ramp may have to be repeated a few times. If the seal does not rupture, try hyperpolarizing the cell to −140 mV until rupture (see Troubleshooting).*

iii. Switch to current clamp and remove negative pressure.

8. Once broken through, assess quality of whole-cell recordings.

> *Typical resting membrane potential and input resistance for visual cortex L5 PCs of postnatal day 14–16 (P14–P16) rats is −65 ± 3.4 mV (mean ± s.d.) and 110 ± 45 MΩ, respectively (n = 325; PJ Sjöström, unpubl.), although the distribution of the latter parameter has a long tail extending beyond 300 MΩ, and both values vary with age (Sjöström et al. 2001). The series resistance should be as low as possible, but is, as a rule of thumb, not less than double the pipette resistance. In practice, series resistances as high as 20–30 MΩ are satisfactory for plasticity experiments performed in current clamp, but voltage clamp is compromised by high or variable series resistance. A typical P14 visual cortex L5 PC will produce a single spike after a 5-msec 1.3-nA current injection. Spikes should typically be 1–2 msec in width at half height—if broader, the series resistance may be too high, which results in artificial spike broadening by temporal filtering.*

Identify Connected Pairs of Neurons

9. Search for connections by evoking spikes in all four cells (Fig. 2A,B, top), staggered by at least 500 msec to avoid accidental spike-timing-dependent plasticity (STDP) induction (Sjöström et al. 2001). Generate spike-triggered averages of 10–40 postsynaptic sweeps to ensure that weak connections are not missed. Verify that connections found are monosynaptic: Response latency and temporal jitter should be submillisecond (Fig. 2C,D).

Induce Plasticity

10. Once connections have been identified, start the STDP protocol.

> *First, a baseline period of 10 min or more should be acquired. Next, STDP is elicited by repeated pre- and postsynaptic spike pairings at the desired frequency and timing. A postpairing period then follows, identical*

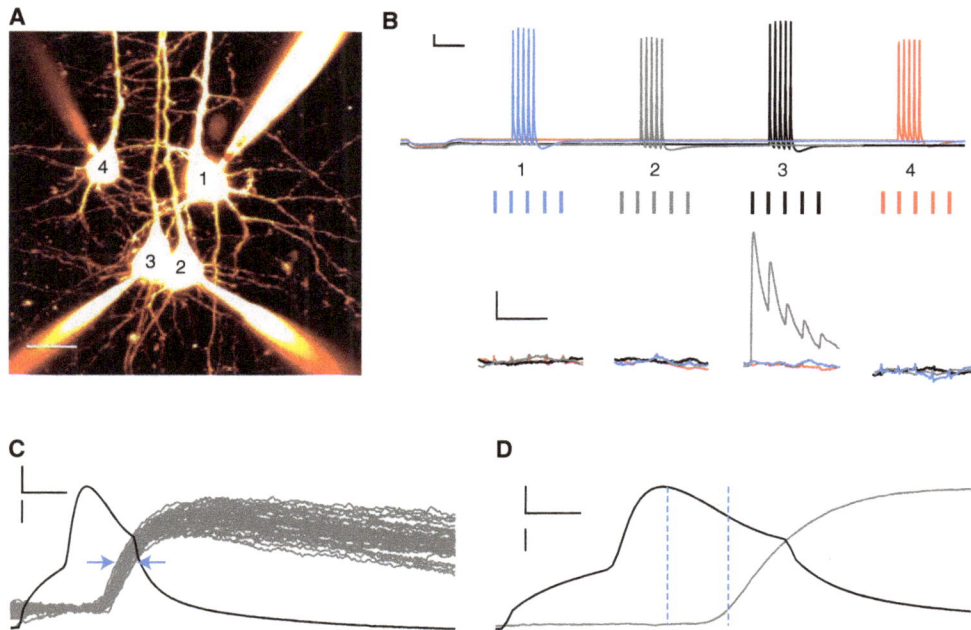

FIGURE 2. Finding connected neurons using quadruple recordings. (*A*) Flattened two-photon-imaging stack of four neighboring L5 PCs filled with Alexa 594 (scale bar: 25 µm). (*B*) Thirty Hertz trains are evoked (*top*, scale bars: 200 msec, 10 mV) to identify responding postsynaptic cells (*bottom*, scale bars: 100 msec, 0.25 mV). Sweeps should be repeated 10–40 times every 10–20 sec, and then averaged. Note short-term depressing connection from cell 3 (black) to cell 2 (gray). (*C*) Monosynaptic connections have a jitter of <1 msec (blue arrows); larger jitter suggests that the responses are polysynaptic. Fifty spike-triggered traces (gray) from cell 2 are represented, whereas the presynaptic action potential (black) is represented by a single sweep (scale bars: 2 msec, 20 mV [black]/300 µV [gray]). (*D*) Monosynaptic connections also have submillisecond latency between presynaptic spike and 10% of EPSP peak (vertical dashed lines). EPSP trace (gray) is a spike-triggered average of 50 sweeps, whereas the presynaptic action potential (black) is represented by a single sweep (scale bars: 1 msec, 20 mV [black]/200 µV [gray]).

to the baseline but maintained for at least 30 min and ideally longer (Fig. 3A,B). At L5 PC connections, repeated pre-before-postsynaptic spike pairings at a timing difference of Δt = +10 msec result in potentiation if the frequency or depolarization is high enough (Fig. 3C), whereas the opposite temporal order may elicit depression (Sjöström et al. 2001).

Analyze Acquired Data

11. Use a dedicated analysis software program, because data analysis is time consuming and highly repetitive.

    Software can be purchased (e.g., pCLAMP, or AxoGraph X), but customization (e.g., in IGOR Pro or MATLAB) maximizes flexibility and speed. Microsoft Excel is also quite adequate for basic analysis.

12. Discard recordings with unstable baseline.

    To avoid bias in the data selection, it is important to apply the same stability criterion to all recordings. One suitable requirement is that the baseline should not change more than, for example, 10% (Markram et al. 1997). An alternative is to apply a t-test to the differences of the means of the two baseline period halves. A stable baseline is indicated by a nonsignificant P-value. Similarly, a two-tailed t-test of Pearson's r for response amplitude versus time during the baseline period should not be significant, as this would violate the null hypothesis that baseline responses have no upward or downward trend.

13. Apply quality-control criteria.

    Recordings should consistently be discarded or truncated if the resting membrane potential, perfusion temperature, or input resistance venture outside bounds. The specifics of these bounds are somewhat arbitrary but can be input resistance should not change >30%, resting membrane potential should not be >8 mV, and temperature should remain within 31°C–34°C (Fig. 3D; Sjöström et al. 2001). In voltage clamp, also monitor series resistance: It should, for example, be <25 MΩ, not change >20%, and be indistinguishable from that in the control experiments (Sjöström et al. 2003).

14. Quantify the magnitude of plasticity. Measure plasticity as the change in EPSP (excitatory postsynaptic potential) amplitude after the induction protocol compared with the amplitude

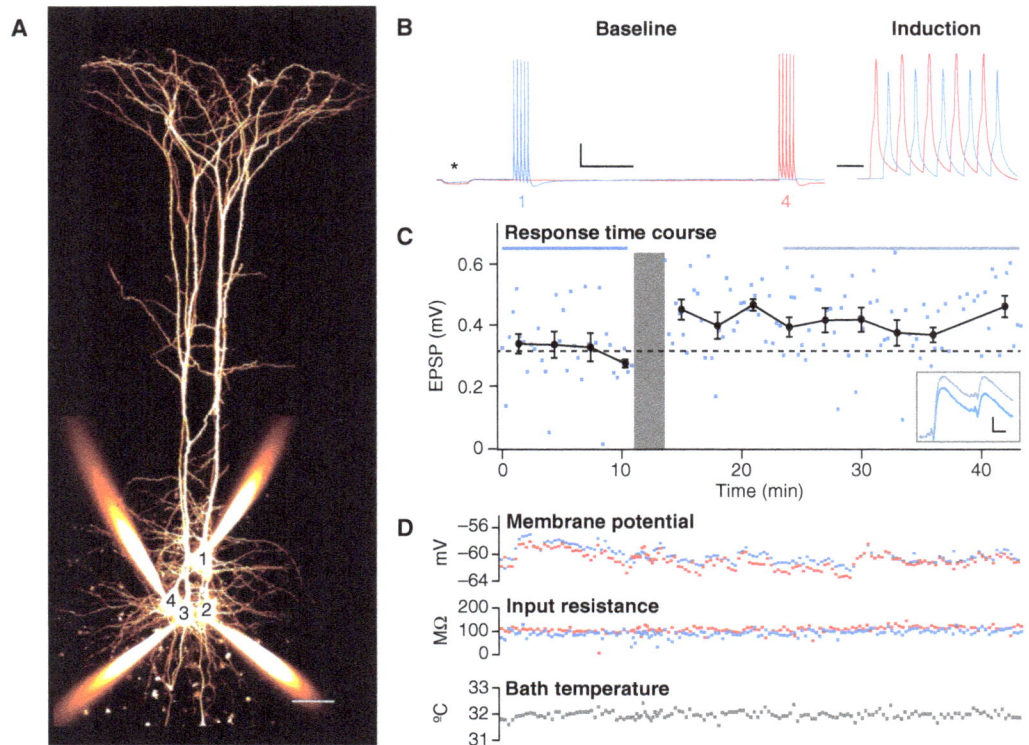

FIGURE 3. Using paired recordings to study plasticity. (*A*) Quadruple whole-cell recording in which PC 4 was connected to PC 1 (scale bar: 50 μm). (*B*) During the 10-min baseline period, 30-Hz trains were repeated every 18 sec in the connected pair of neurons. These bursts were separated by long intervals to avoid induction of long-term plasticity (Sjöström et al. 2001). During the induction, five action potentials at 50 Hz were repeated 15 times every 15 sec. The timing difference was +10 msec. After the induction, the baseline pattern was resumed (scale bars: 500 msec [baseline]/20 msec [induction], 20 mV). Asterisk denotes a 250-msec test pulse of −25 pA, used to monitor input resistance. (*C*) Time course of the first EPSP in the 30-Hz train shows long-term potentiation (LTP). The induction is illustrated by the gray area. Horizontal blue lines (*top*) represent time periods over which averages (*inset*) were taken (scale bars: 10 msec, 0.1 mV). Using such averaged traces, measuring changes in short-term plasticity (Costa et al. 2013) can be used to assess the locus of plasticity expression (Sjöström et al. 2003). (*D*) As a measure of recording quality, resting membrane potential, input resistance, and bath temperature were monitored throughout experiment. Blue and red indicate cell 1 and cell 4, respectively.

before induction, expressed in percentage terms. Ignore the first several minutes after the induction, as other forms of plasticity may be active during this period (e.g. post-tetanic potentiation [Zucker and Regehr 2002]).

> *We typically compare the responses starting 10 min after the induction until the end of the recording to the entire prepairing baseline (Fig. 3C; Sjöström et al. 2001, 2003).*

15. Morphologically classify the recorded cells.

    i. Image the entire volume in which recorded neurons arborize (Fig. 4A). Use software such as Neuromantic, Neurolucida, or Imaris to reconstruct neurons from two-photon laser scanning microscope (2PLSM) image stacks (Blackman et al. 2014), carefully distinguishing dendrites from axons by the presence of dendritic spines characteristic of neocortical pyramidal cells (Fig. 4B).

> *L-Measure provides useful numerical morphometry measurements (Scorcioni et al. 2008).*

    ii. To obtain ensemble averages, create arbor density maps (Buchanan et al. 2012) (Fig. 4C) or carry out Sholl analysis (Fig. 4D; Sholl and Uttley 1953).

> *To save time, Scholl analysis can be directly performed on bitmap images using Fiji (Ferreira et al. 2014).*

16. Repeat experiments. To produce a complete STDP curve or to examine the rate dependence of plasticity, repeat experiments in different paired recordings in the same cell type while varying the

Cite this protocol as *Cold Spring Harb Protoc*; doi:10.1101/pdb.prot091306

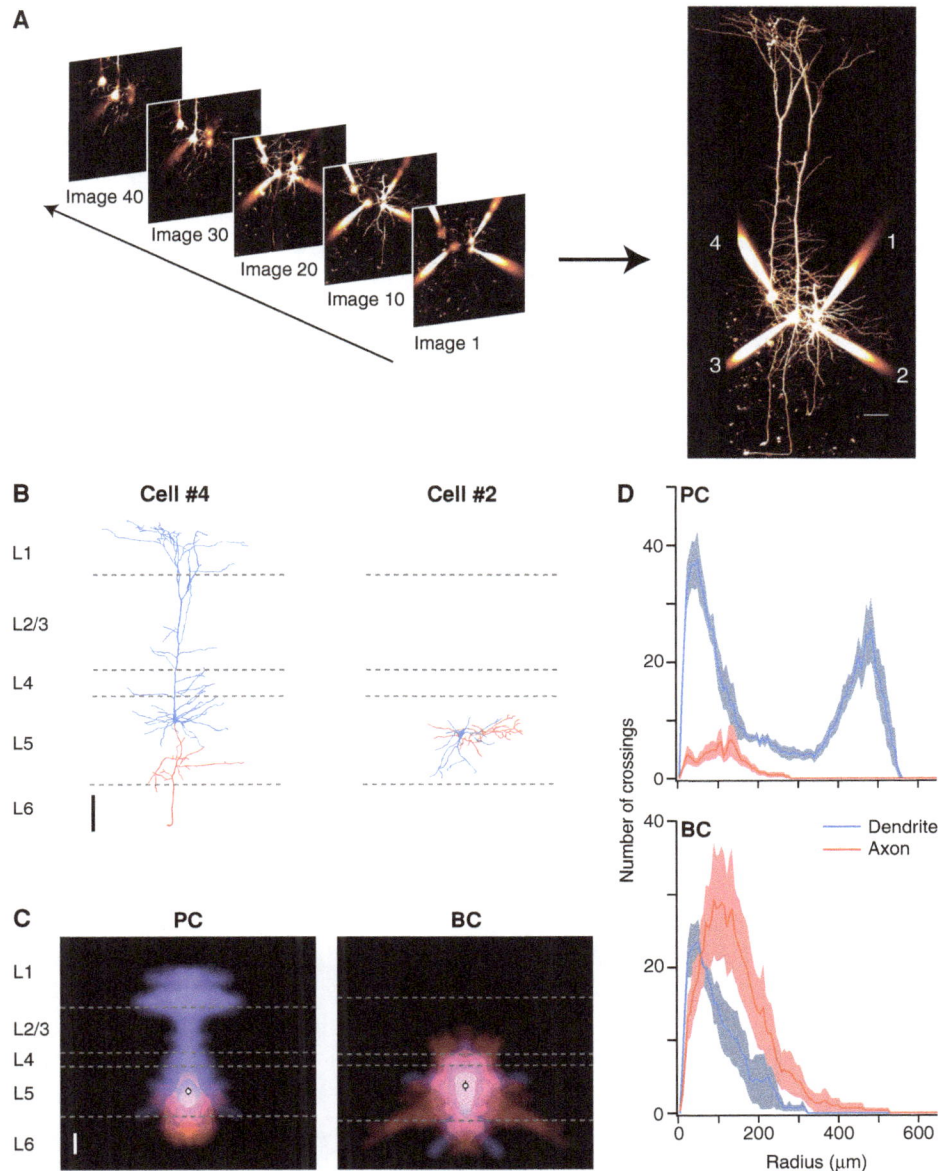

FIGURE 4. Morphological cell classification from two-photon images. (*A*) A stack of two-photon slices provides a 3D representation of recorded neurons filled with Alexa 594. Each slice is an average of two to four 512 × 512-pixel frames. Maximum-intensity projections are assembled for full morphological view (scale bar: 50 μm). (*B*) Digital reconstruction of cell 4 (*left*) shows that this is a neocortical L5 pyramidal cell (PC), whereas the morphology of cell 2 (*right*) is characteristic of a basket cell (BC; scale bar: 100 μm). Reconstructions were performed using Neuromantic (Blackman et al. 2014). Dendrites (blue) and axons (red) were distinguished by the presence of dendritic spines or axonal boutons. Dashed gray lines represent layer boundaries, as determined from simultaneously acquired laser-scanning Dodt contrast images. (*C*) Morphology density map (Buchanan et al. 2012) of six PCs (*left*) highlights the characteristic apical dendrite, with an axonal arborization that remains localized to L5. The corresponding map of six BCs (*right*) shows axonal and dendritic arbors that both remain confined to L5. Average soma location indicated by open circle (scale bar: 100 μm). (*D*) Sholl analysis (Sholl and Uttley 1953) of six PCs (*top*) and six BCs (*bottom*) provides quantitative cell classification criteria from axonal and dendritic branching patterns.

timing (e.g., −120 msec to +40 msec) or rate (e.g., 0.1 Hz to 60 Hz) during the induction (Sjöström et al. 2001).

> It is generally not appropriate to repeat different inductions in sequence in the same connected pair, because depression of previously potentiated synapses is not necessarily the same as depressing a naïve connection (Massey and Bashir 2007).

TROUBLESHOOTING

Problem: Seals and recordings are of poor quality.
Solutions: Consider the following.

- Pipette resistance should be in the right range. Seals form more easily with high pipette resistance (4–6 MΩ), although as a consequence, it may be harder to rupture patches, and the resulting series resistance will be higher. If the patch does not rupture when applying negative pressure, try hyperpolarizing the cell to −140 mV until membrane rupture. With this approach, be ready to quickly switch from voltage to current clamp as soon as whole-cell mode is established, to avoid damaging the cell with the large negative current that results from clamping it to −140 mV. With a GΩ seal that resists rupture even at this point, you can try more aggressive tricks such as applying large negative pressure with a 20-mL syringe or using the amplifier's "buzz" function. These tricks, however, are rarely successful, but represent a last resort. With lower pipette resistance (3–4 MΩ), the pipette tip will be larger, so the series resistance will be lower, but sealing may be more difficult. Once a seal has been established, breaking through to go whole-cell is easier, however. The use of large pipette tips with low pipette resistance is thus recommended for voltage clamp experiments. For quadruple whole-cell recordings in current clamp, however, it is typically more important to establish good seals with high success rate on all four cells, because failure to record from one out of the four cells reduces the number of tested connections from 12 to six—a 50% reduction in yield of connections.

- Pipette tips may be dirty. Remove visible debris with large positive pressure using a 50-mL syringe. If this fails, change the pipette—patching with dirty tips is rarely successful. To reduce clogging of the pipette from the inside, always filter the internal solution with a nylon syringe filter (0.2 μm; e.g., Nalgene #176). Sonicating the filling solution before use may also help.

- Pressure may be too low. Leaky pressure lines should be replaced. Apply positive pressure before the pipette enters the ACSF. In the absence of positive pressure, debris tends to stick to the pipette tip, making patching much more difficult.

- External/internal solutions may not be optimal. The difference in internal and external solution osmolality determines the ease of patching. For rat slices, internal osmolality should be ~294 mOsm (adjusted with sucrose), whereas the ACSF should be ~320 mOsm (adjusted with D-glucose). For mouse slices, osmolalities should be 17 mOsm higher (Bourque 2008).

- Animal age may not be optimal. For rat slices, P13–P16 is the ideal age, whereas for mice it is 1–2 days younger. Although cell health is often inferior in slices from older animals, cells might not have acquired mature properties in slices from young animals. With older animals, cardiac perfusion or high-sucrose dissection solution may improve slice quality (Moyer and Brown 1998).

- Slice quality may have deteriorated. Replace the slice with a new one or dissect a new animal. Slices generally decline faster at 31°C–34°C than at room temperature. In P14 rat slices may be usable up to 10 h after dissection, whereas P18 slices may last no more than 5 h. Slices from mouse brain deteriorate faster than rat slices do.

Problem: Connected pairs of neurons cannot be found.
Solutions: Consider the following.

- The slice may not have been cut at an optimal angle. A slice with L5 PC apical dendrites reaching layer 1 is likely to have well-preserved connectivity.

- Patch cells deeper into the slice. Although visibility and rate of successful patching drop below 80 μm, there are returns in terms of higher connectivity rates (Ko et al. 2011). A reasonable trade-off is to patch 50–80 μm deep.

- Use a train of presynaptic spikes and average more sweeps to better visualize weak postsynaptic responses and synapses with low release probability.

Cite this protocol as *Cold Spring Harb Protoc*; doi:10.1101/pdb.prot091306

Problem: There are electrical problems with the recordings.
Solutions: Consider the following.

- Recordings may be noisy. Make sure headstages and bath chamber are grounded. Do not ground the same device several times, as this creates ground loops and possibly results in more noise. Ground all devices to the same grounding point to minimize grounding loops. Noisy devices should be disconnected, substituted, or moved farther away.

- Voltages may slowly drift because of slow polarization of electrodes. To slow down voltage drift, increase electrode surface area by covering electrodes in chloride. Leave them in bleach (sodium hypochlorite) for 10 min. Alternatively, pass a positive current through the electrode dipped in 1 M NaCl using, for example, a 4.5-V battery until a white layer appears.

- Series resistance may be too high. The GΩ seal may not have been completely ruptured. Apply a light negative pressure. If the problem persists, use pipettes with larger tip diameter.

Problem: The recordings are not stable.
Solutions: Consider the following.

- Prepare new intra- and/or extracellular solutions.

- The slice may not have been cut in the right orientation, and therefore recorded cells are damaged. Pick a new slice from the incubation chamber. Flipping the slice over in the chamber and recording from the other side may also help.

- Baseline stimulation frequency may be too high, hence the synapse may not have enough time to recover from short-term depression. For single EPSPs, use an interstimulus interval of 7–10 sec. A train of five EPSPs at 30–50 Hz requires 15–18 sec for full recovery (Varela et al. 1997; Buchanan et al. 2012).

- Verify that the perfusion temperature is stable and set correctly. Temperature variations cause fluctuations in EPSP amplitude and passive properties. Deterioration can be very rapid at 37°C.

Problem: It is difficult to obtain long-term potentiation (LTP) or long-term depression (LTD).
Solutions: Consider the following.

- Baseline longer than 15 min may result in plasticity washout (Malinow and Tsien 1990).

- Verify that both pre- and postsynaptic cells spike during the induction.

- Average across five to eight paired recordings; do not make too much of the outcome of individual connections.

DISCUSSION

Compared with field recordings, the quadruple patch-clamping technique is expensive to set up and more challenging to master. It does, however, have several advantages such as millisecond temporal precision, specific pharmacological access to pre- and postsynaptic cells, and morphological identification of connected cells. For additional background on both methods, see Introduction: In Vitro Investigation of Synaptic Plasticity (Abrahamsson et al. 2016a).

With multiple whole-cell recordings, the statistics of local connections can also be studied. For example, neocortical pyramidal cells are reciprocally connected to a greater degree than expected from a uniformly random distribution (Song et al. 2005). This may result from Hebbian plasticity, so that cells coding for similar information are wired together more strongly (Ko et al. 2011). The juvenile cerebellar Purkinje cell network, on the other hand, is solely built from chains of unidirectionally connected neurons (Watt et al. 2009). This forms a substrate for traveling waves of activity, which may help wire the circuit up in early development (van Welie et al. 2011). Quadruple recordings provide an excellent means of assessing such network-specific differences in patterns of connectivity.

Multiple whole-cell recordings do represent a financial hurdle, however. A less expensive alternative is to search for presynaptic cells using loose patch while whole-cell recording the postsynaptic cell (Feldmeyer et al. 1999; Barbour and Isope 2000). Unfortunately, this approach has the disadvantage that it may subject the postsynaptic cell to plasticity washout (Malinow and Tsien 1990).

In the end, multiple whole-cell recording is technically similar to patching single cells. With this protocol, an experimentalist capable of reliably patching individual cells should not find quadruple whole-cell recordings too challenging.

RECIPES

Artificial Cerebrospinal Fluid (ACSF) for Synaptic Plasticity Studies

Prepare the following 10× stock solution in double-distilled water (ddH$_2$O). Store it at 4°C for a maximum of 1 wk.

NaCl	1250 mM
KCl	25 mM
NaH$_2$PO$_4$	12.5 mM
NaHCO$_3$	260 mM

On the day of the experiment, dilute the 10× solution 10-fold, bubble for 10 min with carbogen (95% O$_2$/5% CO$_2$), and supplement with the following.

MgCl$_2$	1 mM
CaCl$_2$	2 mM
Glucose	~26 mM

ACSF—in particular Ca^{2+} and Mg^{2+} concentrations—may be varied depending on what is to be studied. For example, higher Ca^{2+} concentration favors long-term potentiation (LTP), whereas lower Ca^{2+} concentration promotes long-term depression (LTD). Adjust osmolality to ~320 mOsm with D-glucose for rat and ~338 mOsm for mouse.

Internal Solution for Whole-Cell Recording

Prepare a 1 M solution of HEPES in ddH$_2$O and adjust the pH to 7 with 3 M KOH. Then prepare 30 mL of the following solution in ddH$_2$O (the final concentrations are for a 50-mL volume). Adjust the volume to 45 mL and verify that pH $\sim$ 7.

KCl	125 mM
K Gluconate	2.5 mM
HEPES (1 M, pH 7)	10 mM

To the previous solution, add the following (the final concentrations are for a 50-mL volume). Readjust the pH to 7.2–7.4 with KOH and correct the osmolality to 294 mOsm (rat) or 311 mOsm (mouse) with sucrose.

MgATP	4 mM
NaGTP	0.3 mM
Na-Phosphocreatine	10 mM

Filter the solution using a nylon filter (pore size 0.2 μm; e.g., Nalgene #176) and make 1-mL aliquots. Store at −20°C for up to 6 mo. Verify the osmolality when thawed.

Cite this protocol as *Cold Spring Harb Protoc*; doi:10.1101/pdb.prot091306

ACKNOWLEDGMENTS

We thank Alanna Watt, Ian Duguid, Karri Lamsa, Elvis Cela, Jérôme Maheux, and Andrew Chung for help and useful discussions. This work was funded by Canada Foundation for Innovation (CFI) LOF 28331 (P.J.S.), Canadian Institutes of Health Research (CIHR) OG 126137 (P.J.S.), CIHR NIA 288936 (P.J.S.), Natural Sciences and Engineering Research Council of Canada (NSERC) DG 418546-2 (P.J.S.), and by an Research Institute of the McGill University Health Centre (RI MUHC) studentship award (T.L.).

REFERENCES

Abrahamsson T, Lalanne T, Watt AJ, Sjöström PJ. 2016a. In vitro investigation of synaptic plasticity. *Cold Spring Harb Protoc* doi: 10.1101/pdb.top087262.

Abrahamsson T, Lalanne T, Watt AJ, Sjöström PJ. 2016b. Long-term potentiation by theta-burst stimulation using extracellular field potential recordings in acute hippocampal slices. *Cold Spring Harb Protoc* doi: 10.1101/pdb.prot091298.

Barbour B, Isope P. 2000. Combining loose cell-attached stimulation and recording. *J Neurosci Methods* 103: 199–208.

Blackman AV, Grabuschnig S, Legenstein R, Sjöström PJ. 2014. A comparison of manual neuronal reconstruction from biocytin histology or 2-photon imaging: Morphometry and computer modeling. *Front Neuroanat* 8: 65.

Bourque CW. 2008. Central mechanisms of osmosensation and systemic osmoregulation. *Nat Rev Neurosci* 9: 519–531.

Buchanan KA, Blackman AV, Moreau AW, Elgar D, Costa RP, Lalanne T, Tudor Jones AA, Oyrer J, Sjöström PJ. 2012. Target-specific expression of presynaptic NMDA receptors in neocortical microcircuits. *Neuron* 75: 451–466.

Costa RP, Sjöström PJ, van Rossum MCW. 2013. Probabilistic inference of short-term synaptic plasticity in neocortical microcircuits. *Front Comput Neurosci* 7. doi: 10.3389/fncom.2013.00075.

Davie JT, Kole MH, Letzkus JJ, Rancz EA, Spruston N, Stuart GJ, Häusser M. 2006. Dendritic patch-clamp recording. *Nat Protoc* 1: 1235–1247.

Feldmeyer D, Egger V, Lübke J, Sakmann B. 1999. Reliable synaptic connections between pairs of excitatory layer 4 neurones within a single 'barrel' of developing rat somatosensory cortex. *J Physiol* 521 Pt 1: 169–190.

Ferreira TA, Blackman AV, Oyrer J, Jayabal S, Chung AJ, Watt AJ, Sjöström PJ, van Meyel DJ. 2014. Neuronal morphometry directly from bitmap images. *Nat methods* 11: 982–984.

Holmgren C, Harkany T, Svennenfors B, Zilberter Y. 2003. Pyramidal cell communication within local networks in layer 2/3 of rat neocortex. *J Physiol* 551: 139–153.

Ko H, Hofer SB, Pichler B, Buchanan KA, Sjöström PJ, Mrsic-Flogel TD. 2011. Functional specificity of local synaptic connections in neocortical networks. *Nature* 473: 87–91.

Malinow R, Tsien RW. 1990. Presynaptic enhancement shown by whole-cell recordings of long-term potentiation in hippocampal slices. *Nature* 346: 177–180.

Markram H, Lübke J, Frotscher M, Sakmann B. 1997. Regulation of synaptic efficacy by coincidence of postsynaptic APs and EPSPs. *Science* 275: 213–215.

Massey PV, Bashir ZI. 2007. Long-term depression: Multiple forms and implications for brain function. *Trends Neurosci* 30: 176–184.

Moyer JR Jr, Brown TH. 1998. Methods for whole-cell recording from visually preselected neurons of perirhinal cortex in brain slices from young and aging rats. *J Neurosci Methods* 86: 35–54.

Perin R, Berger TK, Markram H. 2011. A synaptic organizing principle for cortical neuronal groups. *Proc Natl Acad Sci* 108: 5419–5424.

Scorcioni R, Polavaram S, Ascoli GA. 2008. L-Measure: A web-accessible tool for the analysis, comparison and search of digital reconstructions of neuronal morphologies. *Nat Protoc* 3: 866–876.

Sholl A, Uttley AM. 1953. Pattern discrimination and the visual cortex. *Nature* 171: 387–388.

Sjöström PJ, Turrigiano GG, Nelson SB. 2001. Rate, timing, and cooperativity jointly determine cortical synaptic plasticity. *Neuron* 32: 1149–1164.

Sjöström PJ, Turrigiano GG, Nelson SB. 2003. Neocortical LTD via coincident activation of presynaptic NMDA and cannabinoid receptors. *Neuron* 39: 641–654.

Song S, Sjöström PJ, Reigl M, Nelson S, Chklovskii DB. 2005. Highly nonrandom features of synaptic connectivity in local cortical circuits. *PLoS Biol* 3: e68.

van Welie I, Smith IT, Watt AJ. 2011. The metamorphosis of the developing cerebellar microcircuit. *Curr Opin Neurobiol* 21: 245–253.

Varela JA, Sen K, Gibson J, Fost J, Abbott LF, Nelson SB. 1997. A quantitative description of short-term plasticity at excitatory synapses in layer 2/3 of rat primary visual cortex. *J Neurosci* 17: 7926–7940.

Watt AJ, Cuntz H, Mori M, Nusser Z, Sjöström PJ, Häusser M. 2009. Traveling waves in developing cerebellar cortex mediated by asymmetrical Purkinje cell connectivity. *Nat Neurosci* 12: 463–473.

Zucker RS, Regehr WG. 2002. Short-term synaptic plasticity. *Annu Rev Physiol* 64: 355–405.

Stimulating Neurons with Heterologously Expressed Light-Gated Ion Channels

J. Simon Wiegert, Christine E. Gee, and Thomas G. Oertner[1]

Institute for Synaptic Physiology, Center for Molecular Neurobiology (ZMNH), 20251 Hamburg, Germany

Heterologous expression of ion channels that can be directly gated by light has made it possible to stimulate almost any excitable cell with light. Optogenetic stimulation has been particularly powerful in the neurosciences, as it allows the activation of specific, genetically defined neurons with precise timing. Organotypic hippocampal slice cultures are a favored preparation for optogenetic experiments. They can be cultured for many weeks and, after transfection with optogenetic actuators and sensors, allow the study of individual synapses or small networks. The absence of any electrodes allows multiple imaging sessions over the course of several days and even chronic stimulation inside the incubator. These timescales are not accessible in electrophysiological experiments. Here, we introduce the production of organotypic hippocampal slice cultures and their transduction or transfection with optogenetic tools. We then discuss the options for light stimulation.

OPTOGENETIC TOOLS: DESIGNER CHANNELRHODOPSINS

Channelrhodopsin-2 (ChR2) from the unicellular alga *Chlamydomonas reinhardtii* and its homologs are proteins with seven transmembrane domains. The hydrophobic core regions of ChRs create a binding pocket for retinal and show homology with the light-activated proton pump, bacteriorhodopsin. Nagel et al. discovered in 2002 that expression of this core region from ChR1 (amino acids 76–309 out of 712) is sufficient to form a directly light-gated ion channel in eukaryotic cells (Nagel et al. 2002). One year later, they demonstrated that amino acids 1–315 of ChR2 formed a cation channel that could depolarize cells (Nagel et al. 2003). As a first engineering step, they replaced the carboxy-terminal tail of ChR2 with yellow fluorescent protein (YFP) to identify transfected neurons by their fluorescence (Boyden et al. 2005). Based on the sequence homology with bacteriorhodopsin, they then introduced a point mutation (H134R) that increased stationary photocurrents. This version was used for the first in vivo optogenetic experiments (Nagel et al. 2005).

The structural homology of ChRs to bacteriorhodopsin inspired several laboratories to introduce point mutations at other critical residues (most of which are close to the retinal-binding pocket) to generate novel optogenetic tools with altered properties. It was soon discovered that mutations that accelerate the closing kinetics of channels generally result in smaller photocurrents, whereas "slow" mutants are much more light-sensitive than wild-type ChR2 because, even at low photon flux, a large fraction of channels accumulate in the open state (Berndt et al. 2011; Mattis et al. 2012). An important achievement was the generation of a red-shifted ChR by combining elements from ChR2 and ChR1 to form chimeric ChRs (Prigge et al. 2012; Lin et al. 2013). An even more red-shifted variant with fast kinetics and large currents (ChrimsonR) was discovered in a recent genetic screen (Klapoetke et al.

[1]Correspondence: thomas.oertner@zmnh.uni-hamburg.de

2014). It should be noted that even the most blue- and red-shifted ChRs known today have considerable spectral overlap, meaning that blue light will activate all known ChRs. However, by using wavelengths at the extreme ends of the action spectra (UV/red), two neuronal populations can be activated independently. We have successfully achieved independent activation of neurons expressing the highly potent CheRiff as a UV-sensitive ChR and ChrimsonR as a red-light-sensitive ChR (Hochbaum et al. 2014; Klapoetke et al. 2014).

ChR2 is a nonspecific cation channel with a reversal potential in neurons of ~0 mV; therefore, brief ChR2 activation resembles excitatory synaptic input. Mutating the central gate region of ChR2 has resulted in light-gated Cl^- channels that clamp neurons close to their resting membrane potential and may, therefore, be useful for inhibiting neurons (Berndt et al. 2014; Wietek et al. 2015). Natural anion-conducting ChRs with similar favorable properties have recently been found in cryptophyte algae (Govorunova et al. 2015).

HIPPOCAMPAL SLICE CULTURE PREPARATION

Mammalian neurons can be cultured directly on glass coverslips where they form two-dimensional networks that are spontaneously active. Activity in these dissociated cultures is characterized by highly synchronized bursting patterns that do not resemble natural activity in the brain. As a more physiological alternative, entire slices of brain tissue can be cultured for months under interface conditions on sterile tissue culture inserts. Hippocampal slice cultures establish a pattern of connectivity that resembles in many ways the situation in vivo, including maturation of synaptic properties (De Simoni et al. 2003; Rose et al. 2013). This preparation also combines very good optical and electrophysiological accessibility and can be maintained for several weeks—long enough to address fundamental questions about synaptic plasticity and long-term dynamics of network connectivity (De Roo et al. 2008; Wiegert and Oertner 2013). Furthermore, organotypic slice cultures have proven to be an ideal preparation for testing and characterizing novel optogenetic tools (Schoenenberger et al. 2009; Berndt et al. 2011; Wietek et al. 2014).

Sterility during the preparation, maintenance and experimental interrogation of slice cultures is essential, especially for long-term experiments. The initial description of hippocampal slice cultures (Stoppini et al. 1991) and a more recent protocol (Gogolla et al. 2006) recommend the use of antibiotics or antimycotics to prevent microbial infection. Antibiotics, however, are known to affect many properties of neurons (Amonn et al. 1978; Llobet et al. 2015) and may introduce artifacts. If sterile condtions are maintained during production, maintenance ("feeding"), and handling (e.g., transfection and imaging) of the cultures, antimicrobial drugs can and should be avoided (see Protocol 1: Preparation of Slice Cultures from Rodent Hippocampus [Gee et al. 2016]; Protocol 2: Viral Vector–Based Transduction of Slice Cultures [Wiegert et al. 2016a]).

INTRODUCTION OF TRANSGENES INTO NEURONS OF SLICE CULTURES

Several different delivery methods are available for the introduction of transgenes into cells of organotypic slice cultures, such as biolistic transfection, electroporation, and viral vector–based transduction. These methods give different patterns of transgene expression and, therefore, a suitable delivery method has to be chosen depending on the goals of an experiment. Biolisitic transfection via gene gun is a fast method to achieve a random pattern of transfected neurons (Fig. 1A; Thomas et al. 1998). Drawbacks include poor control of transfection density, resulting in many untransfected cultures, as well as nonspecific tissue damage from the pressure wave and gold particle impact. Transfected cultures with the ideal "Golgi"-like transfection pattern are rare. The impact of gold particles sometimes induces a plasma bridge between two cells, resulting in neurons with two nuclei and pyramidal neurons with two apical dendrites.

Cite this introduction as *Cold Spring Harb Protoc*; doi:10.1101/pdb.top089714

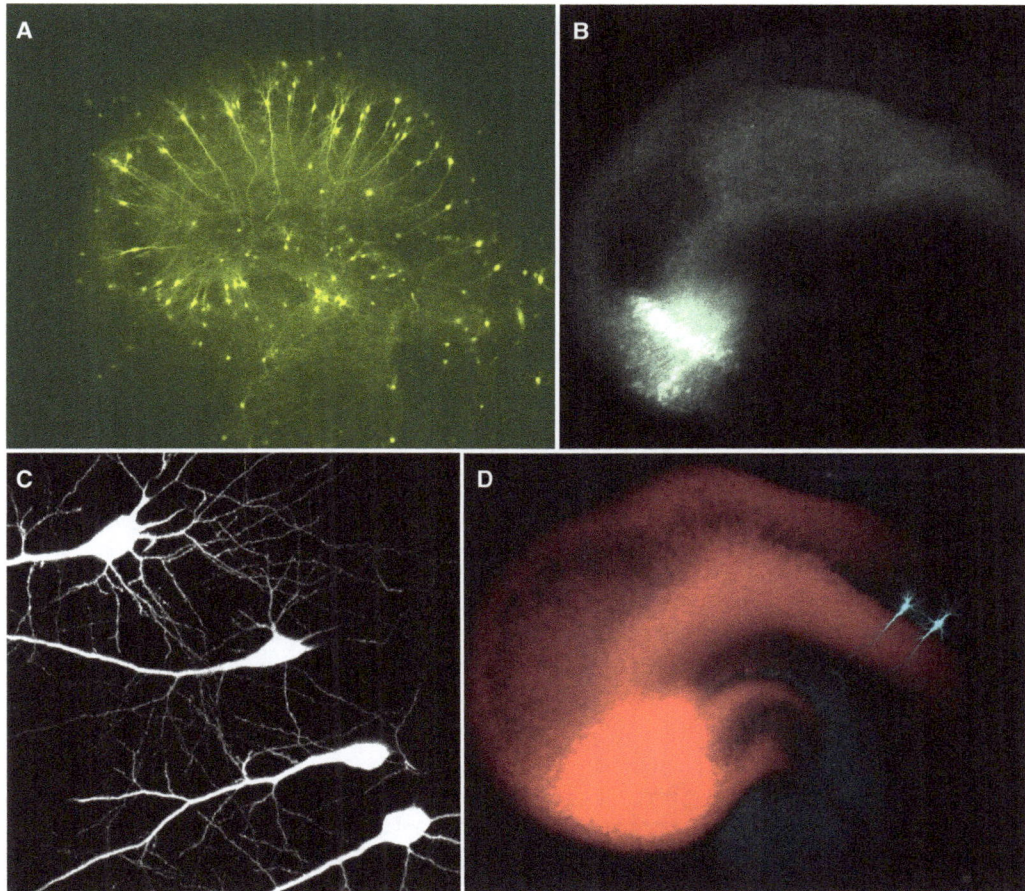

FIGURE 1. Outcome of different gene delivery methods. (*A*) Result of biolistic transfection (BioRad gene gun) of a fluorescent protein driven by the neuron-specific promoter synapsin-1. A sparse, random set of pyramidal cells expresses the fluorescent construct. (*B*) Result of local pressure injection of recombinant adeno-associated virus (rAAV) into area CA3. A dense cluster of neurons expresses the construct at high levels. The axonal projection of these neurons is seen as a faint glow. (*C*) Single-cell electroporation allows the transfection of individual neurons under differential interference contrast (DIC) or Dodt contrast. Electroporation is labor intensive but highly reproducible. (*D*) Combining rAAV injection into CA3 (red) and single-cell electroporation of CA1 pyramidal cells (cyan) allows the expression of different constructs in pre- and postsynaptic cells. (*A*, Adapted from Holbro et al. 2009; *C,D*, adapted from Wiegert and Oertner 2013.)

To produce a small cluster of transgene-expressing neurons, local injection of recombinant adeno-associated virus (rAAV) can be performed (Fig. 1B; Burger et al. 2004; Protocol 2: Viral Vector–Based Transduction of Slice Cultures [Wiegert et al. 2016a]). Transgene expression in individual cells varies with the distance from the injection site, indicating that varying numbers of rAAV particles are taken up by neurons. The serotype of the virus and the transgene promoter also determine levels of expression. After local injection into hippocampal slice cultures, AAV2/9 or AAV2/10 vectors and the synapsin-1 promoter rapidly produce high levels of protein expression in neurons.

If locally defined expression is the goal, single-cell electroporation produces exactly the desired pattern of transfection but is labor intensive (Fig. 1C; Rathenberg et al. 2003; Protocol 3: Single-Cell Electroporation of Neurons [Wiegert et al. 2016b]). The level of expression can even be controlled by the concentration of DNA in the electroporation pipette, and reliable cotransfection of several plasmids (e.g., ChR2, CFP, and synaptophysin-RFP) at defined ratios is possible. Single-cell electroporation can be combined with viral transduction to express different optogenetic constructs in pre- and postsynaptic cells (Fig. 1D). We routinely combine local rAAV transduction of CA3 neurons with single-cell electroporation of CA1 pyramidal neurons to investigate synaptic plasticity at individual

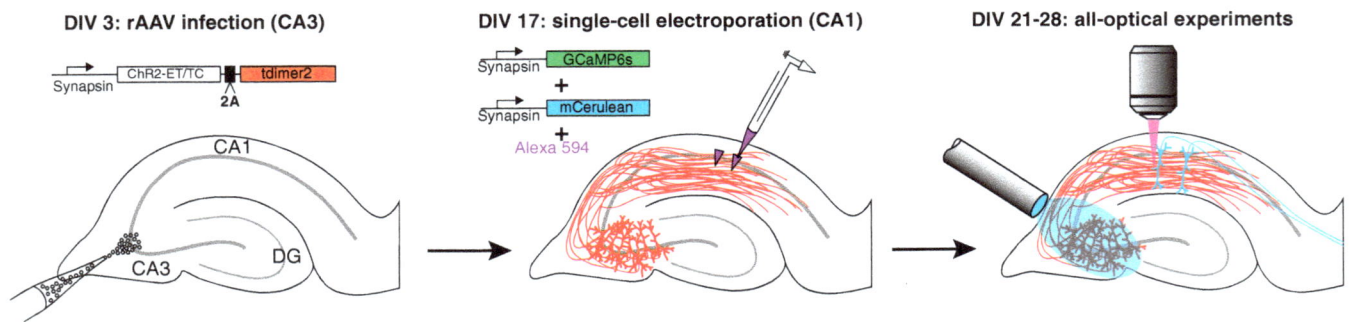

DIV 3: rAAV infection (CA3)

DIV 17: single-cell electroporation (CA1)

DIV 21-28: all-optical experiments

FIGURE 2. Sample timeline of an optogenetic experiment. Example time course for combining rAAV transduction and single-cell electroporation for optogenetic stimulation and readout (adapted from Wiegert and Oertner 2013). At DIV3, CA3 cells are transduced with rAAV2/7 encoding ChR2(ET/TC)-2A-tdimer2. Two weeks after rAAV infection (DIV 17), ChR2 (ET/TC) and tdimer2 have reached stable expression levels. CA1 pyramidal neurons are then transfected with GCaMP6s and mCerulean via single-cell electroporation. Alexa Fluor 594 is included in the pipette to immediately visualize successfully electroporated cells. At DIV 21, expression levels are suitable for optical synaptic plasticity experiments.

Schaffer collateral synapses over many days (Fig. 2; Wiegert and Oertner 2013). Presynaptic ChR2 can be coexpressed with a red label for presynaptic morphology, whereas postsynaptic CA1 neurons can coexpress a green-fluorescent calcium indicator (i.e., GCaMP6) together with the blue mCerulean, which serves as a postsynaptic morphology marker.

OPTOGENETIC STIMULATION OF TRANSFECTED SLICE CULTURES

Optogenetic tools that have been introduced into cells of slice culture can be stimulated with light in a very controlled fashion. Depending on the type of experiment, different illumination methods are available. To calibrate the properties of optogenetic tools and to determine the optimal transfection conditions, it is often necessary to perform patch-clamp recordings from the neuron expressing the tool of interest (Berndt et al. 2011; Wietek et al. 2014).

Light Stimulation through the Epifluorescence Pathway

To characterize the properties of a light-gated channel in a single neuron, the simplest method is to use the epifluorescence pathway of the microscope to deliver light to the cells that are under the objective. Several light-emitting diode (LED)-based light engines are commercially available that provide up to 16 different wavelengths. They can be manually controlled through a graphical user interface (GUI), by transistor-transistor logic (TTL) pulses, or via analog outputs of a data acquisition board. Using a high-pressure Hg lamp with a mechanical shutter is not recommended for optogenetic stimulation because the shutter blades heat up tremendously and pulse timing becomes imprecise. Monochromators based on 75-W xenon lamps are also not ideal because the power output is too low to saturate most channelrhodopsins and pumps (or fluorescent indicators).

To optically stimulate a neuron within the field of view (FoV) of a water immersion objective, the epifluorescence condenser of the microscope can be used to deliver light from high-power LED illumination systems (e.g., Lumencor SPECTRA, CoolLED pE-4000, Mightex, Thorlabs) via brief light pulses under TTL control. To avoid vibrating the microscope by cooling fans, a flexible coupling of the light source through a liquid light guide is recommended. Wavelength and intensity of the LED pulses can be chosen to excite commonly used fluorescent proteins and dyes; in this way a single light source can be used for optogenetic stimulation and visualization of labeled neurons. When upgrading an existing microscope from a high-pressure Hg arc lamp to an LED system, it might be necessary to exchange some of the narrow-band excitation filters in the fluorescence filter cubes. For epifluorescence

Cite this introduction as *Cold Spring Harb Protoc*; doi:10.1101/pdb.top089714

stimulation with multiple wavelengths combined with IR (infrared)-DIC or two-photon imaging, a short-pass dichroic beamsplitter (e.g., Semrock FF705-Di01-25x36) should be mounted in an empty filter cube. Because the epifluorescence condenser is designed to produce an even illumination across the entire FoV, the local intensity can be determined by measuring total light power under the objective and dividing this value by the illuminated area of the objective (see below). Intensities of 10–50 mW/mm^2 are necessary to drive some light-gated channels and pumps to saturation. On the other hand, some optogenetic tools are fully saturated at light levels several orders of magnitude lower. As most LEDs cannot be dimmed below 1%–5% of their maximum light output, neutral density filters must be inserted into the light path to extend the dynamic range of the light stimulation system downward.

To calibrate light intensity in the specimen plane, total power (mW) under the objective (e.g., Newport 1918-R with detector 818-ST2-UV/DB) should be measured and divided by the illuminated field of the objective (mm^2). The illuminated field of an objective is typically larger than its field of view. The illuminated field can be measured by focusing the objective on a calibration slide (e.g., Thorlabs R1L3S1P 10-mm Stage Micrometer) and shining light through the objective (i.e., turn on the epifluorescence light source). A mirror inserted at 45° under the condenser and a focusing lens ($f = 100$ mm) is used to project an image of the calibration slide on a screen mounted sideways. The diameter of the circular illuminated field can be directly read out from the projected image and is used to calculate the illuminated area.

For experiments combining optogenetic stimulation with patch-clamp recordings or imaging, an upright microscope with motorized stage should be used. When slices will be repeatedly stimulated and imaged over multiple days, the chamber and objectives should be sterilized by wiping with 70% ethanol. To avoid constant perfusion and bubbling with oxycarbon (carbogen), sterile HEPES-buffered solution rather than bicarbonate-buffered artificial cerebral spinal fluid can be used. Cultures can easily tolerate 30- to 60-min imaging sessions under a water immersion objective (25°C–30°C) without perfusion as long as the pH and osmolality of the imaging solution is maintained (Wiegert and Oertner 2013).

Light Stimulation through Oblique Light Fiber

When synapses and networks are under investigation (e.g., the CA3–CA1 connection in the hippocampus), the cell bodies of presynaptic neurons are often far outside the field of view of a water immersion objective that is being used to visualize the postsynaptic neurons (Wiegert and Oertner 2013). To stimulate neurons far away from the optical axis of the microscope, a blunt light fiber can be positioned using a micromanipulator just above the area to be stimulated (e.g., CA3). A relatively large-diameter fiber (typically 400 µm) is required to provide enough light from high-powered LEDs (see above) to drive action potentials in ChR2-expressing neurons. Because of the shallow angle of illumination and the divergent cone of light, all parts of the slice culture are illuminated at different intensities. Although this style of light delivery is flexible and economical, it is not possible to exactly reproduce the local light intensity at all cells of interest or to completely avoid illuminating the tissue under the objective.

Off-Center Illumination through Condenser

Certain optogenetic experiments require the illumination of CA3 pyramidal neurons at different wavelengths or with timing different from CA1 illumination. For these experiments, high local light intensities have to be generated at positions up to 1.5 mm off the optical axis with minimal stray light to other parts of the culture. To selectively stimulate neurons far away from the optical axis of the microscope, we steer a collimated laser beam through the oil immersion condenser (Fig. 3). This strategy produces a very restricted (Gaussian) pencil of light with minimal stray light to other areas of the culture, such as the area under the objective, but requires a laser as light source. Measures must also be taken to ensure there is no possibility of accidently looking into the laser through the eyepieces. This could be a notch filter inserted below the tube lens or an interlock switch that enables observation

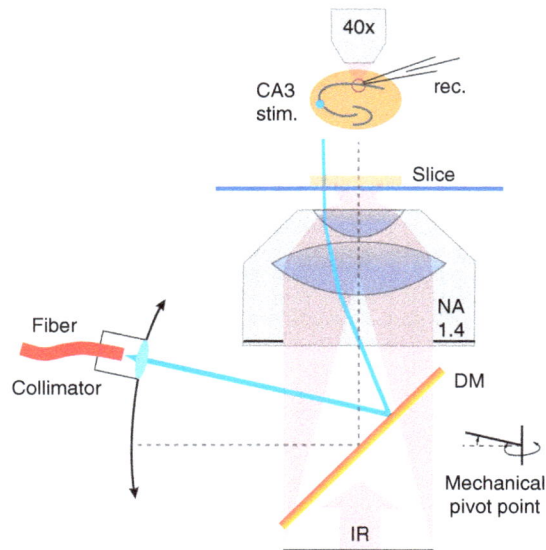

FIGURE 3. Beam steering through condenser for off-center light stimulation. An optical fiber collimator is mounted on a swing arm to point a laser beam at variable angles through the back aperture of a high-NA (numerical aperture) condenser (1.4 NA, oil immersion). Different angles correspond to different positions of the laser spot in the slice. A long-pass dichroic mirror (DM) mounted below the condenser allows simultaneous visualization of the tissue by infrared (IR) DIC or Dodt contrast for patch clamp experiments.

of the laser only with the camera. Because of their high output power, diode-pumped solid-state (DPSS) and diode lasers are well suited for such focal illumination. For two-color stimulation (e.g., 405 and 594 nm), a laser combiner (Omicron LightHub) coupled to a multimode fiber can be used. The far end of the SMA fiber-optic patch cable is connected to a collimator (Thorlabs). The collimator is mounted on a swing arm (lockable, with counter weight) to point the laser beam at different angles through the center of the back aperture of the condenser (Fig. 3). Using this system, high local intensities can be realized away from the optical axis with little scattered light in other hippocampal areas. It is readily combined with patch-camp recordings and optical stimulation of neurons close to the optical axis through the epifluorescence pathway (see above). This simple mechanical system can be refined for multispot illumination, using galvanometric scan mirrors and a scan lens for remote control of the laser spot position.

REFERENCES

Amonn F, Baumann U, Wiesmann UN, Hofmann K, Herschkowitz N. 1978. Effects of antibiotics on the growth and differentiation in dissociated brain cell cultures. *Neuroscience* 3: 465–468.

Berndt A, Schoenenberger P, Mattis J, Tye KM, Deisseroth K, Hegemann P, Oertner TG. 2011. High-efficiency channelrhodopsins for fast neuronal stimulation at low light levels. *Proc Natl Acad Sci* 108: 7595–7600.

Berndt A, Lee SY, Ramakrishnan C, Deisseroth K. 2014. Structure-guided transformation of channelrhodopsin into a light-activated chloride channel. *Science* 344: 420–424.

Boyden ES, Zhang F, Bamberg E, Nagel G, Deisseroth K. 2005. Millisecond-timescale, genetically targeted optical control of neural activity. *Nat Neurosci* 8: 1263–1268.

Burger C, Gorbatyuk OS, Velardo MJ, Peden CS, Williams P, Zolotukhin S, Reier PJ, Mandel RJ, Muzyczka N. 2004. Recombinant AAV viral vectors pseudotyped with viral capsids from serotypes 1, 2, and 5 display differential efficiency and cell tropism after delivery to different regions of the central nervous system. *Mol Ther* 10: 302–317.

De Roo M, Klauser P, Mendez P, Poglia L, Muller D. 2008. Activity-dependent PSD formation and stabilization of newly formed spines in hippocampal slice cultures. *Cereb Cortex* 18: 151–161.

De Simoni A, Griesinger CB, Edwards FA. 2003. Development of rat CA1 neurones in acute versus organotypic slices: Role of experience in synaptic morphology and activity. *J Physiol* 550: 135–147.

Gee CE, Ohmert I, Wiegert JS, Oertner TG. 2016. Preparation of slice cultures from rodent hippocampus. *Cold Spring Harb Protoc* doi: 10.1101/pdb. prot094888.

Gogolla N, Galimberti I, DePaola V, Caroni P. 2006. Preparation of organotypic hippocampal slice cultures for long-term live imaging. *Nat Protoc* 1: 1165–1171.

Govorunova EG, Sineshchekov OA, Janz R, Liu X, Spudich JL. 2015. Natural light-gated anion channels: A family of microbial rhodopsins for advanced optogenetics. *Science* 349: 647–650.

Hochbaum DR, Zhao Y, Farhi SL, Klapoetke N, Werley CA, Kapoor V, Zou P, Kralj JM, Maclaurin D, Smedemark-Margulies N, et al. 2014. All-optical electrophysiology in mammalian neurons using engineered microbial rhodopsins. *Nat Methods* 11: 825–833.

Holbro N, Grunditz A, Oertner TG. 2009. Differential distribution of endoplasmic reticulum controls metabotropic signaling and plasticity at hippocampal synapses. *Proc Natl Acad Sci* 106: 15055–15060.

Klapoetke NC, Murata Y, Kim SS, Pulver SR, Birdsey-Benson A, Cho YK, Morimoto TK, Chuong AS, Carpenter EJ, Tian Z, et al. 2014. Independent optical excitation of distinct neural populations. *Nat Methods* 11: 338–346.

Lin JY, Knutsen PM, Muller A, Kleinfeld D, Tsien RY. 2013. ReaChR: A red-shifted variant of channelrhodopsin enables deep transcranial optogenetic excitation. *Nat Neurosci* 16: 1499–1508.

Llobet L, Montoya J, Lopez-Gallardo E, Ruiz-Pesini E. 2015. Side effects of culture media antibiotics on cell differentiation. *Tissue Eng Part C Methods* 21: 1143–1147.

Mattis J, Tye KM, Ferenczi EA, Ramakrishnan C, O'Shea DJ, Prakash R, Gunaydin LA, Hyun M, Fenno LE, Gradinaru V, et al. 2012. Principles

Cite this introduction as *Cold Spring Harb Protoc*; doi:10.1101/pdb.top089714

for applying optogenetic tools derived from direct comparative analysis of microbial opsins. *Nat Methods* **9**: 159–172.

Nagel G, Ollig D, Fuhrmann M, Kateriya S, Musti AM, Bamberg E, Hegemann P. 2002. Channelrhodopsin-1: A light-gated proton channel in green algae. *Science* **296**: 2395–2398.

Nagel G, Szellas T, Huhn W, Kateriya S, Adeishvili N, Berthold P, Ollig D, Hegemann P, Bamberg E. 2003. Channelrhodopsin-2, a directly light-gated cation-selective membrane channel. *Proc Natl Acad Sci* **100**: 13940–13945.

Nagel G, Brauner M, Liewald JF, Adeishvili N, Bamberg E, Gottschalk A. 2005. Light activation of channelrhodopsin-2 in excitable cells of *Caenorhabditis elegans* triggers rapid behavioral responses. *Curr Biol* **15**: 2279–2284.

Prigge M, Schneider F, Tsunoda SP, Shilyansky C, Wietek J, Deisseroth K, Hegemann P. 2012. Color-tuned channelrhodopsins for multiwavelength optogenetics. *J Biol Chem* **287**: 31804–31812.

Rathenberg J, Nevian T, Witzemann V. 2003. High-efficiency transfection of individual neurons using modified electrophysiology techniques. *J Neurosci Methods* **126**: 91–98.

Rose T, Schoenenberger P, Jezek K, Oertner TG. 2013. Developmental refinement of vesicle cycling at Schaffer collateral synapses. *Neuron* **77**: 1109–1121.

Schoenenberger P, Gerosa D, Oertner TG. 2009. Temporal control of immediate early gene induction by light. *PLoS One* **4**: e8185.

Stoppini L, Buchs PA, Muller D. 1991. A simple method for organotypic cultures of nervous tissue. *J Neurosci Methods* **37**: 173–182.

Thomas A, Kim DS, Fields RL, Chin H, Gainer H. 1998. Quantitative analysis of gene expression in organotypic slice-explant cultures by particle-mediated gene transfer. *J Neurosci Methods* **84**: 181–191.

Wiegert JS, Oertner TG. 2013. Long-term depression triggers the selective elimination of weakly integrated synapses. *Proc Natl Acad Sci* **110**: E4510–E4519.

Wiegert JS, Gee CE, Oertner TG. 2016a. Viral vector–based transduction of slice cultures. *Cold Spring Harb Protoc* doi: 10.1101/pdb.prot094896.

Wiegert JS, Gee CE, Oertner TG. 2016b. Single-cell electroporation of neurons. *Cold Spring Harb Protoc* doi: 10.1101/pdb.prot094904.

Wietek J, Wiegert JS, Adeishvili N, Schneider F, Watanabe H, Tsunoda SP, Vogt A, Elstner M, Oertner TG, Hegemann P. 2014. Conversion of channelrhodopsin into a light-gated chloride channel. *Science* **344**: 409–412.

Wietek J, Beltramo R, Scanziani M, Hegemann P, Oertner TG, Wiegert SJ. 2015. An improved chloride-conducting channelrhodopsin for light-induced inhibition of neuronal activity in vivo. *Sci Reo* **5**: 14807.

Preparation of Slice Cultures from Rodent Hippocampus

Christine E. Gee, Iris Ohmert, J. Simon Wiegert, and Thomas G. Oertner[1]

Institute for Synaptic Physiology, Center for Molecular Neurobiology (ZMNH), 20251 Hamburg, Germany

This protocol describes the preparation of hippocampal slice cultures from rat or mouse pups using sterile conditions that do not require the use of antibiotics or antimycotics. Combining very good optical and electrophysiological accessibility with a lifetime approaching that of the intact animal, many fundamental questions about synaptic plasticity and long-term dynamics of network connectivity can be addressed with this preparation.

MATERIALS

It is essential that you consult the appropriate Material Safety Data Sheets and your institution's Environmental Health and Safety Office for proper handling of equipment and hazardous material used in this protocol.

RECIPES: Please see the end of this protocol for recipes indicated by <R>. Additional recipes can be found online at http://cshprotocols.cshlp.org/site/recipes.

Reagents

Animals (rat [P4–P6] or mouse [P5–P8] pups)
Appropriate anesthesia according to local guidelines
Dissection solution (sterile; 50 mL is sufficient for approximately 10 pups) <R>
Slice culture medium (sterile; 50 mL is sufficient for approximately 10 pups) <R>

Equipment

95% O_2/5% CO_2 regulator and tubing
Beaker (glass; 500-mL)
Cell Saver Tips (1000-µL) (BIOzym 693000)
Culture plates (six-well; e.g., Corning 3516 or Sarstedt 83.1839)
Dissection stereomicroscope with cold platform (e.g., ice-cold metal plate or glass Petri dish filled with ice, prewiped with 70% ethanol)
Filter paper or Whatman paper (sterile, 6-cm diameter; MACHEREY-NAGEL 431005)
Fine paintbrush (e.g., Ted Pella 11810/11812)
Hot bead sterilizer (e.g., Fine Science Tools 18000-45)
Incubator (37°C/5% CO_2 with rapid humidity recovery, copper chamber recommended; e.g., Heracell 150i/160i, Thermo Scientific)
Large shallow ice bucket containing ice scattered with NaCl

---

[1]Correspondence: thomas.oertner@zmnh.uni-hamburg.de

Cite this protocol as *Cold Spring Harb Protoc*; doi:10.1101/pdb.prot094888

Millicell Cell Culture Insert (30-mm, hydrophilic PTFEm [polytetrafluoroethylene], 0.4-µm;
    Millipore PICMORG50)
Paper towels (sterile)
Pasteur pipettes (glass, 9″, sterilized; e.g., Carl Roth 4518.1)
Pasteur pipettes (plastic disposable, sterile; e.g., Sarstedt 86.1171.010)
Pasteur pipettes with tips broken off (glass, sterilized)
    Wear goggles and wrap pipettes before breaking to prevent injury.

Pipette bulbs (two, rubber, 2-mL; e.g., Sigma-Aldrich Z111597)
Pipettor (1000-µL; BIOzym 655070)
Razor blade (two-sided; Fine Science Tools 10050-00)
Small ice bucket containing ice scattered with NaCl
Spray bottle containing 70% ethanol
Stainless steel tray
Sterilized dissection tools
    Bone curette (small, back thinned on sharpening stone; DO608R Lucas sharp spoon, Aesculap)
    Coarse forceps (e.g., Fine Science Tools 11002-16)
    Fine scissors (Carl Roth LT28.1)
    Forceps (two, No. 5 Dumont [fine]; WPI 500342)
    Large scissors (ET140R, Aesculap or Fine Science Tools)
    Scalpel handle and blades (#3 and #10) (e.g., Fine Science Tools 10003-12/10010-00)
    Small spatula or spoon

Tissue Chopper (McIlwain type 10180, Ted Pella)
Tissue culture dishes (60-mm, sterile; Sarstedt 83.1801)
Tissue culture hood
Tubes (sterile, 50-mL)

METHOD

The preparation of slice cultures that are to be made and maintained without antibiotics requires stringent sterility be maintained at all times. Antibiotics affect a myriad of cellular properties (Amonn et al. 1978; Llobet et al. 2015); therefore, it is essential to avoid their use. Aseptic techniques must therefore be strictly observed at all steps and everything coming into contact with brain tissue must be sterile. For a culture to be maintained it must remain uncontaminated. Normally experiments should be performed within 6 wk of culture preparation, but slice cultures can be successfully maintained for more than a year.

Preparation

1. Put a 1-mL slice of culture medium in each well of the six-well plates. Using sterile forceps place one membrane insert into each well. Put plates into 37°C incubator.

    A mouse pup will yield approximately 10 to 15 slices, a rat pup approximately 15 to 20. If two slices are to be placed on each insert, then one six-well plate will hold 12 slice cultures.

2. Next to the tissue culture hood arrange the following.

    - A stainless steel tray with two 60-mm dishes per animal

        One dish should contain a sterile filter paper. Sit the tray in a shallow ice bucket containing ice sprinkled with NaCl. The dishes can be turned upside down and the lids used because the rims are lower.

    - Hot bead sterilizer and beaker filled with water for rinsing tools

    - Large scissors, fine scissors, coarse forceps, scalpel, and spatula arranged between sterile paper towels

    - Sterile paper towel for placing head

    - Container for body

3. Inside the tissue culture hood arrange the following.

- Tissue chopper set to cut 400-μm slices

    The blade must be cleaned of oil/wax before mounting.

- Dissection stereomicroscope with ice-cold platform

- Small bucket of NaCl-sprinkled ice holding a 50-mL tube containing dissection solution (~5 mL per animal), bubbled with 95% O_2/5% CO_2 through a sterile glass Pasteur pipette

    Place a sterile disposable pipette in the dissection solution. When the bubbled dissection solution has turned from magenta to red/orange in color it is ready to use.

- 1000-μL pipettor with Cell Saver Tip

- Sterile glass Pasteur pipette with bulb

- Sterile transfer pipette, made by breaking the tip from a Pasteur pipette and mounting the bulb on the broken end

- Scalpel, two fine forceps (#5 Dumont), bone curette, arranged on sterile paper towel

- Sterile fine paint brush

4. Spray or wipe down the microscope, tissue chopper, ice bucket, and stainless steel tray with 70% ethanol. Spray the blade and stage (with mounted disks) of the tissue chopper and allow to air dry before beginning the procedure.

Dissection

Wear sterile gloves and a mask. Neurons are sensitive to ethanol, including fumes, so if using ethanol to sterilize surfaces, tools, etc., allow thorough drying to prevent tissue exposure.

5. Immediately before beginning the procedure, place 1–2 mL of the cold, bubbled dissection solution into one of the 60-mm dishes containing the filter paper (this can be done while pups are undergoing anesthesia). Place covered dish back on the chilled tray.

6. Anesthetize pups according to local regulations and decapitate using the large scissors.

    Do not put the head in the dissection solution as this will result in contamination of the slice cultures.

7. Place the head on a sterile paper towel and hold (without squeezing) between the fingers of one hand—it is important to note that this hand is now contaminated. With a scalpel, open the skin from the nose to the base of the skull without cutting through the skull. Pull back the skin with the fingers already holding the head, slide the scalpel under the skin on one side and cut downward through the ears, repeat on the other side, and the skin can now easily be held back leaving the skull exposed. Cut the skull with the fine scissors along the midline. Make small cuts at the most rostral end out toward the sides keeping the orientation of the scissors so that the blade closest to the brain is always the same. Put the scissors into the beaker of water. Take the coarse forceps and pull open/remove the skull. The flaps of skull can be held with the fingers holding the head if necessary. Again be sure to keep the orientation of the forceps constant to avoid contaminating the surface of the brain. Put the forceps into the beaker of water and remove the cover from the chilled 6-cm dish with filter paper. Take the spatula/spoon and slide under the brain flipping it out of the skull onto the chilled wetted filter paper.

8. Put the dish with the brain on an ice-cold platform under the dissection microscope in the hood. Drop some of the cold, bubbled dissection solution onto the brain. While the brain chills for ~1 min, quickly dry off the dissection tools and place the tips into the bead sterilizer for 10 sec to sterilize. Replace and/or clean contaminated glove(s) with 70% ethanol. Ensure gloves are dry before moving close to the tissue to avoid exposing the tissue to ethanol vapor.

9. Under the dissection microscope in the hood, dissect the hippocampi. Holding the cerebellum with one pair of the fine forceps, cut along the midline twice with the scalpel without severing the most caudal part of the hemispheres (Fig. 1A). Using the bone curette fold down one hemisphere, which will expose the hippocampus (Fig. 1B). Cut the ends of the hippocampus with the scalpel

Cite this protocol as *Cold Spring Harb Protoc*; doi:10.1101/pdb.prot094888

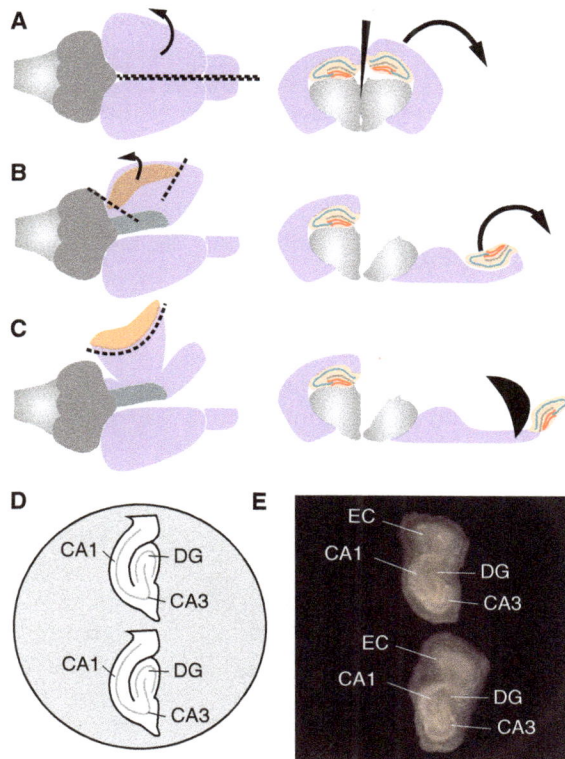

FIGURE 1. Dissecting the hippocampus. (*A*) Cut along the midline with a scalpel, using the cerebellum/brain stem as a handle. (*B*) Fold out the left hemisphere, exposing the hippocampus. Cut free the tips of the hippocampus. (*C*) Flip out the left hippocampus and cut through the entorhinal cortex to completely separate the left hippocampus from the brain. (*D*) Arrangement of slices on a cell culture insert for injection and electroporation. (*E*) Six-week-old Wistar rat hippocampal slice cultures. DG, dentate gyrus; EC entorhinal cortex; CAI and CA3, regions of the hippocampus proper.

and carefully flip out the hippocampus with the curette. Cut away most of the attached cortex (Fig. 1C). Repeat with the other hippocampus.

> *It is often helpful to pull away some of the meninges from the brain with the fine forceps before removing the hippocampi.*

10. Transfer the hippocampi to the stage of the tissue chopper using the transfer pipette. Arrange the hippocampi perpendicular to the blade and remove all surrounding dissection solution with the glass Pasteur pipette. Cut 400-μm thick slices.

> *The sliced hippocampus should remain on the platform. If the slices stick to the blade, there is probably too much dissection solution remaining on the stage.*

11. Take the Teflon chopping disk off the stage and gently rinse the slices into one of the prechilled 60-mm dishes using sterile dissection solution. Quickly clean the chopping disk with 70% ethanol, wipe dry with a sterile paper towel, and replace. If necessary, gently tease apart any slices that are adhering to each other with fine forceps. Do not grasp the slices but slide the forceps between the slices. The slices should be cut through and come apart easily.

12. Transfer the undamaged slices showing nice morphology onto the prewarmed membrane inserts (from Step 1) using the Cell Saver Tip. Two slices are typically placed near the center of each membrane and oriented so that the CA1 regions are pointing to one side using a fine brush (Fig. 1D). This simplifies injection of recombinant adeno-associated virus and electroporation. Carefully suck away any liquid from the top of the membrane using the glass Pasteur pipette and put the slices into the incubator.

> *If slices are to be transfected by gene gun, place three to five slices on each membrane and do not worry about the orientation. From the start of the dissection to placing the slices in the incubator should take 20–30 min per pup. It is more important not to damage the hippocampus and to not contaminate anything that will come in contact with the slices than it is to perform the procedure very quickly.*

13. One or two days after making the cultures, change the medium by aspirating about two-thirds and replacing with 700–800 μL prewarmed fresh medium. Thereafter, change the medium

twice per week (or at least every 5 d) for one to two slices per membrane and at least every 3 d for three to five slices per membrane. Allow 3 d for the slices to firmly adhere to the membrane.

Do not inject recombinant adeno-associated virus or electroporate the cultures within the first 2 d after preparation because they are likely to become detached and are then no longer usable.

RECIPES

Dissection Solution

Reagent	Final concentration	Amount (for 500 mL)
Sucrose (Fluka 84100)	248 mM	40 g
NaHCO$_3$ (Fluka 31437)	26 mM	1.09 g
D-glucose (Fluka 49152)	10 mM	0.9 g
KCl (1 M; Fluka 60129)	4 mM	2 mL
MgCl$_2$ (1 M; Fluka 63020)	5 mM	2.5 mL
CaCl$_2$ (1 M; Fluka 21114)	1 mM	0.5 mL
Phenol red (0.5%; Riedel-De Haen 32662)	0.001%	1 mL
Ultrapure H$_2$O	–	to 500 mL
Kynurenic acid (100 mM; Fluka 61260)	2 mM	1 mL in 50 mL dissection solution

After mixing all ingredients except the kynurenic acid, check the osmolality (it should be 310–320 mOsm/kg) and the color (it should be red-magenta, indicating a pH >8). Filter-sterilize (0.2 µm pore size) and store at 4°C in 50-mL aliquots. Prepare a 100 mM stock solution of kynurenic acid and store at −20°C in 1-mL aliquots. Just before use, thaw and vortex one aliquot of kynurenic acid and add to 50 mL of dissection solution.

Slice Culture Medium

Reagent	Final concentration	Amount (for 500 mL)
MEM (Sigma-Aldrich M7278)		394 mL
Heat-inactivated horse serum[a]	20%	100 mL
L-glutamine (200 mM; Gibco 25030-024)	1 mM	2.5 mL
Insulin (1 mg/mL; Sigma-Aldrich I6634)	0.01 mg/mL	0.5 mL
NaCl (5 M; Sigma-Aldrich S5150)		1.45 mL
MgSO$_4$ (1 M; Fluka 63126)	2 mM	1 mL
CaCl$_2$ (1 M; Fluka 21114)	1.44 mM	0.72 mL
Ascorbic acid (25%; Fluka 11140)	0.00125%	2.4 µL
D-glucose (Fluka 49152)	13 mM	1.16 g

[a]The serum is often a critical factor in slice quality and it is often necessary to test several batches (lots); three products that have been successfully used are Sigma-Aldrich H1138, Gibco 26050070, and Gibco 16050122. Gibco 16050122 must be heat-inactivated for 30 min at 55°C.

After mixing, filter-sterilize (0.2-µm pore size), and store at 4°C in 50-mL aliquots. (The solution should be orange-red [i.e., pH ~7.3] and osmolality should be ~320 mOsm/kg.)

REFERENCES

Amonn F, Baumann U, Wiesmann UN, Hofmann K, Herschkowitz N. 1978. Effects of antibiotics on the growth and differentiation in dissociated brain cell cultures. *Neuroscience* 3: 465–468.

Llobet L, Montoya J, Lopez-Gallardo E, Ruiz-Pesini E. 2015. Side effects of culture media antibiotics on cell differentiation. *Tissue Eng Part C Methods* 21: 1143–1147.

Viral Vector-Based Transduction of Slice Cultures

J. Simon Wiegert, Christine E. Gee, and Thomas G. Oertner[1]

Institute for Synaptic Physiology, Center for Molecular Neurobiology (ZMNH), 20251 Hamburg, Germany

Transgenes can be introduced into the cells of organotypic slice cultures using different delivery methods, such as biolistic transfection, electroporation, and viral vector-based transduction. These methods produce different patterns of transgene expression. Local injection of recombinant adeno-associated virus (rAAV) produces a small cluster of transgene-expressing neurons around the injection site. Expression in individual cells varies with the distance from the injection site, indicating that many neurons take up several rAAV particles. The serotype and promoter also play a role in transgene expression. Here, we present a protocol for the transduction of previously prepared hippocampal slice cultures with rAVV.

MATERIALS

It is essential that you consult the appropriate Material Safety Data Sheets and your institution's Environmental Health and Safety Office for proper handling of equipment and hazardous material used in this protocol.

RECIPES: Please see the end of this protocol for recipes indicated by <R>. Additional recipes can be found online at http://cshprotocols.cshlp.org/site/recipes.

Reagents

Hippocampal slice cultures on inserts (at least 3 d in vitro [DIV])

> *For preparation of hippocampal slice cultures, see Protocol 1: Preparation of Slice Cultures from Rodent Hippocampus (Gee et al. 2016).*

Recombinant adeno-associated virus (rAAV) suspension (concentration: 10^{11}–10^{13} GC/mL)

Slice culture medium (sterile) <R>

Slice culture transduction solution (sterile) <R>

Equipment

Compressed/pressurized air (for Picospritzer)

Culture plates (six-well; e.g., Corning 3516 or Sarstedt 83.1839)

Forceps (coarse; e.g., Fine Science Tools 11002-16, sterile)

Forceps (fine, No. 5 Dumont; e.g., WPI 500342, sterile)

Incubator (37°C/5% CO_2 with rapid humidity recovery, copper chamber recommended; e.g., Heracell 150i/160i, Thermo Scientific)

Microelectrode holder (e.g., WPI MPH6S)

Micromanipulator (e.g., LN Junior 3 or 4 axis; Luigs & Neumann or PatchMan, Eppendorf)

Micropipette puller (e.g., Sutter P-1000)

---

[1]Correspondence: thomas.oertner@zmnh.uni-hamburg.de

Cite this protocol as *Cold Spring Harb Protoc*; doi:10.1101/pdb.prot094896

Microscope (either an upright microscope with camera, wide-field illumination, and 4× objective or a USB microscope [e.g., dnt DigiMicro Profi] and movable stage)
Picospritzer III with foot switch (Parker Hannafin)
Sharps container
Thin-walled borosilicate glass capillaries with filament (WPI TW150F-3)
Tissue culture hood

METHOD

The hippocampal slice cultures used in this protocol must have been prepared under stringent sterile conditions and all steps in this procedure have to be performed under sterile conditions. For precautions to avoid contamination, see Protocol 1: Preparation of Slice Cultures from Rodent Hippocampus (Gee et al. 2016). It is recommended that the injection setup be close to the tissue culture hood. Furthermore, construction of a cabinet around the injection microscope with a fan and HEPA (high-efficiency particulate air) filter to blow clean air down over the setup (Fig. 1C) will reduce the incidence of contamination to almost never. A dedicated viral vector injection setup can be constructed using an inexpensive USB microscope and the required ancillary parts. This setup is shown in Figure 1. Viral vector injections can be performed equally well on an upright microscope using wide-field illumination and a low-magnification objective (i.e., a standard patch-clamp setup).

1. Place a fresh six-well plate containing 1 mL of slice culture medium in each well in the incubator to allow for temperature and pH equilibration.

2. Fabricate injection pipettes using a micropipette puller to pull thin-walled borosilicate capillaries to obtain a long and narrow tip with a shallow taper. Insert an empty injection pipette into the holder attached to the microscope and focus on the tip (Fig. 1A). Break off the tip with fine forceps under visual guidance to achieve a tip diameter of ~10 µm.

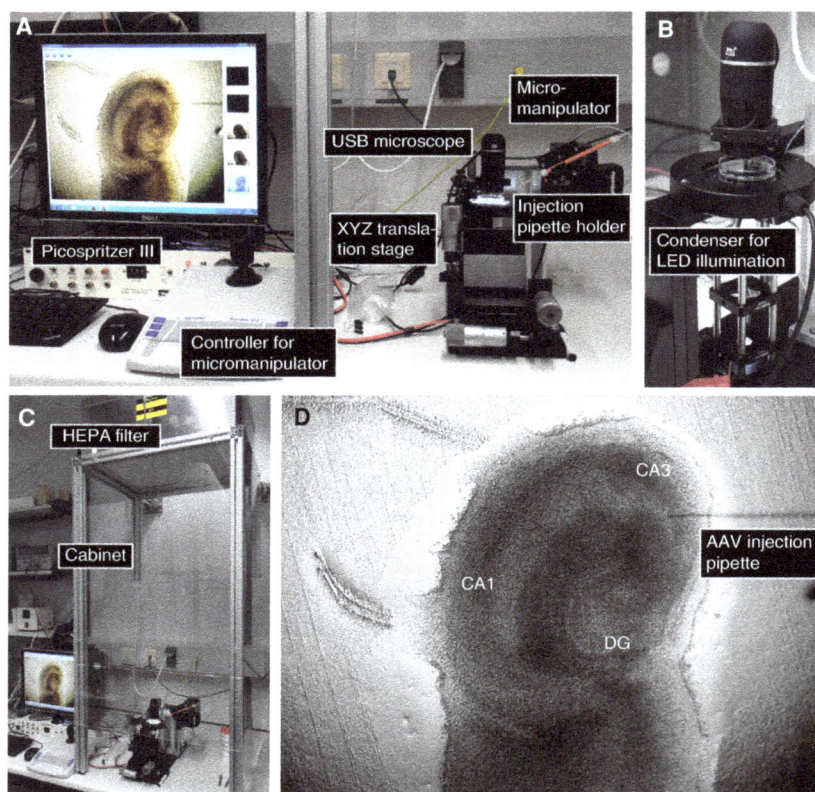

FIGURE 1. Injection of rAAV into CA3. (*A*) Photograph of USB microscope–based injection setup. (*B*) Detail of LED (light-emitting diode) illumination condenser. (*C*) View of setup showing surrounding cabinet. (*D*) Hippocampal slice culture with tip of injection pipette in area CA3.

Cite this protocol as *Cold Spring Harb Protoc*; doi:10.1101/pdb.prot094896

3. Remove the pipette from the holder and back-fill with 1.2 µL rAAV suspension (this typically lasts for 12 slice cultures). Hold the pipette tip down and wait until all liquid has migrated along the filament into the tip. Set the pressure at the Picospritzer to 1.8 bar (25 p.s.i.) and the pulse duration to 50 msec.

4. Insert the pipette into the holder, focus on the tip, and apply a pressure pulse to test whether the rAAV suspension is expelled from the tip. If the tip is open, a small droplet should emerge. This check is most easily and safely done with a membrane insert already in the chamber (see Step 5). If the tip of the pipette is positioned to just touch the membrane (away from the slice cultures to be injected), then the drop will form on the membrane. The pulse duration/pressure should be adjusted if larger/smaller injection volumes are desired.

   Perform all subsequent steps quickly to avoid degradation of the rAAV at room temperature and to minimize drying out of air-exposed slice cultures.

5. Working in a tissue culture hood, fill the microscope chamber with 800 µL transduction solution prewarmed to 37°C and, using coarse forceps, place an insert with slice cultures into the chamber.

   When using a USB camera setup, the lid of a sterile 35-mm tissue culture dish serves as the chamber. For injecting cultures on the stage of an upright microscope a large glass microscope slide (70 × 100 × 1 mm) onto which a Teflon ring (inner diameter 34 mm, 2 mm high) is fixed with silicone aquarium sealant should be used as the chamber. Before use, the chamber should be wiped with 70% ethanol and rinsed with sterilized transduction solution.

6. Transfer the chamber to the microscope and focus on the slice culture. Bring the pipette tip into view and move tip axially into the cell body layer of CA3 (Fig. 1A,D). Depending on the desired infection density, one to three pressure pulses should be given. If desired, this procedure can be repeated several times at neighboring sites until the whole target area is covered. Axially retract the pipette until it no longer touches the culture and move to the second slice culture on the insert to repeat the injection procedure.

7. Transfer the insert to the six-well plate newly prepared in Step 1. Repeat Steps 5 and 6 until all cultures are injected with viral constructs; the same injection pipette may be used for many slice cultures. When finished, remove the injection pipette immediately from the holder and dispose in a sharps container.

   The ideal time between rAAV infection and when cells may be stimulated with light depends on many factors including the promoter, rAAV serotype, and the construct itself. We recommend testing cultures at various intervals before beginning actual experiments.

RECIPES

Slice Culture Medium

Reagent	Final concentration	Amount (for 500 mL)
MEM (Sigma-Aldrich M7278)		394 mL
Heat-inactivated horse serum[a]	20%	100 mL
L-glutamine (200 mM; Gibco 25030-024)	1 mM	2.5 mL
Insulin (1 mg/mL; Sigma-Aldrich I6634)	0.01 mg/mL	0.5 mL
NaCl (5 M; Sigma-Aldrich S5150)		1.45 mL
$MgSO_4$ (1 M; Fluka 63126)	2 mM	1 mL
$CaCl_2$ (1 M; Fluka 21114)	1.44 mM	0.72 mL
Ascorbic acid (25%; Fluka 11140)	0.00125%	2.4 µL
D-glucose (Fluka 49152)	13 mM	1.16 g

[a]The serum is often a critical factor in slice quality and it is often necessary to test several batches (lots); three products that have been successfully used are Sigma-Aldrich H1138, Gibco 26050070, and Gibco 16050122. Gibco 16050122 must be heat-inactivated for 30 min at 55°C.

After mixing, filter-sterilize (0.2-µm pore size), and store at 4°C in 50-mL aliquots. (The solution should be orange-red [i.e., pH ~7.3] and osmolality should be ~320 mOsm/kg.)

Slice Culture Transduction Solution

Reagent	Final concentration	Amount (for 500 mL)
NaCl (Sigma-Aldrich S5150)	145 mM	4.23 g
HEPES (Sigma-Aldrich H4034)	10 mM	1.19 g
D-glucose (Fluka 49152)	25 mM	2.25 g
KCl (1 M; Fluka 60129)	2.5 mM	1.25 mL
MgCl$_2$ (1 M; Fluka 63020)	1 mM	0.5 mL
CaCl$_2$ (1 M; Fluka 21114)	2 mM	1 mL

Adjust pH to 7.4 with NaOH. After mixing all ingredients, check the osmolality—it should be 310–320 mOsm/kg. Filter-sterilize (0.2-μm pore size) and store at 4°C.

REFERENCES

Gee CE, Ohmert I, Wiegert JS, Oertner TG. 2016. Preparation of slice cultures from rodent hippocampus. *Cold Spring Harb Protoc* doi: 10.1101/pdb.prot094888.

Cite this protocol as *Cold Spring Harb Protoc*; doi:10.1101/pdb.prot094896

Single-Cell Electroporation of Neurons

J. Simon Wiegert, Christine E. Gee, and Thomas G. Oertner[1]

Institute for Synaptic Physiology, Center for Molecular Neurobiology (ZMNH), 20251 Hamburg, Germany

Single-cell electroporation allows the transfection of a small number of neurons in an organotypic culture with a single plasmid or a defined mixture of plasmids. Desired protein expression levels can vary depending on the experimental goals (e.g., high expression levels are needed for optogenetic experiments); however, when too much protein is expressed, cellular toxicity and cell death may arise. To a large degree, protein expression can be controlled by adjusting the concentration of plasmid DNA in the electroporation pipette. Here, we present a protocol for transfecting individual neurons in hippocampal slice cultures by electroporation. Essentially, a patch-clamp setup is required that includes an upright microscope with infrared differential interference contrast or Dodt contrast with a camera and a specialized amplifier that is able to deliver large-voltage pulses to the electroporation pipette.

MATERIALS

It is essential that you consult the appropriate Material Safety Data Sheets and your institution's Environmental Health and Safety Office for proper handling of equipment and hazardous material used in this protocol.

RECIPES: Please see the end of this protocol for recipes indicated by <R>. Additional recipes can be found online at http://cshprotocols.cshlp.org/site/recipes.

Reagents

Alexa Fluor 594 (Invitrogen A33082, or other dye compatible with available filter sets)
Hippocampal slice cultures on inserts (at least 5 d in vitro [DIV], optimally 2 wk)

For preparation of hippocampal slice cultures, see Protocol 1: Preparation of Slice Cultures from Rodent Hippocampus (Gee et al. 2016).

K-gluconate-based intracellular solution <R>
Plasmid DNA in TE buffer (pH 7.4)

Prepare using any standard commercial purification kit.

Saturated $FeCl_3$ solution
Slice culture medium <R>
Slice culture transduction solution (sterile, prewarmed to 37°C) <R>

Equipment

Centrifuge (table-top, refrigerated, e.g., Eppendorf 5415 R)
Culture plates (six-well; e.g., Corning 3516 or Sarsted 83.1839)

---

[1]Correspondence: thomas.oertner@zmnh.uni-hamburg.de

Cite this protocol as *Cold Spring Harb Protoc*; doi:10.1101/pdb.prot094904

Electroporation setup

   20×–40× water immersion objective (20× with a variable magnifier tube before the camera)

   Axoporator 800A with pipette holder (Molecular Devices)

   Epifluorescence illumination and filters (optional: see Step 11)

   Headphones or speakers

   Microscope chamber consisting of glass microscope slide (70 × 100 × 1 mm) onto which a Teflon ring (inner diameter ~35 mm, ~2 mm high) is affixed with silicone aquarium sealant

   Motorized micromanipulator (e.g., from Luigs & Neumann or Sutter)

   Silver wire (~0.25 mm diameter)

   Upright microscope with motorized stage, CCD/CMOS (charge-coupled device/complementary metal oxide semiconductor) or video camera and IR-DIC (infrared-differential interference contrast) or Dodt contrast

Forceps (coarse; e.g., Fine Science Tools 11002-16)

Hot bead sterilizer (e.g., Fine Science Tools 18000-45)

Incubator (37°C/5% CO_2 with rapid humidity recovery, copper chamber recommended; e.g., Heracell 150i/160i, Thermo Scientific)

Micropipette puller (e.g., PC-10, Narishige)

Thin-walled borosilicate glass capillaries (WPI TW150F-3)

Tissue culture dishes (60-mm, sterile; Sarstedt 83.1801)

Tissue culture hood

Ultrafree centrifugal filter units (UFC30GV0S, Millipore)

METHOD

Preparation of DNA

1. Dilute Alexa Fluor 594 to a final concentration of 20 μM in K-gluconate-based intracellular solution and filter sterilize the solution through a Millipore Ultrafree centrifugal filter unit by centrifugation at 16,100g for several seconds in a table-top centrifuge at 4°C.

2. Remove the filter insert and add plasmid DNA to the desired final concentration. Centrifuge the solution again for 5 min at 16,100g, 4°C to pellet debris.

   One hundred microliters of solution will be enough for more than 50 slices and can be stored between uses at −20°C. It is important that the DNA-containing solution is not passed through the Millipore Ultrafree centrifugal filter unit. The optimal final DNA concentration must be determined empirically and is usually 1–100 ng/μL.

Single-Cell Electroporation

The hippocampal slice cultures used in this protocol must have been prepared under stringent sterile conditions and all steps in this procedure have to be performed under sterile conditions. For precautions to avoid contamination, see Protocol 1: Preparation of Slice Cultures from Rodent Hippocampus (Gee et al. 2016). It is recommended that the electroporation setup be close to the tissue culture hood. Furthermore, construction of a cabinet around the electroporation microscope with a fan and HEPA (high-efficiency particulate arrestance) filter to blow clean air down over the setup will reduce the incidence of contamination to almost never (for further description of this setup, see Protocol 2: Viral Vector–Based Transduction of Slice Cultures [Wiegert et al. 2016]).

3. Coat the tips of silver wires for the electroporation and grounding electrodes with Cl by bathing them in $FeCl_3$ solution for a few minutes (when scratched) or overnight (when new).

4. Place a fresh six-well plate containing 1 mL of slice culture medium in each well in the incubator for pre-equilibration.

5. Fabricate electroporation pipettes using a micropipette puller to pull thin-walled borosilicate capillaries to a resistance of 10–15 MΩ when filled with the intracellular solution. For each slice to

Cite this protocol as *Cold Spring Harb Protoc*; doi:10.1101/pdb.prot094904

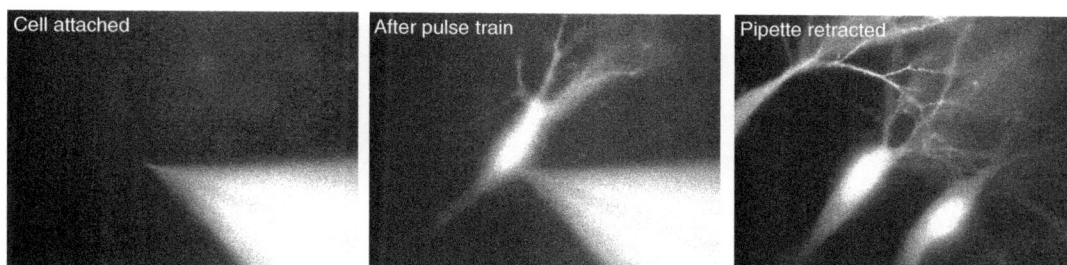

FIGURE 1. Single-cell electroporation. Fluorescence images taken during electroporation procedure. (*Left*) Pipette filled with DNA solution and Alexa Fluor 594 touching target cell. (*Middle*) Immediately after applying pulse train, target cell is filled with DNA solution and Alexa Fluor 594. (*Right*) Successfully electroporated neurons after retraction of the pipette.

be electroporated, back-fill one electroporation pipette with 1.5 µL of DNA/Alexa Fluor 594 solution (from Step 2).

> Several pipettes can be filled at once and kept upright (tip down) for 1–2 h.

6. Working in a tissue culture hood, pipette 1 mL of transduction solution prewarmed to 37°C into the microscope chamber. Use sterile forceps to transfer one slice culture insert into the chamber and add another 2 mL of transduction solution on top of the slice cultures. Place forceps into the hot bead sterilizer for ~10 sec to resterilize between handling of inserts.

7. Cover the microscope chamber with a sterile 60-mm dish and transfer to the microscope.

8. Approach the selected cells with the electroporation pipette while applying positive pressure to the pipette by mouth. Monitor the tip resistance, which should be 10–15 MΩ, by the audio output of the Axoporator 800A amplifier. Make sure the tip is not clogged by monitoring tip resistance and expulsion of Alexa Fluor 594 fluorescence if needed.

9. Move the tip of the electroporation electrode close to a cell of interest while reducing pressure (easily controlled by mouth by blowing less hard). Monitor the tip resistance acoustically.

10. Approach the cell without sealing the electrode with membrane from other cells in the tissue.

11. Lightly touch the plasma membrane, which will cause a rise in pitch and tip resistance.

12. Release the pressure (do not apply suction) and wait for resistance to increase to 25–40 MΩ; however, avoid the formation of a GΩ seal.

13. Apply pulse train (e.g., voltage: −12 V, pulse width: 500 µsec, frequency: 50 Hz, train duration 500 msec; the optimal settings may differ depending on cell type).

14. Slowly retract the pipette and begin applying very light pressure when 2–5 µm away from the soma. Increase pressure at larger distances.

> Do not apply too much pressure while retracting the pipette, otherwise the cell may rupture. Electroporation success can be instantly evaluated by assessing Alexa Fluor 594 fluorescence in electroporated cells (Fig. 1).

15. Repeat Steps 8–14 to electroporate more cells.

16. Change the pipette before electroporating the second culture on insert and when the pipette becomes clogged.

17. Cover the chamber with a 60-mm dish and transfer to the tissue culture hood. Aspirate all solution and return the insert to the slice culture medium in the six-well plate (from Step 4).

> The optimal time between electroporation and starting an experiment has to be determined empirically for each plasmid.

RELATED INFORMATION

This protocol is based upon the electroporation of single neurons in organotypic cultures described by Rathenberg et al. (2003).

RECIPES

K-Gluconate-Based Intracellular Solution

Reagent	Final concentration	Amount (for 50 mL)
Aldrich K-gluconate (Sigma-Aldrich G4500)	135 mM	1.581 g
Aldrich EGTA (Sigma-Aldrich E0396)	0.2 mM	0.004 g
Aldrich HEPES (Sigma-Aldrich H4034)	10 mM	0.119 g
$MgCl_2$ (1 M; Fluka 63020)	4 mM	0.200 mL
Aldrich Na_2-ATP (Sigma-Aldrich A3377)	4 mM	0.121 g
Aldrich Na-GTP (Sigma-Aldrich G8877)	0.4 mM	0.010 g
Aldrich Na_2-phosphocreatine (Sigma-Aldrich P7936)	10 mM	0.128 g
Aldrich Ascorbate (L-ascorbic acid; Sigma-Aldrich A5960)	3 mM	0.026 g

Adjust pH to 7.2 with KOH, and check osmolality (it should be 290–300 mOsm/kg). After mixing, filter-sterilize (0.2-µm pore size) and divide into 100–500-µL aliquots. Store at −20°C.

Slice Culture Medium

Reagent	Final concentration	Amount (for 500 mL)
MEM (Sigma-Aldrich M7278)		394 mL
Heat-inactivated horse serum[a]	20%	100 mL
L-glutamine (200 mM; Gibco 25030-024)	1 mM	2.5 mL
Insulin (1 mg/mL; Sigma-Aldrich I6634)	0.01 mg/mL	0.5 mL
NaCl (5 M; Sigma-Aldrich S5150)		1.45 mL
$MgSO_4$ (1 M; Fluka 63126)	2 mM	1 mL
$CaCl_2$ (1 M; Fluka 21114)	1.44 mM	0.72 mL
Ascorbic acid (25%; Fluka 11140)	0.00125%	2.4 µL
D-glucose (Fluka 49152)	13 mM	1.16 g

[a]The serum is often a critical factor in slice quality and it is often necessary to test several batches (lots); three products that have been successfully used are Sigma-Aldrich H1138, Gibco 26050070, and Gibco 16050122. Gibco 16050122 must be heat-inactivated for 30 min at 55°C.

After mixing, filter-sterilize (0.2-µm pore size), and store at 4°C in 50-mL aliquots. (The solution should be orange-red [i.e., pH ~7.3] and osmolality should be ~320 mOsm/kg.)

Slice Culture Transduction Solution

Reagent	Final concentration	Amount (for 500 mL)
NaCl (Sigma-Aldrich S5150)	145 mM	4.23 g
HEPES (Sigma-Aldrich H4034)	10 mM	1.19 g
D-glucose (Fluka 49152)	25 mM	2.25 g
KCl (1 M; Fluka 60129)	2.5 mM	1.25 mL
$MgCl_2$ (1 M; Fluka 63020)	1 mM	0.5 mL
$CaCl_2$ (1 M; Fluka 21114)	2 mM	1 mL

Adjust pH to 7.4 with NaOH. After mixing all ingredients, check the osmolality—it should be 310–320 mOsm/kg. Filter-sterilize (0.2-µm pore size) and store at 4°C.

REFERENCES

Gee CE, Ohmert I, Wiegert JS, Oertner TG. 2016. Preparation of slice cultures from rodent hippocampus. *Cold Spring Harb Protoc* doi: 10.1101/pdb.prot094888.

Rathenberg J, Nevian T, Witzemann V. 2003. High-efficiency transfection of individual neurons using modified electrophysiology techniques. *J Neurosci Methods* 126: 91–98.

Wiegert JS, Gee CE, Oertner TG. 2016. Viral vector–based transduction of slice cultures. *Cold Spring Harb Protoc* doi: 10.1101/pdb.prot094896.

CHAPTER 13

Combining Optogenetics and Electrophysiology to Analyze Projection Neuron Circuits

Naoki Yamawaki,[1] Benjamin A. Suter,[1] Ian R. Wickersham,[2] and Gordon M.G. Shepherd[1,3]

[1]Department of Physiology, Feinberg School of Medicine, Northwestern University, Chicago, Illinois 60611;
[2]McGovern Institute for Brain Research, Massachusetts Institute of Technology, Cambridge, Massachusetts 02139

A set of methods is described for channelrhodopsin-2 (ChR2)-based synaptic circuit analysis that combines photostimulation of virally transfected presynaptic neurons' axons with whole-cell electrophysiological recordings from retrogradely labeled postsynaptic neurons. The approach exploits the preserved photoexcitability of ChR2-expressing axons in brain slices and can be used to assess either local or long-range functional connections. Stereotaxic injections are used both to express ChR2 selectively in presynaptic axons of interest (using rabies virus [RV] or adeno-associated virus [AAV]) and to label two types of postsynaptic projection neurons of interest with fluorescent retrograde tracers. In brain slices, tracer-labeled postsynaptic neurons are targeted for whole-cell electrophysiological recordings, and synaptic connections are assessed by sampling voltage or current responses to light-emitting diode (LED) photostimulation of ChR2-expressing axons. The data are analyzed to estimate the relative amplitude of synaptic input and other connectivity parameters. Pharmacological and electrophysiological manipulations extend the versatility of the basic approach, allowing the dissection of monosynaptic versus disynaptic responses, excitatory versus inhibitory responses, and more. The method enables rapid, quantitative characterization of synaptic connectivity between defined pre- and postsynaptic classes of neurons.

MATERIALS

It is essential that you consult the appropriate Material Safety Data Sheets and your institution's Environmental Health and Safety Office for proper handling of equipment and hazardous material used in this protocol.

RECIPES: Please see the end of this protocol for recipes indicated by <R>. Additional recipes can be found online at http://cshprotocols.cshlp.org/site/recipes.

Reagents

3-((R)-2-carboxypiperazin-4-yl)-propyl-1-phosphonic acid (CPP) or other N-methyl-D-aspartate (NMDA) antagonist (optional; see Step 4)

Artificial cerebrospinal fluid (ACSF) for LSPS <R>

ACSF is used as the bath solution. For analysis of monosynaptic inputs (see Step 9 and Table 1), add tetrodotoxin (TTX, 1 µM) and 4-aminopyridine (4AP, 100 µM).

Internal (pipette) solution containing desired cation (see Table 1)

Mice of desired genotype and developmental stage

[3]Correspondence: g-shepherd@northwestern.edu

TABLE 1. Recording conditions for isolating different kinds of signals

Drugs in ACSF[a]	Pipette cation	Amplifier mode	Membrane potential[b]	Signals sampled[c]	Notes
None	K⁺	I-clamp	~RMP	EPSP, IPSP, AP	Mono- and disynaptic
None	Cs⁺	V-clamp	−70 mV V_{com}	EPSC	Usually monosynaptic but can be polysynaptic
None	Cs⁺	V-clamp	0 mV V_{com}	IPSC	Disynaptic (assuming presynaptic axons all excitatory)
TTX, 4AP	K⁺ or Cs⁺	V-clamp	−70 mV V_{com}	EPSC	Monosynaptic
TTX, 4AP	Cs⁺	V-clamp	0 mV V_{com}	IPSC	Monosynaptic

[a]Depending on the type of experiment, it may be useful to add blockers of NMDA receptors, HCN channels, or other voltage- or ligand-gated conductances.
[b]RMP: resting membrane potential; V_{com}: command voltage.
[c]EPSP, IPSP, EPSC, IPSC: excitatory/inhibitory postsynaptic potential/current; AP: action potential.

Retrograde tracers (see Step 2)

Examples include fluorescent microspheres (red or green RetroBeads [Lumafluor]) and cholera toxin subunit B conjugated with fluorophores (Molecular Probes).

Viruses encoding ChR2 (see Step 1)

Commercial sources include Penn Vector Core (for AAV-ChR2) and Duke Viral Vector Core (for RV-ChR2); protocols are also available for high-titer preparation of RV (see Protocol: Rabies Viral Vectors for Monosynaptic Tracing and Targeted Transgene Expression in Neurons [Wickersham and Sullivan 2015])

Equipment

Pipette puller and microgrinder

Slice rig equipped for whole-cell recording, photostimulation with a blue LED and/or laser, and epifluorescence microscopy

Software packages (e.g., Ephus [http://scanimage.vidriotechnologies.com]) for hardware control and data acquisition

Stereotaxic frame and micromanipulator

METHOD

See Figure 1 for an outline of the experimental procedures.

In Vivo Labeling of Pre- and Postsynaptic Neurons

Perform in vivo stereotaxic delivery of virus (Step 1) and retrograde tracers (Step 2) into mice following standard procedures (Cetin et al. 2006). For reviews and further descriptions of viral and nonviral tracing methods, see Wickersham et al. (2007), Betley and Sternson (2011), Nassi et al. (2015), and Protocol: Rabies Viral Vectors for Monosynaptic Tracing and Targeted Transgene Expression in Neurons (Wickersham and Sullivan 2015).

1. Inject ChR2-virus following one of the two paradigms described below, depending on the goals of the experiment.

   With either paradigm, the infected neurons' ChR2-expressing axons will be the presynaptic source of input in the subsequent in vitro experiments, as they remain photoexcitable in brain slices (Petreanu et al. 2007). Viral constructs should be chosen to allow coexpression of a fluorescent protein (e.g., Venus, mCherry, etc.) to aid in targeting subsequent recordings to regions containing transfected axons. Choice of virus strain or serotype is also an important consideration, influencing factors such as cytotoxicity in the case of RV, the degree of retrograde labeling in the case of AAV, and more (Nassi et al. 2015; Reardon et al. 2016).

   Paradigm 1: RV-ChR2

   i. Use injection with RV-ChR2 to optogenetically label one or the other of two classes of projection neurons that are intermingled in the same brain region, as a way to study their local interconnections (Kiritani et al. 2012).

      The projection neurons are retrogradely infected by injecting rabies viruses (RV) encoding ChR2 (Osakada et al. 2011; Kiritani et al. 2012) into their axonal projections (i.e., downstream brain area).

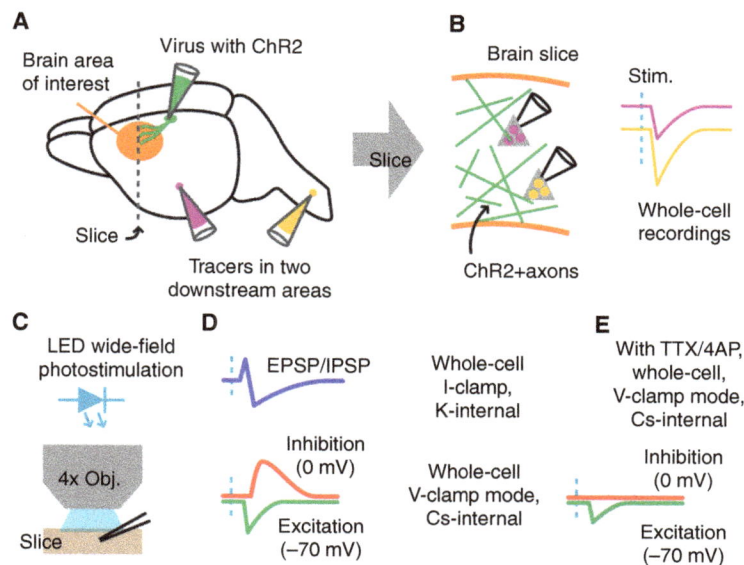

FIGURE 1. General paradigm for optogenetic circuit electroanatomy. (*A*) In vivo injections are made to deliver either rabies or adeno-associated virus carrying ChR2 to one brain region of interest that is either upstream or downstream from the main region of interest (orange area indicates where brain slices will be made), and to deliver retrograde tracers of two different colors in areas that are downstream from this area. (*B*) In brain slices of the main brain area of interest, ChR2-expressing presynaptic axons (green) are photostimulated while recordings are made from the two types of projection neurons identified by fluorescent labeling. (*C*) LED illumination through a low-power objective (Obj.) is used for wide-field photostimulation. (*D*) Electrophysiological manipulations allow different types of responses to be sampled, including mixed excitatory and inhibitory postsynaptic potentials (EPSP/IPSP) in current-clamp (I-clamp) mode, or isolated inhibitory and excitatory currents in voltage-clamp (V-clamp) mode. "K-internal" and "Cs-internal" refer to the cation used in the internal (pipette) solution; see Table 1. (*E*) Pharmacological manipulations allow isolation of purely monosynaptic inputs.

Alternatively, mouse lines are available that express Cre recombinase in various types of projection neurons (Gerfen et al. 2013). In this case, inject AAV carrying Cre-dependent ChR2 into the cortical area of interest to label a small cluster of neurons. A consideration is the potential damage from craniotomy and insertion of a pipette in the same brain region where recordings will subsequently be performed. Grinding the injection pipettes to a sharp beveled edge on a microgrinder and inserting pipettes at an angle to the cortical surface helps to minimize damage.

Paradigm 2: AAV-ChR2

    ii. Use injection with AAV-ChR2 to study long-range afferent innervation to two types of projection neurons that are intermingled in the same brain region.

The long-range axons of projection neurons in an upstream area of interest can be anterogradely labeled by infecting their parent somata with AAV virus encoding ChR2 (e.g., Petreanu et al. 2009; Hooks et al. 2013).

2. Inject two different retrograde tracers.

Stereotaxically inject these as appropriate to label two classes of projection neurons of interest. The somata of the labeled neurons will be targeted as the postsynaptic neurons in the subsequent in vitro experiment. The temporal order of the tracer and viral injections is generally not critical.

Alternatively, injection of only one retrograde tracer suffices for experiments where synaptic inputs will be compared between identified projection neurons at different spatial locations; for example, vertically across layers in a particular cortical area, horizontally along a layer across the cortex, across subcortical nuclei (including over multiple slices), etc. (Suter and Shepherd 2015; Yamawaki and Shepherd 2015).

Analysis of Excitatory and Inhibitory Currents

Analysis of excitatory and inhibitory currents is described in Steps 3–8. For analysis of monosynaptic circuits, see Step 9.

In Vitro Circuit Analysis of Disynaptic Currents

3. After a suitable time for transgene expression, typically ~1 wk for RV-ChR2 and ~3 wk for AAV-ChR2 (Kiritani et al. 2012; Yamawaki and Shepherd 2015), follow standard methods for preparing brain slices for photostimulation experiments (see Protocol: Circuit Mapping by Ultraviolet Uncaging of Glutamate [Shepherd 2012]).

   The wait time is shorter with RV both because transgene expression levels increase rapidly after infection and because cytotoxicity eventually develops and kills infected cells. Control experiments should be performed to assess the health (e.g., intrinsic electrophysiological properties) of RV-infected neurons in the time window of recordings (Kiritani et al. 2012). Other retrograde viruses, such as rabies-pseudotyped lentivirus, may be useful alternatives if longer survival times are needed (see Protocol: Lentiviral Vectors for Retrograde Delivery of Recombinases and Transactivators [Wickersham et al. 2015]).

4. Place a brain slice in the recording chamber perfused with artificial cerebrospinal fluid (ACSF, warmed to 32°C with an in-line heater).

   Pharmacological conditions and other experimental parameters can be set to isolate different kinds of signals of interest for circuit analysis (Table 1). NMDA-receptor blockade (e.g., CPP, 1 µM) is commonly used to prevent synaptic plasticity.

5. Under a high-power objective lens, target postsynaptic neurons based on fluorescent retrograde labeling, and establish a gigaohm seal using a pipette filled with cesium- or potassium-based internal solution (according to the goal of the experiment; see Table 1).

6. Perform recording from the first neuron as follows:

   i. Switch to a low-power objective lens (e.g., 4×, so that photostimuli will broadly excite axons across the neuropil around the neuron's dendrites). Adjust the position of the slice to bring it into an appropriate alignment within the field of view, noting the x–y coordinates so that they can be used for subsequent neurons in the same slice.

   ii. Compensate pipette capacitance and rupture the membrane for whole-cell configuration in voltage-clamp (V-clamp) or current-clamp (I-clamp) mode, depending on the goals of the experiment.

   iii. For routine mapping of excitatory inputs, photostimulate the presynaptic axons while recording from the postsynaptic neuron by driving an LED with a brief (e.g., 5-msec) pulse to deliver a wide-field flash of blue light.

   For the first neuron of the slice, choose one that is expected to receive relatively strong input (based on previous experiments). Start with high intensity (e.g., 1 mW/mm^2), and adjust as needed to obtain (if possible) a response of moderate amplitude (e.g., 100 pA). Then use this LED power setting as the standard level for all following neurons in the same slice. Sampling at additional LED powers can also be useful, particularly in the early stages of a project, to assess how responses scale with stimulus intensity. A common practice is to acquire several traces for online or posthoc averaging, with interstimulus intervals set to allow recovery from desensitization.

   iv. To isolate excitatory currents, apply a command voltage to hold near the GABAergic reversal potential, typically around −70 mV as calculated or determined empirically, such as by GABA uncaging (Wood et al. 2009).

   Inhibitory currents can similarly be sampled at the glutamatergic reversal potential of ~0 mV (usually with cesium-based internal solution, for better voltage control). For cortical slices, GABA antagonists should be avoided as photostimulation of excitatory neurons may generate epileptiform discharges, even at low doses (Weiler et al. 2008).

7. Record from a second, neighboring neuron labeled with the other retrograde tracer, and repeat the photostimulation recording. Use the same x–y coordinates for the slice position, and the same LED power setting(s), so that the same axons are stimulated as for the first neuron. Repeat for additional neurons from the same slice, as needed for the objectives of the experiment.

8. Continue to Step 10 (statistical analysis).

In Vitro Circuit Analysis of Monosynaptic Inputs

9. Perform circuit analysis as in Steps 3–8, except include tetrodotoxin (TTX, 1 μM) and 4-amino-pyridine (4AP, 100 μM) in the bath solution.

   Bath solution containing TTX and 4AP abolishes action potentials, enhances local depolarization of photo-stimulated ChR2-expressing axons, and results in local photoevoked depolarization of presynaptic terminals sufficient to induce neurotransmitter release—isolating purely monosynaptic inputs (Petreanu et al. 2009). The TTX/4AP conditions were originally developed for use with laser scanning photostimulation (subcellular ChR2-assisted circuit mapping, or "sCRACM") (Petreanu et al. 2009), but can be used with LED photo-stimulation to map cellular connections to retrogradely labeled projection neurons (Suter and Shepherd 2015; Yamawaki and Shepherd 2015). The LED and laser optical systems can be implemented in the same microscope.

Statistical Analysis

10. Analyze the traces offline to quantify the responses.

    The relative innervation of two postsynaptic cell types by a single presynaptic axon type can be shown by plotting the response amplitudes (e.g., mean or peak) against each other. Pairwise and other types of comparisons can be made using a variety of statistical measures to analyze inputs to the two types of postsynaptic projection neurons (Yamawaki and Shepherd 2015). For example, for paired observations, use the nonparametric sign test for two medians (Kanji 2006).

    In general, the normalization-based approach provides a measure of the relative amplitude of synaptic input to multiple postsynaptic neurons. It controls for variability in factors that can affect the absolute amplitude, such as the number of infected neurons per animal and expression levels of ChR2 in axons. The method can be used to compare neurons not only in the same slice but also in different slices from the same animal (Yamawaki and Shepherd 2015). Normalized responses can then be pooled across experiments (i.e., animals) for group (population) analyses.

DISCUSSION

Projection neurons, through their interconnections, form the major circuits of the central nervous system. Brain regions typically contain multiple classes of projection neurons, and thus two basic questions about projection neuron circuits are the extents to which different projection classes (1) interconnect locally and (2) receive inputs from upstream regions ("long-range inputs"). The preserved photoexcitability of ChR2-expressing axons in brain slices has ushered in a new era of optogenetic-electrophysiological experiments capable of probing local and long-range connectivity (Petreanu et al. 2007). The presynaptic photostimulation aspect of the protocol described here builds on methods originally described by Petreanu, Svoboda, and colleagues (Petreanu et al. 2007, 2009). A simplification is that connectivity is assessed with an LED light source, which is technically simpler to implement than a laser scanning system and also substantially shortens the recording time per cell, allowing more recordings per slice. An LED and a laser scanning system can both be mounted on the same microscope, enabling the operator to use both methods in the same experiment, including for the same cell. On the postsynaptic side, the key aspect of this protocol is the use of retrograde tracers such as fluorescent microspheres (Katz et al. 1984) to allow identified projection neurons to be targeted for postsynaptic recordings during photostimulation of presynaptic ChR2-expressing axons (Anderson et al. 2010; Mao et al. 2011).

The general paradigm of making pairwise comparisons of electrophysiological responses recorded in two different postsynaptic cell types following photostimulation of presynaptic inputs has been used in a wide variety of experiments. Here, by recording from identified projection neurons, the circuit analysis is extended to include the areas downstream from where the recordings are performed—providing an additional and functionally relevant level of detail in circuit diagrams. For example, detection of excitatory inputs from primary motor cortex (M1) to M1-projecting neurons in secondary somatosensory cortex indicates a pattern of recurrent connectivity between the two areas (Suter and Shepherd 2015).

The basic approach described here should be easy to adapt for many different types of experiments. Using a laser scanning system (and adding TTX and 4AP in the bath solution) allows high-resolution mapping of the dendritic location of presynaptic inputs, as discussed above (Petreanu et al.

2009). Alternatively, with a laser scanning system but with TTX/4AP omitted from the bath, one can generate high-frequency barrages of inputs (from axons at different locations) to study how identified postsynaptic projection neurons translate synaptic input into spiking output; for example, this approach showed a role for hyperpolarization-activated, cyclic nucleotide–gated (HCN) channels in integrating synaptic input from layer 2/3 pyramidal neurons to generate action potentials in corticospinal neurons in motor cortex (Sheets et al. 2011).

While the optogenetic-electrophysiological approach has key advantages due to the ability to selectively activate just the presynaptic axons of interest, it does have several limitations. For one, the approach described here enables one to determine the relative strength of input to different neurons from the same source of input, but unlike paired recordings does not provide a measure of absolute input strength—not only because multiple axons are photoexcited, but because ChR2-related factors (e.g., slow kinetics, Ca^{2+} permeability) can affect release from presynaptic terminals (Schoenenberger et al. 2011). The use of high-frequency repetitive photostimulation (at the same location) is potentially an issue, due to ChR2 desensitization and the slow kinetics of ChR2 (compared to those of potassium channels involved in repolarization), which can distort presynaptic waveforms and affect short-term plasticity (Zhang and Oertner 2007). However, these problems have been reduced in newer variants of ChR2 with faster kinetics (Gunaydin et al. 2010; Berndt et al. 2011; Kleinlogel et al. 2011; Klapoetke et al. 2014), and in some cases may be avoided by photostimulating axons at a short distance away from the recorded postsynaptic neuron (Little and Carter 2013; Jackman et al. 2014).

RECIPE

ACSF for LSPS

1. Prepare the stock solution by combining the following reagents in H_2O to a final volume of 2 L. Store it at 4°C. Its shelf life is up to 1 wk.

Reagents	Quantity
Glucose	9.01 g
KCl	0.373 g
NaCl	14.844 g
$NaH_2PO_4 \cdot H_2O$	0.345 g
$NaHCO_3$	4.2 g

2. Prepare the working solution by adding 1.0 mL of 1 M $CaCl_2$ and 0.5 mL of 1 M $MgCl_2$ to 500 mL of stock solution. This working solution contains standard divalent cation concentrations of 2 mM Ca^{2+} and 1 mM Mg^{2+}. Store it at 4°C. Its shelf life is up to 1 wk.

REFERENCES

Anderson CT, Sheets PL, Kiritani T, Shepherd GMG. 2010. Sublayer-specific microcircuits of corticospinal and corticostriatal neurons in motor cortex. *Nat Neurosci* **13**: 739–744.

Berndt A, Schoenenberger P, Mattis J, Tye KM, Deisseroth K, Hegemann P, Oertner TG. 2011. High-efficiency channelrhodopsins for fast neuronal stimulation at low light levels. *Proc Natl Acad Sci* **108**: 7595–7600.

Betley JN, Sternson SM. 2011. Adeno-associated viral vectors for mapping, monitoring, and manipulating neural circuits. *Hum Gene Ther* **22**: 669–677.

Cetin A, Komai S, Eliava M, Seeburg PH, Osten P. 2006. Stereotaxic gene delivery in the rodent brain. *Nat Protoc* **1**: 3166–3173.

Gerfen CR, Paletzki R, Heintz N. 2013. GENSAT BAC Cre-recombinase driver lines to study the functional organization of cerebral cortical and basal ganglia circuits. *Neuron* **80**: 1368–1383.

Gunaydin LA, Yizhar O, Berndt A, Sohal VS, Deisseroth K, Hegemann P. 2010. Ultrafast optogenetic control. *Nat Neurosci* **13**: 387–392.

Hooks BM, Mao T, Gutnisky D, Yamawaki N, Svoboda K, Shepherd GMG. 2013. Organization of cortical and thalamic input to pyramidal neurons in mouse motor cortex. *J Neurosci* **33**: 748–760.

Jackman SL, Beneduce BM, Drew IR, Regehr WG. 2014. Achieving high-frequency optical control of synaptic transmission. *J Neurosci* **34**: 7704–7714.

Kanji GK. 2006. *100 Statistical Tests.* Sage, London.

Katz LC, Burkhalter A, Dreyer WJ. 1984. Fluorescent latex microspheres as a retrograde neuronal marker for in vivo and in vitro studies of visual cortex. *Nature* **310**: 498–500.

Kiritani T, Wickersham IR, Seung HS, Shepherd GMG. 2012. Hierarchical connectivity and connection-specific dynamics in the corticospinal-corticostriatal microcircuit in mouse motor cortex. *J Neurosci* **32**: 4992–5001.

Klapoetke NC, Murata Y, Kim SS, Pulver SR, Birdsey-Benson A, Cho YK, Morimoto TK, Chuong AS, Carpenter EJ, Tian Z, et al. 2014. Independent optical excitation of distinct neural populations. *Nat Methods* **11**: 338–346.

Kleinlogel S, Feldbauer K, Dempski RE, Fotis H, Wood PG, Bamann C, Bamberg E. 2011. Ultra light-sensitive and fast neuronal activation with the Ca^{2+}-permeable channelrhodopsin CatCh. *Nat Neurosci* **14**: 513–518.

Cite this protocol as *Cold Spring Harb Protoc*; doi:10.1101/pdb.prot090084

Little JP, Carter AG. 2013. Synaptic mechanisms underlying strong reciprocal connectivity between the medial prefrontal cortex and basolateral amygdala. *J Neurosci* **33**: 15333–15342.

Mao T, Kusefoglu D, Hooks BM, Huber D, Petreanu L, Svoboda K. 2011. Long-range neuronal circuits underlying the interaction between sensory and motor cortex. *Neuron* **72**: 111–123.

Nassi JJ, Cepko CL, Born RT, Beier KT. 2015. Neuroanatomy goes viral! *Front Neuroanat* **9**: 80.

Osakada F, Mori T, Cetin AH, Marshel JH, Virgen B, Callaway EM. 2011. New rabies virus variants for monitoring and manipulating activity and gene expression in defined neural circuits. *Neuron* **71**: 617–631.

Petreanu L, Huber D, Sobczyk A, Svoboda K. 2007. Channelrhodopsin-2-assisted circuit mapping of long-range callosal projections. *Nat Neurosci* **10**: 663–668.

Petreanu L, Mao T, Sternson SM, Svoboda K. 2009. The subcellular organization of neocortical excitatory connections. *Nature* **457**: 1142–1145.

Reardon TR, Murray AJ, Turi GF, Wirblich C, Croce KR, Schnell MJ, Jessell TM, Losonczy A. 2016. Rabies Virus CVS-N2c(ΔG) strain enhances retrograde synaptic transfer and neuronal viability. *Neuron* **89**: 711–724.

Schoenenberger P, Scharer YP, Oertner TG. 2011. Channelrhodopsin as a tool to investigate synaptic transmission and plasticity. *Exp Physiol* **96**: 34–39.

Sheets PL, Suter BA, Kiritani T, Chan CS, Surmeier DJ, Shepherd GMG. 2011. Corticospinal-specific HCN expression in mouse motor cortex: I_h-dependent synaptic integration as a candidate microcircuit mechanism involved in motor control. *J Neurophysiol* **106**: 2216–2231.

Shepherd GM. 2012. Circuit mapping by ultraviolet uncaging of glutamate. *Cold Spring Harb Protoc* **2012**: 998–1004.

Suter BA, Shepherd GM. 2015. Reciprocal interareal connections to corticospinal neurons in mouse m1 and s2. *J Neurosci* **35**: 2959–2974.

Weiler N, Wood L, Yu J, Solla SA, Shepherd GMG. 2008. Top-down laminar organization of the excitatory network in motor cortex. *Nat Neurosci* **11**: 360–366.

Wickersham IR, Sullivan HA. 2015. Rabies viral vectors for monosynaptic tracing and targeted transgene expression in neurons. *Cold Spring Harb Protoc* **2015**: 375–385.

Wickersham IR, Finke S, Conzelmann KK, Callaway EM. 2007. Retrograde neuronal tracing with a deletion-mutant rabies virus. *Nat Methods* **4**: 47–49.

Wickersham IR, Sullivan HA, Pao GM, Hamanaka H, Goosens KA, Verma IM, Seung HS. 2015. Lentiviral vectors for retrograde delivery of recombinases and transactivators. *Cold Spring Harb Protoc* **2015**: 368–374.

Wood L, Gray NW, Zhou Z, Greenberg ME, Shepherd GMG. 2009. Synaptic circuit abnormalities of motor-frontal layer 2/3 pyramidal neurons in an RNA interference model of methyl-CpG-binding protein 2 deficiency. *J Neurosci* **29**: 12440–12448.

Yamawaki N, Shepherd GM. 2015. Synaptic circuit organization of motor corticothalamic neurons. *J Neurosci* **35**: 2293–2307.

Zhang YP, Oertner TG. 2007. Optical induction of synaptic plasticity using a light-sensitive channel. *Nat Methods* **4**: 139–141.

CHAPTER 14

Whole-Cell Recording in the Awake Brain

Doyun Lee[1,3] and Albert K. Lee[2,3]

[1]*Center for Cognition and Sociality, Institute for Basic Science, Daejeon 34141, Republic of Korea;* [2]*Howard Hughes Medical Institute, Janelia Research Campus, Ashburn, Virginia 20147*

Intracellular recording is an essential technique for investigating cellular mechanisms underlying complex brain functions. Despite the high sensitivity of the technique to mechanical disturbances, intracellular recording has been applied to awake, behaving, and even freely moving, animals. Here we summarize recent advances in these methods and their application to the measurement and manipulation of membrane potential dynamics for understanding neuronal computations in behaving animals.

INTRODUCTION

Intracellular recording has served as a key technique for investigating the cellular and circuit mechanisms underlying neural activity. Because of its high sensitivity to mechanical disturbances, intracellular recording previously was limited primarily to stable preparations such as brain slices and, more recently, the intact brains of anesthetized animals (Pei et al. 1991; Ferster and Jagadeesh 1992; Metherate et al. 1992; Moore and Nelson 1998; Zhu and Connors 1999; Margrie et al. 2002). Thus, it had not been straightforward to link the accumulated knowledge regarding synaptic and intrinsic mechanisms from experiments conducted mostly in vitro with higher brain functions that generally involve active participation of the animal. The development of intracellular recording methods for awake behaving animals across multiple species (Covey et al. 1996; Fee 2000; Aksay et al. 2001; Margrie et al. 2002; Wilson et al. 2004; Lee et al. 2006, 2014; Harvey et al. 2009; Long et al. 2010; English et al. 2014; Tan et al. 2014) has therefore played and will continue to play a critical role in understanding the neural basis of cognitive processes (for a review, see Long and Lee (2012)).

INTRACELLULAR RECORDINGS IN HEAD-FIXED BEHAVING ANIMALS

Information processing in the brain is affected by vigilance or arousal states. Therefore, some recordings need to be performed during wakefulness to fully understand neuronal computations. Several intracellular recording studies have demonstrated differences in subthreshold membrane potential dynamics during anesthetized or sleep versus awake states (Steriade et al. 2001; Mahon et al. 2006; Constantinople and Bruno 2011; Haider et al. 2013). Furthermore, animals do not simply passively receive sensory input; they actively interact with the environment, which also affects information processing. For instance, whole-cell recordings in the barrel cortex of head-fixed mice revealed membrane potential fluctuations synchronized to active whisker movements. Different

[3]Correspondence: leedoyun@ibs.re.kr; leea@janelia.hhmi.org

Cite this introduction as *Cold Spring Harb Protoc*; doi:10.1101/pdb.top087304

neurons had different phase offsets, which could contribute to position coding (Crochet and Petersen 2006).

Intracellular recording has also been performed in head-fixed animals engaged in activities involving much larger bodily motions such as walking, running, or wing-beating. It had been shown previously that locomotion affects the firing rate of orientation-tuned visual neurons (Niell and Stryker 2010). Cellular mechanisms for such gain modulation have been investigated using whole-cell recordings in head-fixed mice that alternated between periods of running and sitting still on a spherical treadmill while being presented with visual stimuli (Bennett et al. 2013; Polack et al. 2013). Whole-cell recordings obtained from flies fixed in place at the head but free to move their wings were used to study flight-dependent gain modulation in visual neurons from the lobula plate (Maimon et al. 2010).

Head-fixed preparations that allow intracellular recordings to be performed during locomotion on a treadmill have been combined with visually simulated spatial environments where visual scenes are updated based on the animal's movements. Such virtual reality systems allow the study of the brain's navigation system. Whole-cell recordings from mouse hippocampal CA1 place cells revealed intracellular features such as a ramping subthreshold depolarization of the membrane potential during traversals through place fields and phase precession of the intracellular theta-band oscillation with respect to the extracellular local field potential theta rhythm, providing deeper understanding into place cell mechanisms (Harvey et al. 2009). Similar approaches have been used for measuring intracellular features of entorhinal grid cells, allowing investigation into the mechanisms underlying their periodic firing pattern as a function of the animal's location in space (Domnisoru et al. 2013; Schmidt-Hieber and Häusser 2013), as well as for assessing the role of large calcium-based events in place field plasticity in CA1 (Bittner et al. 2015).

Protocol 1: In Vivo Patch-Clamp Recording in Awake Head-Fixed Rodents (Lee and Lee 2016a) describes a detailed training procedure and an efficient patching method for obtaining whole-cell recordings from awake head-fixed animals (Lee et al. 2014).

INTRACELLULAR RECORDINGS IN FREELY MOVING ANIMALS

Although many questions regarding input–output transformations within individual neurons in different behavioral paradigms can be addressed using head-fixed animals, other topics require animals to be non-head-fixed—for example, processing that depends on vestibular input or requires rapid head movements (Monaco et al. 2014). Lee et al. (2006) were the first to demonstrate that intracellular recording (in this case using the whole-cell method) was possible in freely moving animals. These recordings were obtained in anesthetized animals using a head-mounted linear microdrive to advance the pipette electrode; the pipette was subsequently fixed (i.e., "anchored") in place with dental acrylic and the animal awakened for unrestrained behavior. The recordings could survive severe mechanical disturbances and remain stable for ∼20 min while rats explored their environment freely. Other groups have shown that sharp intracellular recordings could be obtained and maintained in freely moving birds (Long et al. 2010; Hamaguchi et al. 2014; Kosche et al. 2015; Vallentin and Long 2015) and mice (English et al. 2014; Schneider et al. 2014) using a head-mountable linear microdrive. The first such recordings were made from bird premotor HVC neurons to measure the subthreshold membrane potential dynamics during singing, and the resulting recordings supported a synaptic chain-like mechanism for birdsong generation (Long et al. 2010). Using this method in freely moving mice, Schneider et al. (2014) demonstrated that the membrane potential of excitatory neurons in auditory cortex depolarizes and decreases its variability immediately before and during bodily movements including locomotion, head movements, and grooming. Additional intracellular recordings in head-fixed mice running in place in concert with optogenetic manipulations revealed that these membrane potential dynamics reflect motor-related corollary discharge originating from secondary motor cortex.

An alternate version of the freely moving whole-cell recording technique uses pipette anchoring combined with a conventional micromanipulator located to the side of the animal (Lee et al. 2009) instead of a head-mounted microdrive. Conventional micromanipulators not only allow the recording pipette to be held in place more strongly during the anchoring process but also allow the pipette to move horizontally, thus providing a higher probability of obtaining high-quality whole-cell recordings (see Lee and Lee [2016a] for a description of how horizontal movement is used for in vivo patching). This technique has been applied to the study of hippocampal place cells in freely moving rats exploring spatial environments. One such study focused on spikelets, which are all-or-none events with similarities to the action potential but much smaller in amplitude and thus not easily detectable extracellularly. Recordings of hippocampal CA1 pyramidal neurons showed that spikelets occur in the behaving animal and can contribute directly to place cell firing during spatial exploration (Epsztein et al. 2010). In another study, the intracellular features of place cells and silent cells were compared in rats exploring a novel environment. Contrary to expectations from the basic model of neuronal integration, silent cells—unlike place cells—did not display large, spatially tuned inputs at the soma. Furthermore, the excitability of these cells measured even before the rats explored the environment (i.e., for the first time) could partially predict which cells would become place cells, suggesting that intrinsic excitability plays a significant role in place field origin and memory allocation (Epsztein et al. 2011). This method using conventional micromanipulators initially used dental acrylic to fix pipettes in place. However, a limitation of this pipette stabilization technique was that the anchoring process took a relatively long time (~10 min) and thus required that the whole-cell recording be obtained in an initially anesthetized animal that was subsequently awakened for free behavior (Lee et al. 2009). In addition, recording losses during the anchoring process limited the overall success rate of the method (Lee et al. 2009).

To improve this method, the use of dental acrylic was replaced with an ultraviolet (UV)-cured adhesive and UV-transparent collar to anchor the whole-cell pipette in place. This provides highly efficient pipette stabilization with no recording losses. Furthermore, it anchors a pipette in 15 sec, thus allowing the whole-cell recording to be obtained in both anesthetized and awake animals (Lee et al. 2014). This technique has also been applied to the study of cellular mechanisms underlying place cell activity. In particular, injecting a spatially uniform holding current into a silent cell reversibly converted it into a place cell during spatial exploration. Injecting different amounts of current into the silent cell revealed the sudden emergence of large, spatially tuned inputs and place cell spiking above a cell-specific threshold, supporting the idea that nonlinear dendritic mechanisms underlie place cell firing by gating the propagation of spatial inputs to the soma (Lee et al. 2012). These observations were reproduced in drug-free, awake-patched recordings in freely moving animals (Lee et al. 2014).

Protocol 2: Efficient Method for Whole-Cell Recording in Freely Moving Rodents Using UV-Cured Collar-Based Pipette Stabilization (Lee and Lee 2016b) describes a detailed step-by-step procedure for awake-patched whole-cell recording in freely moving, drug-free rats and mice.

We end this introduction with a comment on whole-cell and sharp intracellular recording methods for freely moving animals. High-quality and long-duration recordings of both types have been obtained successfully. There is, at present, no indication of differences in overall success rates when considering the latest versions of the methods applied to the same brain region (e.g., hippocampal subregion CA1) of the same species (e.g., mice) (English et al. 2014; Lee et al. 2014), although differences might emerge with increased use. However, there are some technical differences inherent to each approach. The ability to achieve low series resistance recordings with the whole-cell configuration makes voltage-clamp experiments possible. On the other hand, multiple recordings can be obtained with the same sharp microelectrode, whereas whole-cell pipettes must be replaced after each recording, as well as each attempt in which negative pressure has been applied.

ACKNOWLEDGMENTS

Work in our laboratories is supported by the Howard Hughes Medical Institute and the Institute for Basic Science, Republic of Korea.

REFERENCES

Aksay E, Gamkrelidze G, Seung HS, Baker R, Tank DW. 2001. In vivo intracellular recording and perturbation of persistent activity in a neural integrator. *Nat Neurosci* **4:** 184–193.

Bennett C, Arroyo S, Hestrin S. 2013. Subthreshold mechanisms underlying state-dependent modulation of visual responses. *Neuron* **80:** 350–357.

Bittner KC, Grienberger C, Vaidya SP, Milstein AD, Macklin JJ, Suh J, Tonegawa S, Magee JC. 2015. Conjunctive input processing drives feature selectivity in hippocampal CA1 neurons. *Nat Neurosci* **18:** 1133–1142.

Constantinople CM, Bruno RM. 2011. Effects and mechanisms of wakefulness on local cortical networks. *Neuron* **69:** 1061–1068.

Covey E, Kauer JA, Casseday JH. 1996. Whole-cell patch-clamp recording reveals subthreshold sound-evoked postsynaptic currents in the inferior colliculus of awake bats. *J Neurosci* **16:** 3009–3018.

Crochet S, Petersen CCH. 2006. Correlating whisker behavior with membrane potential in barrel cortex of awake mice. *Nat Neurosci* **9:** 608–610.

Domnisoru C, Kinkhabwala AA, Tank DW. 2013. Membrane potential dynamics of grid cells. *Nature* **495:** 199–204.

English DF, Peyrache A, Stark E, Roux L, Vallentin D, Long MA, Buzsáki G. 2014. Excitation and inhibition compete to control spiking during hippocampal ripples: Intracellular study in behaving mice. *J Neurosci* **34:** 16509–16517.

Epsztein J, Lee AK, Chorev E, Brecht M. 2010. Impact of spikelets on hippocampal CA1 pyramidal cell activity during spatial exploration. *Science* **327:** 474–477.

Epsztein J, Brecht M, Lee AK. 2011. Intracellular determinants of hippocampal CA1 place and silent cell activity in a novel environment. *Neuron* **70:** 109–120.

Fee MS. 2000. Active stabilization of electrodes for intracellular recording in awake behaving animals. *Neuron* **27:** 461–468.

Ferster D, Jagadeesh B. 1992. EPSP-IPSP interactions in cat visual cortex studied with in vivo whole-cell patch recording. *J Neurosci* **12:** 1262–1274.

Haider B, Häusser M, Carandini M. 2013. Inhibition dominates sensory responses in the awake cortex. *Nature* **493:** 97–100.

Hamaguchi K, Tschida KA, Yoon I, Donald BR, Mooney R. 2014. Auditory synapses to song premotor neurons are gated off during vocalization in zebra finches. *Elife* **3:** e01833.

Harvey CD, Collman F, Dombeck DA, Tank DW. 2009. Intracellular dynamics of hippocampal place cells during virtual navigation. *Nature* **461:** 941–946.

Kosche G, Vallentin D, Long MA. 2015. Interplay of inhibition and excitation shapes a premotor neural sequence. *J Neurosci* **35:** 1217–1227.

Lee D, Lee AK. 2016a. In vivo patch-clamp recording in awake head-fixed rodents. *Cold Spring Harb Protoc* doi: 10.1101/pdb.prot095802.

Lee D, Lee AK. 2016b. Efficient method for whole-cell recording in freely moving rodents using UV-cured collar-based pipette stabilization. *Cold Spring Harb Protoc* doi: 10.1101/pdb.prot095810.

Lee AK, Manns ID, Sakmann B, Brecht M. 2006. Whole-cell recordings in freely moving rats. *Neuron* **51:** 399–407.

Lee AK, Epsztein J, Brecht M. 2009. Head-anchored whole-cell recordings in freely moving rats. *Nat Protoc* **4:** 385–392.

Lee D, Lin B-J, Lee AK. 2012. Hippocampal place fields emerge upon single-cell manipulation of excitability during behavior. *Science* **337:** 849–853.

Lee D, Shtengel G, Osborne JE, Lee AK. 2014. Anesthetized- and awake-patched whole-cell recordings in freely moving rats using UV-cured collar-based electrode stabilization. *Nat Protoc* **9:** 2784–2795.

Long MA, Lee AK. 2012. Intracellular recording in behaving animals. *Curr Opin Neurobiol* **22:** 34–44.

Long MA, Jin DZ, Fee MS. 2010. Support for a synaptic chain model of neuronal sequence generation. *Nature* **468:** 394–399.

Mahon S, Vautrelle N, Pezard L, Slaght SJ, Deniau J-M, Chouvet G, Charpier S. 2006. Distinct patterns of striatal medium spiny neuron activity during the natural sleep–wake cycle. *J Neurosci* **26:** 12587–12595.

Maimon G, Straw AD, Dickinson MH. 2010. Active flight increases the gain of visual motion processing in *Drosophila*. *Nat Neurosci* **13:** 393–399.

Margrie T, Brecht M, Sakmann B. 2002. In vivo, low-resistance, whole-cell recordings from neurons in the anaesthetized and awake mammalian brain. *Pflügers Archiv* **444:** 491–498.

Metherate R, Cox CL, Ashe JH. 1992. Cellular bases of neocortical activation: Modulation of neural oscillations by the nucleus basalis and endogenous acetylcholine. *J Neurosci* **12:** 4701–4711.

Monaco JD, Rao G, Roth ED, Knierim JJ. 2014. Attentive scanning behavior drives one-trial potentiation of hippocampal place fields. *Nat Neurosci* **17:** 725–731.

Moore CI, Nelson SB. 1998. Spatio-temporal subthreshold receptive fields in the vibrissa representation of rat primary somatosensory cortex. *J Neurophysiol* **80:** 2882–2892.

Niell CM, Stryker MP. 2010. Modulation of visual responses by behavioral state in mouse visual cortex. *Neuron* **65:** 472–479.

Pei X, Volgushev M, Vidyasagar TR, Creutzfeldt OD. 1991. Whole cell recording and conductance measurements in cat visual cortex in-vivo. *NeuroReport* **2:** 485–488.

Polack P-O, Friedman J, Golshani P. 2013. Cellular mechanisms of brain state-dependent gain modulation in visual cortex. *Nat Neurosci* **16:** 1331–1339.

Schmidt-Hieber C, Häusser M. 2013. Cellular mechanisms of spatial navigation in the medial entorhinal cortex. *Nat Neurosci* **16:** 325–331.

Schneider DM, Nelson A, Mooney R. 2014. A synaptic and circuit basis for corollary discharge in the auditory cortex. *Nature* **513:** 189–194.

Steriade M, Timofeev I, Grenier F. 2001. Natural waking and sleep states: A view from inside neocortical neurons. *J Neurophysiol* **85:** 1969–1985.

Tan AYY, Chen Y, Scholl B, Seidemann E, Priebe NJ. 2014. Sensory stimulation shifts visual cortex from synchronous to asynchronous states. *Nature* **509:** 226–229.

Vallentin D, Long MA. 2015. Motor origin of precise synaptic inputs onto forebrain neurons driving a skilled behavior. *J Neurosci* **35:** 299–307.

Wilson RI, Turner GC, Laurent G. 2004. Transformation of olfactory representations in the *Drosophila* antennal lobe. *Science* **303:** 366–370.

Zhu JJ, Connors BW. 1999. Intrinsic firing patterns and whisker-evoked synaptic responses of neurons in the rat barrel cortex. *J Neurophysiol* **81:** 1171–1183.

Cite this introduction as *Cold Spring Harb Protoc*; doi:10.1101/pdb.top087304

In Vivo Patch-Clamp Recording in Awake Head-Fixed Rodents

Doyun Lee[1,3] and Albert K. Lee[2,3]

[1]Center for Cognition and Sociality, Institute for Basic Science, Daejeon 34141, Republic of Korea; [2]Howard Hughes Medical Institute, Janelia Research Campus, Ashburn, Virginia 20147

Whole-cell recording has been used to measure and manipulate a neuron's spiking and subthreshold membrane potential, allowing assessment of the cell's inputs and outputs as well as its intrinsic membrane properties. This technique has also been combined with pharmacology and optogenetics as well as morphological reconstruction to address critical questions concerning neuronal integration, plasticity, and connectivity. This protocol describes a technique for obtaining whole-cell recordings in awake head-fixed animals, allowing such questions to be investigated within the context of an intact network and natural behavioral states. First, animals are habituated to sit quietly with their heads fixed in place. Then, a whole-cell recording is obtained using an efficient, blind patching protocol. We have successfully applied this technique to rats and mice.

MATERIALS

It is essential that you consult the appropriate Material Safety Data Sheets and your institution's Environmental Health and Safety Office for proper handling of equipment and hazardous material used in this protocol.

RECIPES: Please see the end of this protocol for recipes indicated by <R>. Additional recipes can be found online at http://cshprotocols.cshlp.org/site/recipes.

Reagents

Agarose (2% [w/v] in physiological saline; Sigma-Aldrich A9539)
Analgesic (e.g., Buprenex [Reckitt Benckiser Pharmaceuticals])
Bupivacaine hydrochloride monohydrate (Sigma-Aldrich B5274)
Dental acrylic (e.g., Jet Denture Repair Package [Lang Dental])
Dental adhesive, light-cured (e.g., OptiBond FL [Kerr Dental])
Dental composite, light-cured (e.g., Charisma [Heraeus Kulzer])
Experimental animals

This protocol has been successfully used with juvenile rats (age of 4–6 wk) and adult mice (age of ~3 mo).

Intracellular pipette solution <R>
Isoflurane (e.g., IsoSol [Vedco])
Ketamine hydrochloride/xylazine hydrochloride (Sigma-Aldrich K4138)
Paraformaldehyde (4%)
Phosphate-buffered saline (PBS) (0.1 M)
Physiological saline (i.e., 0.9% NaCl)

[3]Correspondence: leedoyun@ibs.re.kr; leea@janelia.hhmi.org

Silicone elastomer (e.g., Kwik-Cast Sealant [World Precision Instruments])
Veterinary ophthalmic ointment (e.g., Puralube [Dechra Veterinary Products])

Equipment

Acquisition software (e.g., Patchmaster [HEKA Elektronik])
Acquisition system, analog-to-digital (e.g., InstruTECH [HEKA Elektronik])
Ag/AgCl pipette with reference electrode wire
Amplifier (e.g., npi electronic ELC-03XS)
Anesthesia system (e.g., VetEquip 901806)
Audio monitor (e.g., Grass Technologies AM10)
Curing unit (e.g., Translux Power Blue [Heraeus Kulzer])
Faraday cage
Glass capillaries, borosilicate, with filament, 1.5 mm OD, 0.87 mm ID (Hilgenberg)
Glass capillary, reference electrode-holding (~15 mm long, 1 mm OD)
Headplate, custom-designed (Fig. 1A)
Light source, halogen (Schott KL1500 HAL)
Manometer (Sigmann Elektronik)
Microdrill (Foredom)
Microdrill bit (0.45 mm)
Micromanipulator, motorized, 3- or 4-axis (e.g., Luigs & Neumann)
Micropipette storage jar (e.g., World Precision Instruments)
Patching station
   This custom-designed unit consists of a base plate to which a headplate holder (Fig. 1B) is attached.

Pipette puller (e.g., Narishige PC-10 or Sutter Instrument P-97)
Reference electrode-holder cap
   An ~5-mm-long piece of Tygon tubing (1/32 in. ID) with one end blocked is suitable for this purpose.

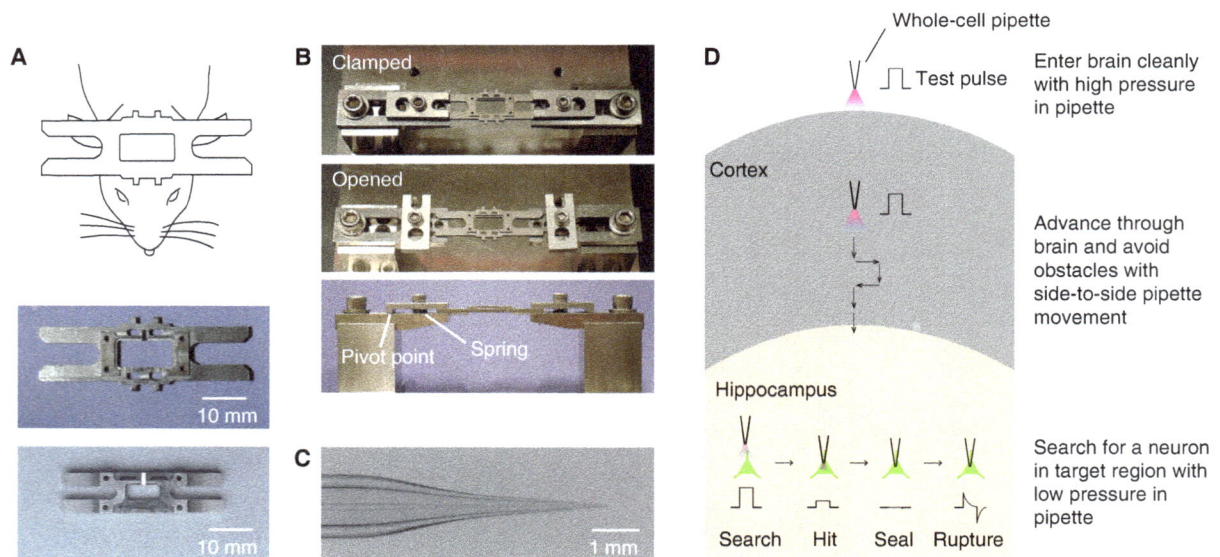

FIGURE 1. Whole-cell recordings in awake head-fixed animals. (A) Drawing showing position of headplate attached to a rat's head (top). Headplate designs for rats (middle) and mice (bottom). (B) Rapidly releasable headplate holder. (C) An example of a recording pipette. (D) Whole-cell patching procedure for awake head-fixed animals. (B, reprinted, with permission, from Lee et al. 2014.)

Stereoscope (e.g., Olympus SZ61)
Stereotaxic apparatus (e.g., Narishige SR-6)
Surgical instruments (Fine Science Tools)
> *Clean and sterilize instruments thoroughly before use.*

Temperature control system (e.g., FHC)
Vibration isolation table (Newport)

METHOD

Headplate Implantation Surgery

> *Use a temperature control system to maintain the body temperature of the experimental animal at ~37°C during surgery.*

1. Using an anesthesia system, sedate the experimental animal using 1.5%–2.0% isoflurane.

2. Affix the animal's head in the stereotaxic apparatus.

3. Apply ophthalmic ointment to the subject's eyes. Cover with small pieces of black paper.
   > *This protects the subject's eyes during surgery.*

4. Subcutaneously inject bupivacaine (2 mg/kg body weight) into the subject's scalp.

5. Make an incision along the scalp midline. Open the skin to expose the skull. Scrape away the connective tissue covering the bone.

6. Mark the coordinates overlying the region of the brain of interest.
   > *For example, if the dorsal part of the rat hippocampal subregion CA1 is to be studied, mark a point 3.5 mm posterior and 2.5 mm lateral to the bregma.*

7. Taking care to avoid the target site, apply three to four drops of dental adhesive to the skull. Light-cure the adhesive.

8. Apply a 0.5- to 1-mm thick layer of dental composite on top of the cured adhesive. Light-cure the composite.

9. Affix the headplate to the cured composite on the subject's skull with a layer of dental acrylic. Cover the inner wall of the headplate completely with dental acrylic such that the area inside the headplate is isolated electrically from the rest of the recording station.
   > *Make sure there is no gap between the skull and the headplate.*

10. Using dental acrylic, attach a reference electrode-holding capillary to the headplate.

11. After surgery, proceed as follows:

    i. Rehydrate the experimental subject by injecting it subcutaneously with ~2 mL warmed saline/100 g body weight.

    ii. Administer an analgesic (e.g., 0.05 mg Buprenex/kg body weight) immediately after surgery and for the next 2 d.

    iii. House experimental animals individually until they are fully recovered (at least 2 d).

Habituation to Head Fixation

12. Affix the headplate attached to the experimental subject's head to the patching station's headplate holder. Leave the animal undisturbed for ~5 min on the first day. Over the course of 4 d, gradually increase the time of habituation to ~1 h.

13. Once the animal is habituated to head fixation, expose it to experiment-related noises during each training session.

14. Continue fixation training daily with 1 h-sessions until the animal is sufficiently accustomed to being head-fixed and can sit quietly for several minutes at a time without movement.

> *Obtaining a giga-seal is difficult if an animal moves during patching. Proceed to Step 15 only with animals that are sufficiently habituated to head fixation. In our hands, >50% of animals proceed to patching.*

Craniotomy and Target Depth Determination

15. On the first day of patching, anesthetize the animal with 1.5%–2.0% isoflurane. Fix its head into the patching station.

16. Drill a 1–2 mm diameter hole in the skull at the previously marked target coordinates.

17. Optionally (required for rats; for mice this step can be omitted), use a ≥26-gauge needle and/or microscissors to cut a small opening in the dura (∼0.5 mm diameter).

> *Be careful not to make the dura opening larger than ∼1 mm, as this can otherwise lead to tissue swelling. Be especially careful not to damage the brain tissue below when targeting cortical areas.*

18. Insert an Ag/AgCl reference electrode through the reference electrode-holding capillary.

19. Insert an extracellular glass electrode (1–3 MΩ). Search for characteristic activity by listening to the audio monitor.

> *This step might not be necessary for shallow cortical target areas but is essential for deeper targets such as the hippocampus (for which the characteristic activity of the CA1 pyramidal cell layer is intermittent bursting).*

20. Once the target depth has been established, protect the brain surface by covering it with a layer of 2% agarose, then cover the agarose with a layer of silicone elastomer. Remove the reference electrode, and cap the reference electrode-holding capillary.

21. Return the animal to its home cage for at least 2 h of recovery before starting a whole-cell recording session.

Awake Whole-Cell Recording

22. Head-fix the animal in the patching station.

23. Insert an Ag/AgCl reference electrode through the reference electrode-holding capillary.

24. Fill a glass pipette (prepared in advance; 4–7 MΩ) with ∼5 μL of the intracellular pipette solution.

> *See Fig. 1C for an example of a glass pipette. However, other pipette shapes have also been successfully used in our hands and those of others.*

25. Using the micromanipulator onto which the headstage is mounted, position the pipette tip a few hundred microns above the brain surface. Apply high pressure to the inside of the pipette (∼800 mbar for a deep target such as the hippocampus, ∼300 mbar for a superficial target). Lower the pipette onto the brain surface (Fig. 1D).

> *It is critical to enter the brain surface through a clean spot. Especially avoid spots with clearly visible blood. When the dura is not removed in mice, the electrode passes through the dura, but patching through dura does not otherwise require a different technique.*

26. While maintaining high-positive pipette pressure, slowly lower the pipette in 5-μm increments into the brain to the search depth (e.g., in the case of the dorsal CA1 pyramidal cell layer, 100–150 μm above the estimated target depth).

> *It is critical to keep the pipette tip clean until attempting giga-seal formation. Carefully monitor the pipette resistance (a good indicator of the cleanness of the pipette tip) while advancing the pipette.*
> *See Troubleshooting.*

27. When the pipette reaches the search depth:
   i. Reduce the positive pressure applied to the pipette to 25–35 mbar (i.e., "search pressure").

ii. Advance the pipette using 1-μm steps to search for a target neuron (Fig. 1D).

iii. When there is an increase in the pipette resistance (by ~20% or more) for 4–5 consecutive steps, attempt to form a giga-seal by removing the positive pressure and applying gentle negative pressure (<10 mbar).

> *At this point, hyperpolarizing the pipette holding potential to −65 mV can facilitate giga-seal formation.*
>
> *See Troubleshooting.*

28. Rupture the membrane within the patch pipette by applying a brief and strong pulse of negative pressure (>100 mbar for ~0.2 sec) to achieve the whole-cell configuration.

> *If no giga-seal is attained or the break-in to whole-cell configuration is unsuccessful, retract the pipette and repeat Steps 24–28 with a new pipette.*

29. At the end of each recording, withdraw the recording pipette slowly to facilitate resealing of the membrane for histological recovery.

30. When a recording session is finished:

i. Cover the brain surface with a layer of 2% agarose, followed by a layer of silicone elastomer.

ii. Remove the reference electrode. Cap the reference electrode-holding capillary.

iii. Return the animal to its home cage.

> *If the same subject undergoes two or more recording sessions on the same day, permit the animal to recuperate in its home cage for 1–2 h between sessions. (We have obtained up to five whole-cell recordings from hippocampal area CA1 in the same hemisphere and have not, in this or any other cases, observed any noticeable impairment in the network activity.)*

Histology

31. After the last recording session, deeply anesthetize the animal by administrating ketamine (200 mg/kg body weight)/xylazine (10 mg/kg body weight).

32. Perfuse the animal transcardially with 0.1 M PBS, then with 4% paraformaldehyde.

33. Extract the brain. Fix overnight in paraformaldehyde. Transfer to PBS.

34. Slice the brain into 150-μm thick sections. Visualize neuronal morphology using an avidin–biotin–peroxidase method (Horikawa and Armstrong 1988).

> *When multiple neurons are recorded from the same hemisphere, unambiguously identifying each neuron is usually difficult. Therefore, for unambiguous identification, filling only one neuron per hemisphere is recommended.*
>
> *See Troubleshooting.*

TROUBLESHOOTING

Problem (Step 26): Resistance increases when the pipette is inserted into the brain.

Solution: If the dura is dissected (e.g., as for rats), pipette resistance should not change on entry. Repeat the dissection of the dura, or target the pipette to a clean spot on the dissected area. If inserting the pipette through an intact dura (e.g., as in mice), pipette resistance might increase temporarily, but it should return to the original value as soon as the pipette penetrates the dura. If the resistance does not recover, increase the positive pressure applied to the inside of the pipette.

Problem (Step 26): The pipette encounters an obstacle (e.g., a cell ahead of the target depth, a blood vessel, etc.) during its advance to the search depth.

Solution: Generally, an increase in resistance by 10% or more is indicative of an obstruction. Retract the pipette a few tens of microns until the resistance returns to its original value, then advance it again. In many cases, this will allow one to bypass the obstacle without any increase in resistance.

If this remains unsuccessful after several attempts, retract the pipette until the resistance returns to its original value, reposition it to the side by up to 50 µm (usually <10 µm), then advance the pipette again in an attempt to bypass the obstacle (Fig. 1D). Repeat as necessary to avoid all obstacles on the way to the search depth.

Problem (Steps 26–27): Pipette resistance cannot be maintained until giga-seal formation is attempted.
Solution: Excessive movement while the subject is restrained can translate to large brain movements. Wait until the animal stops moving, extend the habituation period, or reject the subject in favor of a calmer animal. In practice, fewer than half of the animals need to be rejected for this reason.

Problem (Step 27): Gigaohm seal cannot be established.
Solution: Using a microscope, examine the pipette tip for clogging; in such cases, the positive pressure applied to the inside of the pipette cannot effectively push away thin dendrites or the extracellular matrix. If the same issue occurs repeatedly using different pipettes, try the following steps to eliminate the source of the clog: filter the internal solution; clean the inside of the glass capillaries before pulling pipettes; rechloride the recording wire. If no clog is evident, advance the pipette further into the membrane after the increase in resistance before removing positive pressure (e.g., instead of 4–5 steps of 1 µm, use 6–8 steps of 1 µm).

Problem (Step 27): The pipette does not hit the somata of neurons.
Solution: If the pipette hits dendrites, glia, or other structures in the brain, estimation of the target depth might be incorrect. Repeat Step 19 to find the correct depth of the target brain region. If the pipette hits dendrites, the search pressure might be too low. If the pipette does not hit any structures during the search for a neuron, the search pressure might be too high.

Problem (Step 34): The recorded cells cannot be observed histologically.
Solution: If a recording ends because of a loss of the whole-cell configuration, there is a greater chance that the biocytin will leak out, resulting in poor visualization of the recorded neuron. If histological recovery is required in such cases, perfuse the brain as soon as possible. If the whole-cell configuration is still intact at the end of a recording, slowly withdraw the recording pipette to allow the membrane to reseal, thus limiting biocytin leakage.

DISCUSSION

Because of the high sensitivity of whole-cell recording to mechanical perturbations, in vivo whole-cell recording has been performed more often in anesthetized animals. Recently, an increasing number of studies have performed whole-cell recording in awake head-fixed animals (Margrie et al. 2002; Crochet and Petersen 2006; Harvey et al. 2009; Lee et al. 2014; Tan et al. 2014). Under optimal conditions, the success rate, recording duration, and recording quality rely largely on the extent of brain movement. Thus, habituating animals to sit quietly for long enough periods of time is important for successful recording. If an appropriately habituated (i.e., calm) animal is used, the patching method described here can usually achieve more than one recording per every three attempts (i.e., three different pipettes). Because it can be particularly challenging to obtain a giga-seal while the animal is moving, one should attempt giga-seal formation only while the animal is sitting quietly; once a tight seal is formed, recordings can tolerate gentle-to-moderate brain movement. In many experiments, active movement of the animal is required, such as for comparing processing between a quiet period and a period with locomotion (Bennett et al. 2013; Polack et al. 2013; Schiemann et al. 2015), or running on a treadmill to navigate virtual environments (Harvey et al. 2009; Domnisoru et al. 2013; Schmidt-Hieber and Häusser 2013; Bittner et al. 2015). In these cases, selecting animals that walk or run smoothly—which cause less-sudden brain motions—is critical for long-lasting recordings.

Cite this protocol as *Cold Spring Harb Protoc*; doi:10.1101/pdb.prot095802

RECIPE

Intracellular Pipette Solution

Reagent	Final concentration
Potassium gluconate	135 mM
HEPES	10 mM
Na$_2$-phosphocreatine	10 mM
KCl	4 mM
MgATP	4 mM
Na$_3$GTP	0.3 mM
Biocytin	~0.05% (w/v)

Prepare in distilled water. Adjust pH to 7.3 using KOH (target osmolarity is 295 mOsm). Store the solution for up to 1 yr at −20°C.

ACKNOWLEDGMENTS

We thank Jason Osborne for assistance in designing the custom equipment. Work in our laboratories is supported by the Howard Hughes Medical Institute and the Institute for Basic Science, Republic of Korea.

REFERENCES

Bennett C, Arroyo S, Hestrin S. 2013. Subthreshold mechanisms underlying state-dependent modulation of visual responses. *Neuron* **80:** 350–357.

Bittner KC, Grienberger C, Vaidya SP, Milstein AD, Macklin JJ, Suh J, Tonegawa S, Magee JC. 2015. Conjunctive input processing drives feature selectivity in hippocampal CA1 neurons. *Nat Neurosci* **18:** 1133–1142.

Crochet S, Petersen CCH. 2006. Correlating whisker behavior with membrane potential in barrel cortex of awake mice. *Nat Neurosci* **9:** 608–610.

Domnisoru C, Kinkhabwala AA, Tank DW. 2013. Membrane potential dynamics of grid cells. *Nature* **495:** 199–204.

Harvey CD, Collman F, Dombeck DA, Tank DW. 2009. Intracellular dynamics of hippocampal place cells during virtual navigation. *Nature* **461:** 941–946.

Horikawa K, Armstrong WE. 1988. A versatile means of intracellular labeling: Injection of biocytin and its detection with avidin conjugates. *J Neurosci Methods* **25:** 1–11.

Lee D, Shtengel G, Osborne JE, Lee AK. 2014. Anesthetized- and awake-patched whole-cell recordings in freely moving rats using UV-cured collar-based electrode stabilization. *Nat Protoc* **9:** 2784–2795.

Margrie T, Brecht M, Sakmann B. 2002. In vivo, low-resistance, whole-cell recordings from neurons in the anaesthetized and awake mammalian brain. *Pflügers Archiv* **444:** 491–498.

Polack P-O, Friedman J, Golshani P. 2013. Cellular mechanisms of brain state–dependent gain modulation in visual cortex. *Nat Neurosci* **16:** 1331–1339.

Schiemann J, Puggioni P, Dacre J, Pelko M, Domanski A, van Rossum MCW, Duguid I. 2015. Cellular mechanisms underlying behavioral state-dependent bidirectional modulation of motor cortex output. *Cell Rep* **11:** 1319–1330.

Schmidt-Hieber C, Häusser M. 2013. Cellular mechanisms of spatial navigation in the medial entorhinal cortex. *Nat Neurosci* **16:** 325–331.

Tan AYY, Chen Y, Scholl B, Seidemann E, Priebe NJ. 2014. Sensory stimulation shifts visual cortex from synchronous to asynchronous states. *Nature* **509:** 226–229.

Efficient Method for Whole-Cell Recording in Freely Moving Rodents Using Ultraviolet-Cured Collar-Based Pipette Stabilization

Doyun Lee[1,3] and Albert K. Lee[2,3]

[1]Center for Cognition and Sociality, Institute for Basic Science, Daejeon 34141, Republic of Korea; [2]Howard Hughes Medical Institute, Janelia Research Campus, Ashburn, Virginia 20147

Whole-cell recording is a key technique for investigating synaptic and cellular mechanisms underlying various brain functions. However, because of its high sensitivity to mechanical disturbances, applying the whole-cell recording method to freely moving animals has been challenging. Here, we describe a technique for obtaining such recordings in freely moving, drug-free animals with a high success rate. This technique involves three major steps: obtaining a whole-cell recording from awake head-fixed animals, reliable and efficient stabilization of the pipette with respect to the animal's head using an ultraviolet (UV)-transparent collar and UV-cured adhesive, and rapid release of the animal from head fixation without loss of the recording. This technique has been successfully applied to obtain intracellular recordings from the hippocampus of freely moving rats and mice exploring a spatial environment, and should be generally applicable to other brain areas in animals engaged in a variety of natural behaviors.

MATERIALS

It is essential that you consult the appropriate Material Safety Data Sheets and your institution's Environmental Health and Safety Office for proper handling of equipment and hazardous material used in this protocol.

RECIPES: Please see the end of this protocol for recipes indicated by <R>. Additional recipes can be found online at http://cshprotocols.cshlp.org/site/recipes.

Reagents

Acrylic paint, black (e.g., Crafter's acrylic [DecoArt])
Adhesive, UV-cured (Norland NOA63)
Agarose (2% [w/v] in physiological saline; Sigma-Aldrich A9539)
Agarose, darkened (4% black acrylic paint in 2% agarose)
Analgesic (e.g., Buprenex [Reckitt Benckiser Pharmaceuticals])
Bupivacaine hydrochloride monohydrate (Sigma-Aldrich B5274)
Dental acrylic (e.g., Jet Denture Repair Package [Lang Dental 1234PNK])
Dental adhesive, light-cured (e.g., OptiBond FL [Kerr Dental])
Dental composite, light-cured (e.g., Charisma [Heraeus Kulzer])
Experimental animals
> *This protocol has been successfully used with juvenile rats (age of 4–6 wk) and adult mice (age of ~3 mo).*

[3]Correspondence: leedoyun@ibs.re.kr; leea@janelia.hhmi.org

Intracellular pipette solution <R>
Isoflurane (e.g., IsoSol [Vedco])
Ketamine hydrochloride/xylazine hydrochloride (Sigma-Aldrich K4138)
Physiological saline (i.e., 0.9% NaCl)
Silicone elastomer (e.g., Kwik-Cast Sealant [World Precision Instruments])
Veterinary ophthalmic ointment (e.g., Puralube [Dechra Veterinary Products])

Equipment

The procedure for whole-cell recording in freely moving rodents requires several custom-designed parts that have been described previously (Lee et al. 2014). Contact the authors for further details.

Acquisition software (e.g., Patchmaster [HEKA Elektronik])
Acquisition system, analog-to-digital (e.g., InstruTECH [HEKA Elektronik])
Acrylic paint marker, black (e.g., DecoColor)
Air tubing and tubing connectors (World Precision Instruments)

> *Attach a three-way tubing connector, a pipette electrode wire, and an air tube as shown in Figure 1D.*

Amplifier, with miniature headstage (e.g., npi electronic ELC-03XS; see Fig. 1B)
Anesthesia system (e.g., VetEquip 901806)
Audio monitor (e.g., Grass Technologies AM10)
Behavioral arena, electrically shielded and grounded
Cable counterweighting system

> *The weight of the headstage/light-emitting diode (LED) assembly is relieved by counterweights attached to a few locations on the recording wires coming from the headstage via pulleys. Carefully adjust the weights such that the head-mounted device floats at the level of animal's head when not attached to the animal.*

Camera, charge-coupled device (CCD) (optional; see Step 25.i)
Collar, synthetic sapphire, 3 mm tall, with a 1.55-mm diam (± 0.005 mm) hole at the center (Rayotek Scientific; see Fig. 1A, inset)

> *The collar must be custom-designed and fabricated precisely, as the hole size determines the amount of UV adhesive required around the pipette.*

Curing unit (e.g., Translux Power Blue [Heraeus Kulzer])
Eye cover, hinged, custom-made (see Fig. 1E, top)

> *The hinged eye cover—made from a piece of lightweight black paper placed in front of the headplate holder— is designed to protect the animal's eyes from the lights and the experimenter's movements. Because of its low weight, the animal can easily lift the cover with its nose when the animal's head is released for free behavior.*

Faraday cage
Glass capillaries, borosilicate, with filament, 1.5 mm OD, 0.87 mm ID (Hilgenberg)
Glass capillary, reference electrode-holding, ~15 mm long, 1 mm OD
Gripper jaws, custom-made (Fig. 1C)
Headplate and rapidly releasable headplate holder, custom-designed (Fig. 1A,D)

> *It is essential to make the headplate holder quickly releasable. Unscrewing by a half-turn allows the headplate clamp to rotate because the spring beneath it lifts it, thus allowing the headplate to be released by pulling it up (by the wires on the headstage).*

Headstage, miniaturized, equipped with custom LED assembly (npi electronic; see Fig. 1B)
Heating plate, equipped with magnetic stirrer
Light source, halogen (Schott KL1500 HAL)
Manometer (Sigmann Elektronik)
Microdrill (Foredom)
Microdrill bit (0.45 mm)
Micromanipulator, motorized, 3- or 4-axis (e.g., Luigs & Neumann)
Micropipette storage jar (e.g., World Precision Instruments)

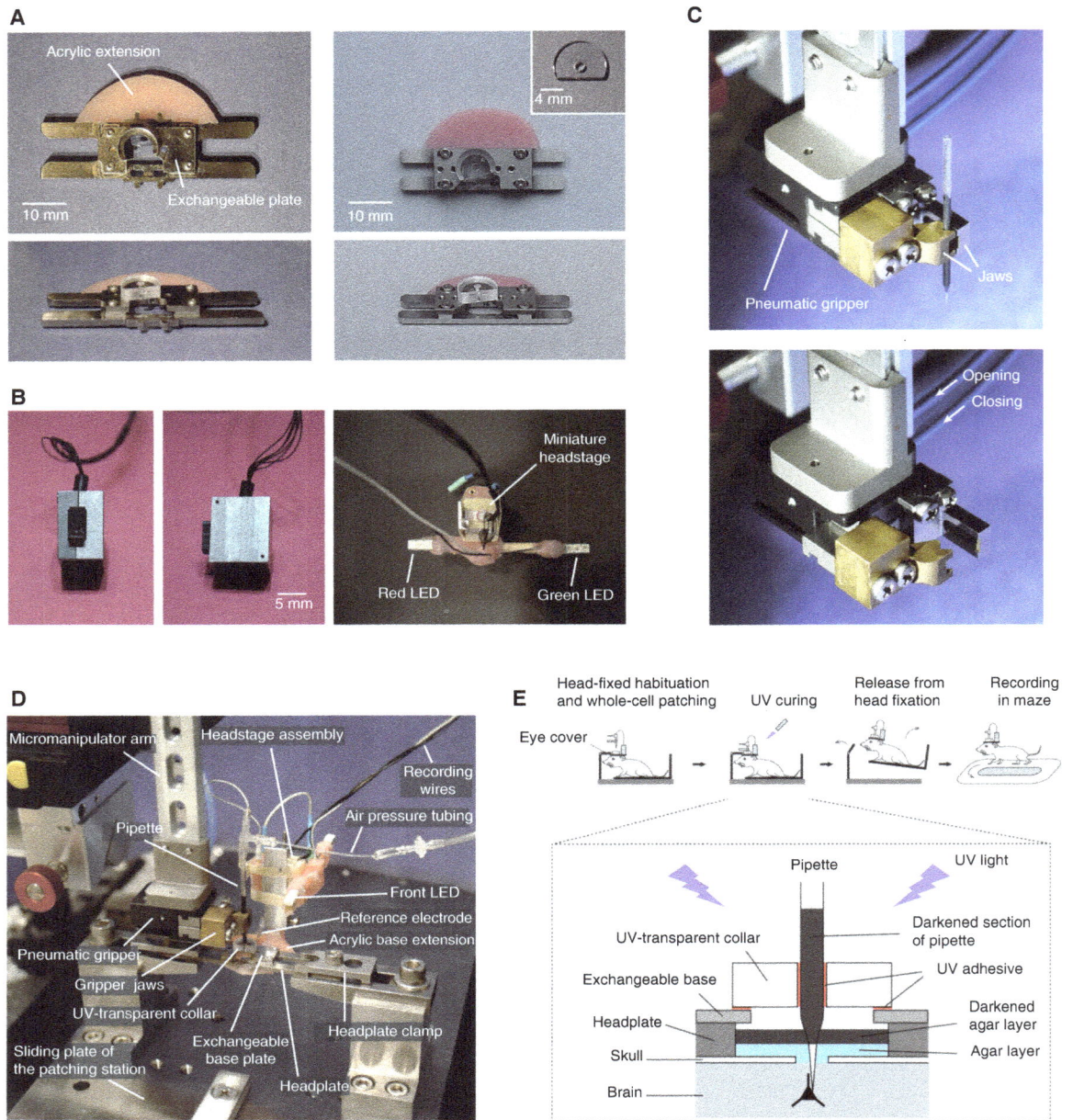

FIGURE 1. Awake-patched whole-cell recordings in freely moving animals. (*A*) Headplate for rats (*left*) and mice (*right*) with acrylic extension and an exchangeable base with a UV-transparent collar placed on it. A close-up of the collar (*inset*). (*B*) Miniaturized headstage (*left* and *center*) and a top view of a headstage assembled with LEDs (*right*). (*C*) Pneumatic pipette gripper. Custom-made jaws hold (*top*) and release (*bottom*) a pipette. (*D*) Arrangement of head-mounted equipment and head-fixed patching station with a rapidly releasable headplate holder. (*E*) Procedure for awake-patched whole-cell recordings in freely moving animals (*top*). Note that the animal is sitting inside the transfer box until it is released for recording in the maze. Arrangement of UV-transparent collar/UV-cured adhesive-based pipette stabilization (*bottom*). (Modified, with permission, from Lee et al. 2014.)

Patching station

This custom-designed unit consists of a sliding base plate to which a headplate holder is attached (Fig. 1D).

Photodiode power sensor (Thorlabs S120VC)
Pipette electrode wire, Ag/AgCl

Solder a 7- to 8-mm-long silver wire to an insulated connecting wire with a gold pin at the other end. Chloride the silver wire from the tip to near the soldered end.

Pipette puller (Sutter Instrument P-97)

Pneumatic grippers, 2-jaw (Zimmer Group MGP801N or MGP803N)

Position tracking system (e.g., Neuralynx)

Power meter console (Thorlabs PM100D)

Reference electrode, Ag/AgCl

Prepare a 30-mm-long silver wire and a 3-mm-long piece of thin tubing (with an OD slightly larger than 1/32 in.). Insert the wire into the tubing. Glue the tube at the middle of the wire. Chloride the wire to one side of the tube. Solder a gold pin to the other side. Insert the wire into a 6- to 7-mm-long piece of plastic (e.g., Tygon) tubing (1/32-in. ID) such that the piece of thin tubing glued at the middle of the wire wedges into the longer tube.

Reference electrode-holder cap

An ~5-mm-long piece of Tygon tubing (1/32 in. ID) with one end blocked is suitable for this purpose.

Stereoscope (e.g., Olympus SZ61)

Stereotaxic apparatus (e.g., Narishige SR-6R)

Surgical instruments (Fine Science Tools)

Clean and sterilize instruments thoroughly before use.

Syringe, equipped with blunt-ended 18-gauge needle

This is used for applying agarose.

Temperature control system (e.g., FHC)

Transfer box with urine absorber, custom-made

Cover the floor of the box with laboratory tissue and then a sintered aluminum plate such that urine will be absorbed through it, allowing the animal to remain more comfortable.

Ultraviolet curing system with bifurcated light guide (e.g., Hypercure 200 [Hologenix])

Vibration isolation table (Newport)

METHOD

All the training and patching procedures are performed while animals are head-fixed and sitting inside the transfer box. Once a whole-cell recording has been obtained and stabilized on the head, the animal can be released from head fixation and transferred to the behavioral arena. Wear gloves and a face shield to protect exposed skin from the UV light.

Headplate Implantation Surgery

1. Prepare the headplate with dental acrylic such that it extends horizontally to form a flat, semi-circular acrylic "base" (see Fig. 1A).

   This base will serve as the site of attachment for the headstage/LED assembly (see Step 11).

2. Attach the headplate to the skull of the experimental subject as described in Steps 1–11 of Protocol 1: In Vivo Patch-Clamp Recording in Awake Head-Fixed Rodents (Lee and Lee 2016).

Habituation to Head Fixation and Release

3. Place the experimental animal in the transfer box. Fix the animal's head into the custom-designed rapidly releasable headplate holders (see Fig. 1E, top left).

4. Close the hinged eye cover to protect the animal's eyes from the lights and experimenter movements (see Fig. 1E, top left).

5. Habituate the animal to head fixation as described in Steps 12–14 of Protocol 1: In Vivo Patch-Clamp Recording in Awake Head-Fixed Rodents (Lee and Lee 2016).

6. Habituate the animal to the release process (i.e., being rapidly released from head fixation and having its head lifted while remaining in the transfer box):

   i. Using the sliding plate, move the animal away from the gripper and light guides.

ii. While pressing down firmly on the headplate with a metal bar, unscrew the plate from the holder.

iii. Remove the bar. Simultaneously, swiftly and gently pull up the recording wires coming from the headstage/LED assembly to lift the animal's head.

Craniotomy and Target Depth Determination

7. Perform craniotomy and target depth determination as described in Steps 15–21 of Protocol 1: In Vivo Patch-Clamp Recording in Awake Head-Fixed Rodents (Lee and Lee 2016).

Prepare Painted Patch Pipettes

8. Pull patch pipettes (4–7 MΩ) before each patching session.

9. Paint the pipettes with a black acrylic paint marker to prevent a voltage offset caused by direct UV illumination of the internal solution.

   Paint the pipettes under a stereoscope. If the paint layer is too thick, the pipette will get stuck in the hole of the UV-transparent collar. If it is too thin, UV illumination will cause a voltage offset. Painting two-thirds of the pipette length is generally enough to protect 5 μL of internal solution. Leave ∼4 mm of length from the pipette tip unpainted to allow one to inspect the tip for trapped air bubbles (the unpainted area will be protected by a layer of agar containing black paint).

Awake Patching and Pipette Stabilization

10. Fix the animal's head in the patching station.

11. Attach the miniaturized headstage/LEDs assembly to the headplate (Fig. 1B,D).

    This attachment should be reversible such that the assembly can be easily detached at the end of each patching session.

12. Insert an Ag/AgCl reference electrode through the reference electrode-holding capillary.

13. Apply UV-cured adhesive to the bottom of the UV-transparent collar, then place the collar on top of the exchangeable base plate (which is itself attached to the headplate with screws; Fig. 1A,D).

    Giga-seal formation will not occur if the pipette tips are contaminated with adhesive. If the same collar is used for two or more pipettes, verify that the hole in the collar is not blocked by adhesive. If necessary, clean the hole with tissue before inserting a fresh pipette.

14. Fill a painted glass pipette with ∼5 μL of the intracellular pipette solution.

15. Insert the prepared pipette into the air-pressure-controlled gripper attached to the micromanipulator.

16. Apply a small amount of UV adhesive around the pipette. Insert the pipette into the hole of the collar (Fig. 1E).

    As the pipette is inserted, the adhesive fills the gap between the pipette and the inner surface of the hole.

17. Place UV light guides in front and back of the animal's head. Point them at the collar at ∼45° with respect to vertical.

    Perform the following calibration before the patching session. Use the photodiode power sensor and power meter console to determine the power at a given distance from the end of each light guide. The UV intensity from each light guide should be ∼42 mW/cm² at the site of curing. Adjust the distance between the end of the light guide and the collar to control the intensity.

18. Obtain a whole-cell recording:

    i. Use the micromanipulator to position the pipette tip a few hundred microns above the brain surface.

    ii. Apply high pressure to the inside of the pipette (∼800 mbar for a deep target such as the hippocampus, ∼300 mbar for a superficial target). Lower the pipette onto the brain surface.

       It is critical to enter the brain surface through a clean spot. Especially avoid spots with clearly visible blood.

iii. While maintaining high-positive pipette pressure, slowly lower the pipette in 5-μm increments into the brain to the search depth (e.g., in the case of the dorsal CA1 pyramidal cell layer, 100–150 μm above the estimated target depth).

It is critical to keep the pipette tip clean until attempting giga-seal formation. Carefully monitor the pipette resistance (a good indicator of the cleanness of the pipette tip) while advancing the pipette.

iv. When the pipette reaches the search depth, reduce the positive pressure applied to the pipette to 25–35 mbar (i.e., "search pressure").

v. Apply a layer of 2% agarose around the pipette and covering the craniotomy. Add a second layer of 2% darkened agarose on top of the first layer such that both the craniotomy and the unpainted tip of the pipette will be protected from UV illumination.

vi. Advance the pipette using 1-μm steps to search for a target neuron.

vii. When there is an increase in the pipette resistance (by ∼20% or more) for four to five consecutive steps, attempt to form a giga-seal by removing the positive pressure and applying gentle negative pressure (<10 mbar).

At this point, hyperpolarizing the pipette holding potential to −65 mV can facilitate giga-seal formation.

viii. Rupture the membrane within the patch pipette by applying a brief and strong pulse of negative pressure (>100 mbar for ∼0.2 sec) to achieve the whole-cell configuration.

19. Observe the quality and stability of the recording for 2–5 min.

Monitor both the resting membrane potential and the series resistance carefully; proceed to Step 20 only if they are stable. Otherwise, retract the pipette, remove the collar, remove the agar, wash the craniotomy surface with physiological saline, and repeat the process from Step 13 with a new pipette.

20. Cure the UV adhesive by illuminating the collar with UV light using two 5-sec pulses, separated by a 5-sec gap.

This step anchors the pipette rigidly to the collar and the collar rigidly to the exchangeable base plate.
See Troubleshooting.

Release and Transfer Animal to the Arena

21. Carefully cut the air pressure tube to the pipette.

22. Carefully open the air-pressure-controlled gripper that holds the recording pipette.

23. Release the animal from head fixation as described in Step 6.i–iii.

Make sure that the headplate does not hit the headplate holder below; otherwise, recordings can be easily lost.

24. Move the transfer box containing the animal adjacent to the behavioral arena. Allow the animal to walk out of the box into the arena.

Recording

25. Record the membrane potential in current-clamp mode:

i. If necessary, track the animal's position using a CCD camera that captures the LEDs attached to the headstage/LED assembly.

ii. Continue recording until the baseline membrane potential depolarizes above physiological levels (either suddenly or gradually) or the series resistance increases significantly.

If another recording is to be attempted in the same recording session, fix the animal's head into the patching station, hold the pipette with the gripper, detach the exchangeable base plate from the headplate, retract the pipette (with exchangeable plate and collar attached), remove the agar, wash the craniotomy surface with physiological saline, then repeat from Step 13 with a new pipette.

26. After each recording session, cover the brain surface with a layer of 2% agarose, followed by a layer of silicone elastomer. Remove the reference electrode. Cap the reference electrode-holding capillary.

If the same animal is used for two or more recording sessions on the same day, provide a 1–2 h rest period in its home cage between sessions. We have obtained multiple recordings from hippocampal area CA1 in the same hemisphere and have not in any cases observed any noticeable impairment in the network activity.

Histology

27. Recover the morphology of the recorded cell as described in Steps 31–34 of Protocol 1: In Vivo Patch-Clamp Recording in Awake Head-Fixed Rodents (Lee and Lee 2016).

    See Troubleshooting.

TROUBLESHOOTING

Problem (Step 20): Pipette fails to anchor properly when cured using UV illumination.

Solution: Establish that the UV light intensity is sufficiently high; check the UV intensity regularly and adjust it to the recommended value. Apply enough UV adhesive such that it completely fills the gaps between the collar and the pipette and between the collar and the exchangeable base plate.

Problem (Step 20): Illumination with UV light changes the membrane potential permanently.

Solution: This can occur when the paint on the pipette is not thick enough and thus does not effectively prevent the UV light from illuminating the internal solution directly. Paint pipettes under a stereoscope and ensure there are no unpainted gaps. A higher-than-recommended UV intensity could also be a problem. Check UV intensity and adjust it if too high.

Problem (Step 27): The recorded cells cannot be observed histologically.

Solution: Because the pipette cannot be gently retracted while the animal is not head-fixed, recordings generally end as a result of the loss of the whole-cell configuration. This can lead to biocytin leaking from the cell interior. Therefore, if histological recovery is required, the brain should be perfused immediately after the end of recording.

DISCUSSION

Different approaches involving both whole cell (Lee et al. 2006, 2009, 2012, 2014; Epsztein et al. 2010, 2011) and sharp (Long et al. 2010; English et al. 2014; Hamaguchi et al. 2014; Schneider et al. 2014; Kosche et al. 2015; Vallentin and Long 2015) intracellular methods have been attempted to overcome the high sensitivity of intracellular recordings to mechanical instability so that the technique can be applied more readily to freely moving animals (Long and Lee 2012). Rigid fixation of the recording pipette to the animal's head can yield stable, long-lasting, low series resistance whole-cell recordings in freely moving animals (Lee et al. 2006, 2009, 2014). Success is attributed to a high rate of giga-seal formation facilitated by the use of a three-axis micromanipulator (see Protocol 1: In Vivo Patch-Clamp Recording in Awake Head-Fixed Rodents [Lee and Lee 2016]), together with a strong, air-pressure-controlled gripper, as well as virtually 100%-efficient UV-based pipette stabilization (Lee et al. 2014). In particular, the nearly instantaneous stabilization of the pipette using the UV-cured, collar-based method allows whole-cell recordings to be obtained and stabilized in awake, drug-free animals. The protocol described here has been used in juvenile rats and adult mice by multiple experimenters with success rates of 40% or higher, where the success rate is defined as the percentage of all whole-cell recordings in which one has attempted UV-based pipette stabilization that ultimately leads to a good-quality intracellular recording in the freely moving animal (irrespective of the duration of the recording). Using this protocol, we have obtained 19 successful recordings from 16 rats; after release from head fixation, recordings lasted from 11 sec to 8.4 min (mean = 2.1 min), during which the rats traveled 1163 ± 232 cm on a linear track.

RECIPE

Intracellular Pipette Solution

Reagent	Final concentration
Potassium gluconate	135 mM
HEPES	10 mM
Na$_2$-phosphocreatine	10 mM
KCl	4 mM
MgATP	4 mM
Na$_3$GTP	0.3 mM
Biocytin	~0.05% (w/v)

Prepare in distilled water. Adjust pH to 7.3 using KOH (target osmolarity is 295 mOsm). Store the solution for up to 1 yr at $-20°$C.

ACKNOWLEDGMENTS

We thank Gleb Shtengel for providing the idea for the UV curing-aided pipette anchoring method and Jason Osborne for assistance in designing the custom equipment. Work in our laboratories is supported by the Howard Hughes Medical Institute and the Institute for Basic Science, Republic of Korea.

REFERENCES

English DF, Peyrache A, Stark E, Roux L, Vallentin D, Long MA, Buzsáki G. 2014. Excitation and inhibition compete to control spiking during hippocampal ripples: Intracellular study in behaving mice. *J Neurosci* **34**: 16509–16517.

Epsztein J, Lee AK, Chorev E, Brecht M. 2010. Impact of spikelets on hippocampal CA1 pyramidal cell activity during spatial exploration. *Science* **327**: 474–477.

Epsztein J, Brecht M, Lee AK. 2011. Intracellular determinants of hippocampal CA1 place and silent cell activity in a novel environment. *Neuron* **70**: 109–120.

Hamaguchi K, Tschida KA, Yoon I, Donald BR, Mooney R. 2014. Auditory synapses to song premotor neurons are gated off during vocalization in zebra finches. *eLife* **3**: e01833.

Kosche G, Vallentin D, Long MA. 2015. Interplay of inhibition and excitation shapes a premotor neural sequence. *J Neurosci* **35**: 1217–1227.

Lee D, Lee AK. 2016. In vivo patch-clamp recording in awake head-fixed rodents. *Cold Spring Harb Protoc* doi: 10.1101/pdb.prot095802.

Lee AK, Manns ID, Sakmann B, Brecht M. 2006. Whole-cell recordings in freely moving rats. *Neuron* **51**: 399–407.

Lee AK, Epsztein J, Brecht M. 2009. Head-anchored whole-cell recordings in freely moving rats. *Nat Protoc* **4**: 385–392.

Lee D, Lin B-J, Lee AK. 2012. Hippocampal place fields emerge upon single-cell manipulation of excitability during behavior. *Science* **337**: 849–853.

Lee D, Shtengel G, Osborne JE, Lee AK. 2014. Anesthetized- and awake-patched whole-cell recordings in freely moving rats using UV-cured collar-based electrode stabilization. *Nat Protoc* **9**: 2784–2795.

Long MA, Lee AK. 2012. Intracellular recording in behaving animals. *Curr Opin Neurobiol* **22**: 34–44.

Long MA, Jin DZ, Fee MS. 2010. Support for a synaptic chain model of neuronal sequence generation. *Nature* **468**: 394–399.

Schneider DM, Nelson A, Mooney R. 2014. A synaptic and circuit basis for corollary discharge in the auditory cortex. *Nature* **513**: 189–194.

Vallentin D, Long MA. 2015. Motor origin of precise synaptic inputs onto forebrain neurons driving a skilled behavior. *J Neurosci* **35**: 299–307.

General Safety and Hazardous Material Information

This manual should be used by laboratory personnel with experience in laboratory and chemical safety or students under the supervision of such trained personnel. The procedures, chemicals, and equipment referenced in this manual are hazardous and can cause serious injury unless performed, handled, and used with care and in a manner consistent with safe laboratory practices. Students and researchers using the procedures in this manual do so at their own risk. It is essential for your safety that you consult the appropriate Material Safety Data Sheets, the manufacturers' manuals accompanying equipment, and your institution's Environmental Health and Safety Office, as well as the General Safety and Disposal Cautions in this appendix for proper handling of hazardous materials in this manual. Cold Spring Harbor Laboratory makes no representations or warranties with respect to the material set forth in this manual and has no liability in connection with the use of these materials.

All registered trademarks, trade names, and brand names mentioned in this book are the property of the respective owners. Readers should please consult individual manufacturers and other resources for current and specific product information.

Users should always consult individual manufacturers, the manufacturers' safety guidelines and other resources, including local safety offices, for current and specific product information and for guidance regarding the use and disposal of hazardous materials.

PRIMARY SAFETY INFORMATION RESOURCES FOR LABORATORY PERSONNEL

Institutional Safety Office. The best source of toxicity, hazard, storage, and disposal information is your institutional safety office, which maintains and makes available the most current information. Always consult this office for proper use and disposal procedures.

Post the phone numbers for your local safety office, security office, poison control center, and laboratory emergency personnel in an obvious place in your laboratory.

Material Safety Data Sheets (MSDSs). The Occupational Safety and Health Administration (OSHA) requires that MSDSs accompany all hazardous products that are shipped. These data sheets contain detailed safety information. MSDSs should be filed in the laboratory in a central location as a reference guide.

GENERAL SAFETY AND DISPOSAL CAUTIONS

The guidance offered here is intended to be generally applicable. However, proper waste disposal procedures vary among institutions; therefore, always consult your local safety office for specific instructions. All chemically constituted waste must be disposed of in a suitable container clearly labeled with the type of material it contains and the date the waste was initiated.

It is essential for laboratory workers to be familiar with the potential hazards of materials used in laboratory experiments and to follow recommended procedures for their use, handling, storage, and disposal.

The following general cautions should always be observed.

- **Before beginning the procedure,** become completely familiar with the properties of substances to be used.

- **The absence of a warning** does not necessarily mean that the material is safe, because information may not always be complete or available.

- **If exposed** to toxic substances, contact your local safety office immediately for instructions.

- **Use proper disposal procedures** for all chemical, biological, and radioactive waste.

- **For specific guidelines on appropriate gloves to use,** consult your local safety office.

- **Handle concentrated acids and bases** with great care. Wear goggles and appropriate gloves. A face shield should be worn when handling large quantities.

   Do not mix strong acids with organic solvents because they may react. Sulfuric acid and nitric acid especially may react highly exothermically and cause fires and explosions.

   Do not mix strong bases with halogenated solvents because they may form reactive carbenes that can lead to explosions.

- **Handle and store pressurized gas containers** with caution because they may contain flammable, toxic, or corrosive gases; asphyxiants; or oxidizers. For proper procedures, consult the Material Safety Data Sheet that is required to be provided by your vendor.

- **Never pipette** solutions using mouth suction. This method is not sterile and can be dangerous. Always use a pipette aid or bulb.

- **Keep halogenated and nonhalogenated** solvents separately (e.g., mixing chloroform and acetone can cause unexpected reactions in the presence of bases). Halogenated solvents are organic solvents such as chloroform, dichloromethane, trichlorotrifluoroethane, and dichloroethane. Non-halogenated solvents include pentane, heptane, ethanol, methanol, benzene, toluene, *N,N*-dimethylformamide (DMF), dimethylsulfoxide (DMSO), and acetonitrile.

- **Laser radiation,** visible or invisible, can cause severe damage to the eyes and skin. Take proper precautions to prevent exposure to direct and reflected beams. Always follow the manufacturer's safety guidelines and consult your local safety office. See caution below for more detailed information.

- **Flash lamps,** because of their light intensity, can be harmful to the eyes. They also may explode on occasion. Wear appropriate eye protection and follow the manufacturer's guidelines.

- **Photographic fixatives, developers, and photoresists** also contain chemicals that can be harmful. Handle them with care and follow the manufacturer's directions.

- **Power supplies and electrophoresis equipment** pose serious fire hazard and electrical shock hazards if not used properly.

- **Microwave ovens and autoclaves** in the laboratory require certain precautions. Accidents have occurred involving their use (e.g., when melting agar or Bacto Agar stored in bottles or when sterilizing). If the screw top is not completely removed and there is inadequate space for the steam to vent, the bottles can explode and cause severe injury when the containers are removed from the microwave or autoclave. Always completely remove bottle caps before microwaving or autoclaving. An alternative method for routine agarose gels that do not require sterile agar is to weigh out the agar and place the solution in a flask.

- **Ultrasonicators** use high-frequency sound waves (16–100 kHz) for cell disruption and other purposes. This "ultrasound," conducted through air, does not pose a direct hazard to humans, but the associated high volumes of audible sound can cause a variety of effects, including headache, nausea, and tinnitus. Direct contact of the body with high-intensity ultrasound (not medical imaging equipment) should be avoided. Use appropriate ear protection and display signs on the door(s) of laboratories where the units are used.

- **Use extreme caution when handling cutting devices,** such as microtome blades, scalpels, razor blades, or needles. Microtome blades are extremely sharp! Use care when sectioning. If unfamiliar with their use, have an experienced user demonstrate proper procedures. For proper disposal, use the "sharps" disposal container in your laboratory. Discard used needles *unshielded*, with the syringe still attached. This prevents injuries and possible infections when manipulating used needles because many accidents occur while trying to replace the needle shield. Injuries may also be caused by broken Pasteur pipettes, coverslips, or slides.

- **Procedures for the humane treatment of animals** must be observed at all times. Consult your local animal facility for guidelines. Animals, such as rats, are known to induce allergies that can increase in intensity with repeated exposure. Always wear a lab coat and gloves when handling these animals. If allergies to dander or saliva are known, wear a mask.

DISPOSAL OF LABORATORY WASTE

There are specific regulatory requirements for the disposal of all medical waste and biological samples mandated by the U.S. Environmental Protection Agency (see http://www.epa.gov/epa waste/hazard/tsd/index.htm) and regulated by the individual states and territories (see http://www.epa.gov/epawaste/wyl/stateprograms.htm). Medical and biological samples that require special handling and disposal are generally termed Medical Pathological Waste (MPW), and medical, veterinary, and biological facilities will have programs for the collection of MPW and its disposal. Restrictions on how radioactive waste can be disposed of as regulated by the U.S. Nuclear Regulatory Commission can be found in 10 CFR 20.2001, General requirements for waste disposal (see http://www.nrc.gov/reading-rm/doc-collections/cfr/part020/part020-2001.html) or the individual Agreement States. The preferred method for the disposal of radioactively contaminated MPW'is decay-in-storage (see http://www.nrc.gov/reading-rm/doc-collections/cfr/part035/part035-0092.html).

Waste and any materials contaminated with biohazardous materials must be decontaminated and disposed of as regulated medical waste. No harmful substances should be released into the environment in an uncontrolled manner. This includes all tissue samples, needles, syringes, scalpels, etc. Be sure to contact your institution's safety office concerning the proper practices associated with the handling and disposal of biohazardous waste.

Some basic rules are outlined below. For treatment of radioactive and biological waste, see sections on Radioactive Safety Procedures and Biological Safety Procedures.

- In practice, only **neutral aqueous solutions** without heavy metal ions and without organic solvents can be poured down the drain (e.g., most buffers). Acid and basic aqueous solutions need to be neutralized cautiously before their disposal by this method.

- For proper disposal of **strong acids and bases,** dilute them by placing the acid or base onto ice and neutralize them. Do not pour water into them. If the solution does not contain any other toxic compound, the salts can be flushed down the drain.

- For disposal of **other liquid waste,** similar chemicals can be collected and disposed of together, whereas chemically different wastes should be collected separately. This avoids chemical reactions between components of the mixture (see above). Collect at least inorganic aqueous waste, non-halogenated solvents, and halogenated solvents separately.

- Waste **from photo processing and automatic developers** should be collected separately to recycle the silver traces found in it.

RADIOACTIVE SAFETY PROCEDURES

In the United States and other countries, the access to radioactive substances is strictly controlled. You may be required to become a registered user (e.g., by attending a mandatory seminar and receiving a personal dosimeter). A convenient calculator to perform routine radioactivity calculations can also be found at http://www.graphpad.com/quickcalcs/ChemMenu.cfm.

If you have never worked with radioactivity before, follow the steps below.

- *Try to avoid it!* Many experiments that are traditionally performed with the help of radioactivity can now be done using alternatives based on fluorescence or chemiluminescence and colorimetric assays, including, for example, DNA sequencing, Southern and northern blots, and protein kinase assays. However, in other cases (e.g., metabolic labeling of cells), use of radioactivity cannot be avoided.

- **Be informed.** While planning an experiment that involves the use of radioactivity, include the physicochemical properties of the isotope (half-life, emission type, and energy), the chemical form of the radioactivity, its radioactive concentration (specific activity), total amount, and its chemical concentration. Order and use only as much as is really needed.

- **Familiarize yourself** with the designated working area. Perform a mental and practical dry run (replacing radioactivity with a colored solution) to make sure that all equipment needed is available and to get used to working behind a shield. Handle your samples as if sterility would be required to avoid contamination.

- **Always wear appropriate gloves**, lab coat, and safety goggles when handling radioactive material.

- **Check the work area** for contamination before, during, and after your experiment (including your lab coat, hands, and shoes).

- **Localize your radioactivity.** Avoid formation of aerosols or contamination of large volumes of buffers.

- **Liquid scintillation cocktails** are often used to quantitate radioactivity. They contain organic solvents and small amounts of organic compounds. Try to avoid contact with the skin. After use, they should be regarded as radioactive waste; the filled vials are usually collected in designated containers, separate from other (aqueous) liquid radioactive waste.

- **Dispose of radioactive waste** only into designated, shielded containers (separated by isotope, physical form [dry/liquid], and chemical form [aqueous/organic solvent phase]). Always consult your safety office for further guidance in the appropriate disposal of radioactive materials.

- Among the experiments requiring **special precautions** are those that use [^{35}S]methionine and ^{125}I, because of the dangers of airborne radioactivity. [^{35}S]methionine decomposes during storage into sulfoxide gases, which are released when the vial is opened. The isotope ^{125}I accumulates in the thyroid and is a potential health hazard. ^{125}I is used for the preparation of Bolton–Hunter reagent to radioiodinate proteins. Consult your local safety office for further guidance in the appropriate use and disposal of these radioactive materials before initiating any experiments. Wear appropriate gloves when handling potentially volatile radioactive substances, and work only in a radioiodine fume hood.

BIOLOGICAL SAFETY PROCEDURES

Biological safety fulfills three purposes: to avoid contamination of your biological sample with other species; to avoid exposure of the researcher to the sample; and to avoid release of living material into the environment. Biological safety begins with the receipt of the living sample; continues with its storage, handling, and propagation; and ends only with the proper disposal of all contaminated materials. A catalog of operations known as "sterile handling" is usually employed in manipulating living matter. However, the actual manner of treatment largely depends on the actual sample, which can be quite diverse: *Escherichia coli* and other bacterial strains, yeasts, tissues of animal or plant origin, cultures of mammalian cells, or even derivatives from human blood are routinely handled in a biological laboratory. Two of these, bacteria and human blood products, are discussed in more detail below.

The Department of Health, Education, and Welfare (HEW) has classified various bacteria into different categories with regard to shipping requirements (see Sanderson and Zeigler 1991). Non-pathogenic strains of *E.coli* (such as K12) and *Bacillus subtilis* are in Class 1 and are considered to present no or minimal hazard under normal shipping conditions. However, *Salmonella*, *Haemophilus*, and certain strains of *Streptomyces* and *Pseudomonas* are in Class 2. Class 2 bacteria are "[a]gents of ordinary potential hazard: agents which produce disease of varying degrees of severity... but which are contained by ordinary laboratory techniques." Contact your institution's safety office concerning shipping biological material.

Human blood, blood products, and tissues may contain occult infectious materials such as hepatitis B virus and human immunodeficiency virus (HIV) that may result in laboratory-acquired infections. Investigators working with lymphoblast cell lines transformed by Epstein–Barr virus (EBV) are also at risk of EBV infection. Any human blood, blood products, or tissues should be considered a biohazard and should be handled accordingly until proved otherwise. Wear appropriate disposable gloves, use mechanical pipetting devices, work in a biological safety cabinet, protect against the possibility of aerosol generation, and disinfect all waste materials before disposal. Autoclave contaminated plasticware before disposal; autoclave contaminated liquids or treat with bleach (10% [v/v] final concentration) for at least 30 minutes before disposal (this is valid also for used bacterial media).

Always consult your local institutional safety officer for specific handling and disposal procedures of your samples. Further information can be found in the Frequently Asked Questions of the ATCC homepage (http://www.atcc.org) and is also available from the National Institute of Environmental Health and Human Services, Biological Safety (http://www.niehs.nih.gov/about/stewardship).

GENERAL PROPERTIES OF COMMON HAZARDOUS CHEMICALS

The hazardous materials list can be summarized in the following categories.

- **Inorganic acids**, such as hydrochloric, sulfuric, nitric, or phosphoric, are colorless liquids with stinging vapors. Avoid spills on skin or clothing. Spills should be diluted with large amounts of water. The concentrated forms of these acids can destroy paper, textiles, and skin and cause serious injury to the eyes.

- **Inorganic bases**, such as sodium hydroxide, are white solids that dissolve in water and under heat development. Concentrated solutions will slowly dissolve skin and even fingernails.

- **Salts of heavy metals** are usually colored, powdered solids that dissolve in water. Many of them are potent enzyme inhibitors and therefore toxic to humans and the environment (e.g., fish and algae).

- Most **organic solvents** are flammable volatile liquids. Avoid breathing the vapors, which can cause nausea or dizziness. Also avoid skin contact.

- **Other organic compounds** including organosulfur compounds, such as mercaptoethanol or organic amines, can have very unpleasant odors. Others are highly reactive and should be handled with appropriate care.

- If improperly handled, **dyes and their solutions** can stain not only your sample but also your skin and clothing. Some are also mutagenic (e.g., ethidium bromide), carcinogenic, and toxic.

- **Nearly all names ending with "ase"** (e.g., catalase, β-glucuronidase, or zymolyase) refer to enzymes. There are also other enzymes with nonsystematic names such as pepsin. Many of them are provided by manufacturers in preparations containing buffering substances, etc. Be aware of the individual properties of materials contained in these substances.

- **Toxic compounds** are often used to manipulate cells. They can be dangerous and should be handled appropriately.

- Be aware that several of the compounds listed have not been thoroughly studied with respect to their toxicological properties. Handle each chemical with appropriate respect. Although the toxic effects of a compound can be quantified (e.g., LD_{50} values), this is not possible for carcinogens or mutagens where one single exposure can have an effect. Also realize that dangers related to a given compound may also depend on its physical state (fine powder vs. large crystals/diethyl ether vs. glycerol/dry ice vs. carbon dioxide under pressure in a gas bomb). Anticipate under which circumstances during an experiment exposure is most likely to occur and how best to protect yourself and your environment.

Cold Spring Harbor Laboratory Press (CSHLP) has used its best efforts in collecting and preparing the material contained herein but does not assume, and hereby disclaims, any liability for any loss or damage caused by errors and omissions in the publication, whether such errors and omissions result from negligence, accident, or any other cause. CSHLP does not assume responsibility for the user's failure to consult more complete information regarding the hazardous substances listed in this publication.

REFERENCE

Sanderson KE, Zeigler DR. 1991. Storing, shipping, and maintaining records on bacterial strains. *Methods Enzymol* **204**: 248–264.

WWW RESOURCES

ATCC Home page http://www.atcc.org

ATCC, for Sample Handling (in Frequently Asked Questions) http://www.atcc.org/CulturesandProducts/TechnicalSupport/FrequentlyAskedQuestions/tabid/469/Default.aspx

GraphPad Software, Radioactivity Calculations http://www.graphpad.com/quickcalcs/ChemMenu.cfm

National Institute of Environmental Health and Human Services, Biological Safety (NIEHS) http://www.niehs.nih.gov/about/stewardship

U.S. Environmental Protection Agency (EPA), Federal waste disposal regulations, Laboratory http://www.epa.gov/epawaste/hazard/tsd/index.htm

U.S. Environmental Protection Agency (EPA), Individual States and Territories http://www.epa.gov/epawaste/wyl/stateprograms.htm

U.S. Nuclear Regulatory Commission (NRC), Medical Pathological Radioactively Contaminated Waste (Decay-in-Storage) http://www.nrc.gov/reading-rm/doc-collections/cfr/part035/part035-0092.html

U.S. Nuclear Regulatory Commission (NRC), Radioactive Waste Disposal Regulations: General Requirements http://www.nrc.gov/reading-rm/doc-collections/cfr/part020/part020-2001.html

Index